D1620792

Global Legislation for Food Packaging Materials

Edited by
Rinus Rijk and Rob Veraart

Related Titles

Brennan, J. G. / Grandison, A. S. (eds.)

Food Processing Handbook

Second Edition

2010

ISBN: 978-3-527-32468-2

Wintgens, J. N. (ed.)

Coffee: Growing, Processing, Sustainable Production

A Guidebook for Growers, Processors, Traders, and Researchers

Second, Updated Edition

2009

ISBN: 978-3-527-32286-2

Piringer, O. G., Baner, A. L. (eds.)

Plastic Packaging

Interactions with Food and Pharmaceuticals

2008

ISBN: 978-3-527-31455-3

Ziegler, H. (ed.)

Flavourings

Production, Composition, Applications, Regulations

Second, Completely Revised Edition

2007

ISBN: 978-3-527-31406-5

Global Legislation for Food Packaging Materials

Edited by
Rinus Rijk and Rob Veraart

WILEY-VCH Verlag GmbH & Co. KGaA

The Editors

Rinus Rijk
Keller and Heckman LLP and
Advisor
Bendienlaan 18
3431 RA Nieuwegein
The Netherlands

Dr. Rob Veraart
Keller and Heckman LLP
Avenue Louise 523
1050 Brussels
Belgium

Library of Congress Card No.: applied for

British Library Cataloguing-in-Publication Data
A catalogue record for this book is available from the British Library.

Bibliographic information published by the Deutsche Nationalbibliothek
The Deutsche Nationalbibliothek lists this publication in the Deutsche Nationalbibliografie; detailed bibliographic data are available on the Internet at http://dnb.d-nb.de.

Typesetting Thomson Digital, Noida, India
Printing and Binding Strauss GmbH, Mörlenbach
Cover Design Anne Christine Keßler, Karlsruhe

Printed in the Federal Republic of Germany
Printed on acid-free paper

ISBN: 978-3-527-31912-1

Contents

Preface *XXI*
List of Contributors *XXIII*

1 **EU Legislation** *1*
Annette Schäfer
1.1 Introduction *1*
1.2 Community Legislation *2*
1.2.1 Horizontal Legislation *2*
1.2.1.1 Framework Regulation *2*
1.2.1.2 Good Manufacturing Practice *5*
1.2.2 Specific Measures *6*
1.2.2.1 Plastics *6*
1.2.2.2 Recycled Plastics *10*
1.2.3 Other Materials *11*
1.2.3.1 Ceramic Articles *12*
1.2.3.2 Regenerated Cellulose Film (Cellophane) *12*
1.2.3.3 Rubber Teats and Soothers *12*
1.2.3.4 BADGE, BFDGE, and NOGE in Coated Materials, Plastics, and Adhesives *13*
1.2.3.5 Active and Intelligent Materials and Articles *13*
1.2.4 Control of Food Contact Materials in the EU *17*
1.2.4.1 Role of the Business Operators: Food Industry and Packaging/Contact Material Industry *17*
1.2.4.2 Role of the Member States *17*
1.2.4.3 Role of the European Commission *18*
1.2.4.4 Methods for Sampling and Analysis in the Official Control *18*
1.3 Specific National Legislation *19*
1.4 Future Trends *20*
1.4.1 Plastic *20*

Global Legislation for Food Packaging Materials. Edited by Rinus Rijk and Rob Veraart

ISBN: 978-3-527-31912-1

1.4.2 Nanomaterials *20*
1.4.3 Risk Assessment *24*
1.4.4 Other Materials *24*
References *24*

2 Petitioning Requirements and Safety Assessment in Europe *27*
Paul Tobback and Rinus Rijk
2.1 Introduction *27*
2.2 EFSA and its Role in Safety Evaluation of Food Contact Materials *28*
2.3 Data Requirement on a Substance for its Safety Assessment by the EFSA *30*
2.3.1 Introduction *30*
2.3.2 Data to be Supplied within a Submission *31*
2.3.2.1 General Information on the Substance *32*
2.3.2.2 Information on Physical and Chemical Properties of the Substance *33*
2.3.2.3 Information on the Intended Application of the Substance *35*
2.3.2.4 Information on Authorization of the Substance *36*
2.3.2.5 Information on the Migration of the Substance *36*
2.3.2.6 Information on the Residual Content *39*
2.3.2.7 Antimicrobial Substances *39*
2.3.2.8 Toxicological Information *40*
2.4 Evaluation Process of a Food Contact Substance *42*
2.4.1 Re-Evaluation of Substances *44*
2.5 Public Access to Data *46*
References *47*

3 Council of Europe Resolutions *49*
Luigi Rossi
3.1 Introduction *49*
3.2 Activity of CoE in the 1960s–1970s *50*
3.2.1 Resolution on Plastics *50*
3.2.2 Guidelines for the Evaluation of Substances *50*
3.2.3 Estimation of Exposure *51*
3.2.4 Conventional Classification of Food *52*
3.3 EU Activity and Relationship Between EU and CoE *52*
3.4 Resolutions of the CoE *52*
3.4.1 Procedure for the Adoption of a Resolution and Guidelines and Technical Documents *52*
3.4.2 CoE Resolutions, Guidelines, and TDs *53*
3.5 Status of the Packages (Resolutions, Guidelines, and Technical Documents) *54*
3.5.1 Resolutions on Colorants in Plastics *54*
3.5.1.1 Inventory of the Documents *54*

3.5.1.2 Chronological Development 54
3.5.1.3 Content of the Resolution 54
3.5.2 Resolutions on Control of Aids to Polymerization for Plastic 54
3.5.2.1 Inventory of the Documents 54
3.5.2.2 Chronological Development 55
3.5.2.3 Content of the Resolution 55
3.5.3 Resolution on Silicones 55
3.5.3.1 Inventory of the Documents 55
3.5.3.2 Chronological Development 55
3.5.3.3 Content of the Resolution 55
3.5.4 Resolution on Paper 56
3.5.4.1 Inventory of the Documents 56
3.5.4.2 Chronological Development 56
3.5.4.3 Content of Package 56
3.5.5 Resolution on Coatings 57
3.5.5.1 Inventory of the Documents 57
3.5.5.2 Chronological Development 57
3.5.5.3 Content of the Resolution 57
3.5.6 Resolution on Cork Stoppers and Other Cork Materials 58
3.5.6.1 Inventory of the Documents 58
3.5.6.2 Chronological Development 58
3.5.6.3 Content of the Resolution 58
3.5.7 Resolution on Ion Exchange and Adsorbant Resins Used in the Processing of Foodstuffs 58
3.5.7.1 Inventory of the Documents 58
3.5.7.2 Chronological Development 59
3.5.7.3 Content of the Resolution 59
3.5.8 Resolution on Rubber Products 59
3.5.8.1 Inventory of the Documents 59
3.5.8.2 Chronological Development 59
3.5.8.3 Content of the Resolution 59
3.5.9 Resolution on Packaging Inks 60
3.5.9.1 Inventory of the Documents 60
3.5.9.2 Chronological Development 60
3.5.9.3 Content of the Resolution 60
3.5.10 Guidelines on Metals and Alloys 61
3.5.10.1 Inventory of the Documents 61
3.5.10.2 Chronological Development 61
3.5.10.3 Content of the Guidelines 61
3.5.11 Guidelines on Glass 62
3.5.11.1 Inventory of the Documents 62
3.5.11.2 Chronological Development 62
3.5.11.3 Content of the Resolution 62

3.5.12 Guidelines on Tissue Paper Kitchen Towels *62*
3.5.12.1 Inventory of the Documents *62*
3.5.12.2 Chronological Development *62*
3.5.12.3 Content of the Guidelines *62*
3.6 Transposition of the Resolutions, Guidelines, and Technical Documents *63*
3.7 The Future of the CoE *63*
3.7.1 Procedure of Authorization of a New Substance at National/CoE Level *64*
3.8 Conclusions *64*
References *65*

4 National Legislation in Germany *67*
Wichard Pump
4.1 Introduction *67*
4.1.1 Commodities Defined in LFGB *67*
4.1.2 Basic Requirements on Commodities *68*
4.1.3 BfR Recommendations *69*
4.1.4 Further Requirements *71*
4.2 Legal Assessment of Different Materials Classes in Food Contact *71*
4.2.1 Plastic Materials *71*
4.2.2 Colorants *74*
4.2.3 Plastics Dispersions (BfR Recommendation XIV) *74*
4.2.4 Silicones (BfR-Recommendation XV) *75*
4.2.5 Rubber and Elastomers (BfR Recommendation XXI) *75*
4.2.6 Hard Paraffins, Microcrystalline Waxes and Mixtures of these with Waxes, Resins, and Plastics (BfR Recommendation XXV) *77*
4.2.7 Conveyor Belts Made from Gutta-Percha and Balata (BfR Recommendation XXX) *77*
4.2.8 Paper, Carton, and Board *77*
4.2.8.1 XXXVI: Paper, Board for Food Contact *78*
4.2.8.2 XXXVI/1: Cooking and Hot-filter Paper and Filter Layers *78*
4.2.8.3 XXXVI/2: Papers, Cartons, and Boards for Baking Purposes *79*
4.2.8.4 XXXVI/3: Absorber Pads based on Cellulosic Fibers for Food Packaging *79*
4.2.9 Artificial Sausage Casings (BfR Recommendation XLIV) *79*
4.2.10 Materials for Coating the Outside of Hollow Glassware (BfR Recommendation XLVIII) *80*
4.2.11 Soft Polyurethane Foams as Cushion Packaging for Fruit (BfR Recommendation IL) *80*
4.2.12 Temperature-Resistant Polymer Coating Systems for Frying, Cooking, and Baking Utensils (BfR Recommendation LI) *80*
4.2.13 Fillers for Commodities Made of Plastic (BfR Recommendation LII) *80*

4.2.14 Absorber Pads and Packagings with Absorbing Function, in which Absorbent Materials Based on Cross-Linked Polyacrylates are Used, for Foodstuffs (BfR Recommendation LIII) *81*
References *81*

5 The French Regulation on Food Contact Materials *83*
Jean Gauducheau and Alexandre Feigenbaum
5.1 Introduction *83*
5.1.1 Basic Principles *84*
5.1.2 Categories of Reference Binding Texts *84*
5.1.2.1 Binding Texts: Lois (Laws), Décrets (Decrees), and Arrêtés (Orders) *84*
5.1.2.2 Additional Information on the Texts *85*
5.1.2.3 The Consumption Code (Code de la Consommation) *86*
5.2 Integration of European Directives and Regulations on Food Contact Materials into the French Regulation *86*
5.2.1 Main EU Regulations on Materials Intended for Contact with Food *87*
5.2.2 Transposition of the Directives on Materials Intended to Come in Contact with Foodstuffs *87*
5.2.2.1 Regenerated Cellulose Films *87*
5.2.2.2 Plastic Materials *88*
5.2.2.3 Ceramics *88*
5.2.2.4 Rubber *88*
5.3 Specific French Legislation on Plastic Materials Intended to Come in Contact with Food *89*
5.3.1 Reference Texts on Coloring Agents for Plastic Materials *89*
5.3.2 Reference Texts on Additives for Plastic Materials *89*
5.3.2.1 Information Note of DGCCRF 2003-27 of 24 March 2003 *89*
5.3.3 Reference Texts on Recycled Materials *89*
5.3.3.1 Avis of CSHPF Dated 1 June 1993 *89*
5.3.3.2 Avis of CSHPF Dated 7 September 1993 *90*
5.3.3.3 Avis of AFSSA of 27 November 2006 *90*
5.4 Supplementary French Legislation on Materials other than Plastics Intended to Come in Contact With Food *90*
5.4.1 Introduction *90*
5.4.2 Application of Regulations: the Decree 2007-766 of 10 May 2007 *91*
5.4.3 French Regulation on Materials in Paper and Board Intended to Come in Contact with Food *91*
5.4.4 French Regulation on Materials and Articles in Rubber Intended to Come in Contact with Food *93*
5.4.4.1 Note of Information 2004-64 of DGCCRF, Chapter "Rubber" *93*
5.4.5 French Regulation on Materials and Articles in Silicone Elastomers Intended to Come in Contact with Food *93*
5.4.6 French Regulation on Materials and Articles in Glass and Ceramics Intended to Come in Contact with Food *94*

5.4.7 French Regulation on Materials and Articles in Metals Intended to Come in Contact with Food *94*
5.4.7.1 General Case: Metallic Materials and Objects, Packaging *94*
5.4.7.2 Household Metallic Articles and Parts of Industrial Materials *96*
5.4.7.3 French Regulation on Materials and Objects in Wood Intended to Come in Contact with Food *99*
5.4.7.4 French Regulation on Materials and Objects as Complexes Intended to Come in Contact with Food *100*
5.5 French Regulation on Coatings Coming into Contact with Foodstuffs *100*
5.6 French Regulation on Coloring Matters Used on or Within Food Packaging in Contact with Food *101*
5.6.1 Preliminary Remarks *101*
5.6.2 Basic Principles *101*
5.6.3 French Regulation on Colorants Used in the Mass of Paper or Plastic Materials *102*
5.6.3.1 Texts and Regulations *102*
5.6.3.2 Instruction of 2 August 1993 *102*
5.6.4 French Regulation on Inks Used to Print Food Packaging *103*
5.7 French Regulation on Requests for Authorization of Use of Constituents of Materials and Articles in Contact with Food *104*
5.7.1 Advice of CSHPF of 9 December 1997 *105*
5.8 Conclusion *106*

6 Dutch Legislation on Food Contact Materials *109*
Rob Veraart
6.1 Introduction *109*
6.2 Plastics *111*
6.2.1 Nonepoxy Plastics *111*
6.2.2 Epoxy Plastics *112*
6.3 Paper and Board *112*
6.3.1 Paper and Board for General Purpose *112*
6.3.1.1 Overall Migration *113*
6.3.1.2 Specific Migration *113*
6.3.2 Paper for Filtering and Cooking Above 80 °C *114*
6.3.2.1 Overall Migration *114*
6.3.2.2 Specific Migration *114*
6.4 Rubber *114*
6.4.1 Categories *114*
6.4.2 Positive List *117*
6.4.3 Other Restrictions *118*
6.5 Metals *118*
6.5.1 Metals Used for the Application of Packaging Materials *118*
6.5.2 Metals Used for the Application of Utensils *119*
6.5.3 Restrictions *119*

6.6 Glass and Glass Ceramics *119*
6.7 Ceramics and Enamels *119*
6.7.1 Ceramics *119*
6.7.2 Enamels *120*
6.8 Textiles *120*
6.9 Regenerated Cellulose *121*
6.10 Wood and Cork *121*
6.11 Coatings *121*
6.12 Colored Materials *122*
6.13 Calculations *122*
References *123*

7 National Legislation in Italy *125*
Maria Rosaria Milana
7.1 Introduction *125*
7.2 Decrees on General Principles *126*
7.3 Decrees on Specific Materials *127*
7.3.1 The Ministerial Decree 21 March 1973 and its Amendments *127*
7.3.1.1 General Part *127*
7.3.1.2 Specific Part *129*
7.3.2 Ceramic *135*
7.3.3 Tin Plate *135*
7.3.4 Tin-Free Steel *136*
7.3.5 Aluminum *137*
7.3.6 Materials Without Specific Regulation *138*
7.4 How to Get the List and the Text of the Italian Legislation *138*
References *138*

8 Switzerland *141*
Roger Meuwly and Vincent Dudler
8.1 Legislative System *141*
8.1.1 Availability of Legal Texts and Official Documents *142*
8.1.2 Attestation of Conformity (Letter of No Objection (LNOs) *143*
8.1.3 Council of Europe *143*
8.2 Food Contact Materials and Articles *144*
8.2.1 Plastic Materials and Articles *144*
8.2.1.1 Limits on Migration *145*
8.2.1.2 Recycled Plastic Materials *146*
8.2.2 Regenerated Cellulose *146*
8.2.3 Materials and Articles in Ceramic, Glass, Enamel, or Other Analogue Materials *147*
8.2.4 Metals and Alloys *147*
8.2.5 Materials and Articles in Paper and Board *147*
8.2.6 Active and Intelligent Materials and Articles *147*
8.2.7 Paraffin, Waxes, and Colorants *148*

8.2.8 Silicone Materials and Articles *148*
8.2.9 Inks *148*
8.3 Conclusions *149*
References *149*

9 Legislation on Food Contact Materials in the Scandinavian Countries and Finland *151*
Bente Fabech, Pirkko Kostamo, Per Fjeldal, and Kristina Salmén
9.1 Introduction *151*
9.2 Legislation in Denmark *151*
9.2.1 Public Control and In-House Documentation *152*
9.3 Legislation in Finland *152*
9.4 Legislation in Norway *153*
9.4.1 The Packaging Convention of Norway and the EK Declaration *153*
9.4.2 Paper and Board Food Contact Materials *154*
9.4.3 Control of Critical Points *154*
9.4.4 Metals in Ceramics, Glass, Metalwares, and Nonceramic Materials Without Enamel *155*
9.5 Legislation in Sweden *157*
9.5.1 Voluntary Agreement *157*

10 Code of Practice for Coatings in Direct Contact with Food *161*
Peter Oldring
10.1 Introduction *161*
10.2 Contents of the Code of Practice *163*
10.3 Main Points of the Code of Practice *164*
10.3.1 Scope of the Code of Practice: Article 1 *164*
10.3.2 Good Manufacturing Practice: Article 2 *165*
10.3.3 Substance Lists: Articles 3, 4, and 5 *166*
10.3.4 Dual Use Additives – Article 6 *167*
10.3.5 Restrictions of Substances and Testing – Articles 7, 8 and 11 *168*
10.3.6 Multilayer Coatings – Article 9 *168*
10.3.7 Declaration of Conformity – Article 10 *168*
10.3.8 Risk Assessment: Annex VI *171*
10.4 How can the Code of Practice be Applied? *171*
10.5 Conclusions *173*
References *173*

11 Estimating Risks Posed by Migrants from Food Contact Materials *175*
Peter Oldring *175*
11.1 Introduction *175*
11.2 Hazard Assessment *176*
11.2.1 Cramer Class I *177*
11.2.2 Cramer Class II *177*

11.2.3 Cramer Class III 177
11.3 Exposure Assessment 178
11.3.1 Introduction 178
11.3.2 What is Exposure? 179
11.3.3 What Data are Needed to Estimate Exposure? 180
11.3.4 Who Should be Protected in an Exposure Assessment? 180
11.3.5 Factors to Consider in an Exposure Assessment 181
11.3.6 Estimating Exposure to Migrants from Food Contact Articles 181
11.3.6.1 Simplistic 182
11.3.6.2 Deterministic 182
11.3.6.3 Probabilistic (Stochastic) Modeling 183
11.3.7 The US FDA Approach to Estimating Consumer Exposure to Migrants from Food Contact Materials 184
11.3.8 Approaches to Determining Exposure to Migrants from Packaging of Foodstuffs that could be Used in the EU 185
11.3.8.1 Overview of Dietary Surveys 185
11.3.8.2 Concentration Data 187
11.3.8.3 Packaging of Foodstuffs Containing the Migrant(s) of Interest 189
11.4 FACET: Overview of Food Contact Material Work Package and Interactions with Other Work Packages 190
11.5 Conclusions 193
References 194

12 Compliance Testing, Declaration of Compliance, and Supporting Documentation in the EU 197
Rob Veraart
12.1 Introduction 197
12.2 Composition of the FCM 198
12.2.1 How to Demonstrate Compliance of Composition 198
12.3 Selection of Contact Conditions 199
12.3.1 Food Simulants 199
12.3.1.1 Substitute Fatty Food Simulants 201
12.3.1.2 Alternative Fatty Food Simulants 201
12.3.1.3 Reduction Factors 201
12.3.1.4 Future Developments 203
12.3.2 Selection of Time and Temperature Conditions 203
12.4 Migration Experiments 205
12.4.1 Contact Methods 205
12.4.2 Overall Migration 208
12.4.2.1 Overall Migration into Aqueous Food Simulants 208
12.4.2.2 Overall Migration into Fatty Food Simulants 208
12.4.2.3 Overall Migration into MPPO 210
12.4.3 Specific Migration 210
12.4.4 Mathematic Modeling 211
12.5 Residual Content 212

12.5.1 Worst-Case Calculations 212
12.5.2 Analytical Determination 212
12.6 Miscellaneous 213
12.6.1 Not Intentionally Added Substances 213
12.6.2 Organoleptic Testing 214
12.6.3 Other Tests 215
12.6.3.1 Demonstrating the Absence of PAHs 215
12.6.3.2 Color Release 216
12.6.3.3 Colorants Purity 217
12.7 Declaration of Compliance and Supporting Documentation 217
12.7.1 Ceramics 218
12.7.2 Plastic Materials 218
12.7.3 Materials with BADGE 219
12.7.4 Other Materials 219
References 220

13 Food Packaging Law in the United States 223
Joan Sylvain Baughan and Deborah Attwood
13.1 Introduction 223
13.2 FDA Rules and Regulations 224
13.2.1 Definition of a Food Additive 224
13.2.2 Suitable Purity, the Delaney Clause, and FDA's Constituents Policy 225
13.3 Exemptions 226
13.3.1 "No Migration" 226
13.3.2 Functional Barrier 228
13.3.3 Prior Sanction 228
13.3.4 GRAS 229
13.3.5 Threshold of Regulation 229
13.3.6 Housewares 230
13.3.7 Basic Polymer/Resin Doctrine 232
13.3.8 Mixture Doctrine 233
13.4 The Food Contact Notification System: An Outgrowth of the Food Additive Regulation System for Food Contact Substances 233
13.4.1 Introduction 233
13.4.2 Prenotification Consultations 235
13.4.3 Requirements for an FCN 235
13.4.3.1 Comprehensive Summary 236
13.4.3.2 Chemical Identity 236
13.4.3.3 Intended Conditions of Use 236
13.4.3.4 Intended Technical Effect 236
13.4.3.5 Estimation of Dietary Intake 237
13.4.3.6 Toxicity Information 237
13.4.3.7 Environmental Information 237
13.5 The Food Additive Petition Process 238

13.6 Conclusions 239
References 239

14 Food Packaging Law in Canada – DRAFT 243
Anastase Rulibikiye and Catherine R. Nielsen
14.1 Introduction 243
14.2 The Legal Structure 244
14.2.1 The Agencies Involved 244
14.2.2 The Laws Involved 244
14.3 The Regulatory Scheme 245
14.3.1 Mandatory Requirements: Federally Registered Establishments 245
14.3.2 Not so Voluntary Requirements 246
14.3.2.1 The NOL Process 247
14.3.2.2 The Result 251
14.4 Enforcement 252
14.5 Conclusions 252
References 253

15 Food Packaging Legislation in South and Central America 255
Marisa Padula
15.1 South America 255
15.1.1 MERCOSUR 255
15.1.1.1 Plastic Materials 256
15.1.1.2 Elastomeric Materials 262
15.1.1.3 Adhesives 263
15.1.1.4 Waxes and Paraffins 264
15.1.1.5 Cellulosic Materials 264
15.1.1.6 Regenerated Cellulose Films 265
15.1.1.7 Regenerated Cellulose Casings 266
15.1.1.8 Metallic Materials 266
15.1.1.9 Glass 267
15.1.1.10 Legislation Update 268
15.1.1.11 Implementation of GMC Resolutions in the MERCOSUR Member States' National Legislations 268
15.1.2 Venezuela 269
15.1.3 Chile 271
15.1.4 Andean Community 271
15.2 Mexico 274
15.3 Central America 275
15.3.1 Costa Rica 276
15.3.2 El Salvador 276
15.3.3 Guatemala 276
15.3.4 Honduras 277
15.3.5 Belize 277
15.4 Cuba 277

15.5 Conclusions 277
References 278

16 Israel's Legislation for Food Contact Materials: Set for the Global Markets 283
Haim H. Alcalay
16.1 Introduction 283
16.2 The Standards: Legislative Process in Israel 284
16.3 Technical and Expert Forums at SII 284
16.4 Voluntary Standards and "Compulsory – Official Standards" 285
16.5 Food Contact SI-5113 Provisions 286
16.6 Documentation Requirements 287
16.7 Approved Test Laboratories in Israel 287
16.8 Introducing the New Standard in Israeli 288
16.9 Imports of Packaging Materials into Israel 288
16.10 Global Israeli Approach to Food Contact Legislation 289
16.11 Kosher Regulations 289
16.12 Implementation Issues (FAQ) 289
16.13 Guiding Principles 290
References 290

17 Rules on Food Contact Materials and Articles in Japan 291
Yasuji Mori
17.1 Introduction 291
17.2 Thc Food Sanitation Law 292
17.2.1 Articles of the Food Sanitation Law 292
17.3 Specifications and Standards for Food and Food Additives (Notification No. 370) 295
17.3.1 Contents 295
17.3.2 Abstract of Restrictions for Packages/Containers 296
17.3.2.1 General Restrictions (Abstracts) 296
17.3.2.2 Specifications for each Material (Abstracts and Summary) 296
17.3.2.3 Specifications for Packages/Containers for Specific Food Type (Abstract) 304
17.3.2.4 Standards for Manufacturing of Apparatus and Packages/Containers (Abstract) 306
17.4 The Ordinance of Specifications and Standards for Milk and Milk Products 306
17.4.1 Articles of the Ordinance of Specifications and Standards for Milk and Milk Products (Abstracts) 306
17.4.2 Attached List (Specification and Standards for Milk and Milk Products) 307
17.4.3 Specifications and Standards for Apparatus, Packages/Containers, Their Materials, and Methods of Manufacturing in Attached List (Abstract) 307

17.4.3.1 Specifications and Standards for Apparatus, Packages/Containers, Their Materials and Methods of Manufacturing for Milk, Special Milk, Sterilized Goat Milk, Milk Making Adjustment for Constituent Part of Milk, Low-Fat Milk, No-Fat Milk, Processed Milk, and Cream (Group 1) *308*
17.4.3.2 Specifications and Standards for Apparatus, Packages/Containers, Their Materials and Methods of Manufacturing for Fermented Milk, Lactic Acid Bacteria Drink, Milk Drink (Group 2) *310*
17.4.3.3 Specifications and Standards for Apparatus, Package/Container, Their Materials and Methods of Manufacturing for Formulated Powder Milk (Group 3) *312*
17.5 The Food Safety Basic Law and Relationship with the Food Sanitation Law *313*
17.6 Industrial Voluntary Rules *314*
17.7 Sheets of Confirmation with Compliance *316*
17.8 Conclusions *316*
References *316*

18 China Food Contact Chemical Legislation Summary *319*
Caroline Li and Sam Bian
18.1 Introduction *319*
18.2 Current Status *319*
18.3 China Food Safety Regulatory Entity *320*
18.4 Regulations and Rules under the China Food Hygiene (Sanitary) Law *320*
18.5 Hygiene Standards on Food Container and Packing Material *321*
18.6 Hygiene Standards for Food Contact Substance (Indirect Additive) *323*
18.7 Update in China Food Contact Chemical Regulation and Forecast on Future *326*
18.7.1 Update of China Food Contact Chemical Legislation *326*
18.7.2 Forecast on the New Food Contact Regulation in China *326*
18.7.2.1 Expanded Food Contact Additive Positive List *327*
18.7.2.2 The New Food Contact Substance Notification Procedure *327*
18.7.2.3 Future Steps *328*
18.8 Other Food Packaging Material-Related Regulation in China: Requirements for Local Manufacturing and Import of Food Contact Materials *328*
18.8.1 Requirements for Local Manufacturing of Food Contact Materials *328*
18.8.2 Requirements for Importing Food Contact Materials *329*
References *335*

19 Principal Issues in Global Food Contact: Indian Perspective *337*
Sameer Mehendale
19.1 The Indian Subcontinent: A Study in Contrast *337*
19.2 Food Contact Legislation *337*

19.3 General Guidelines 337
19.4 Condition for Sale and License 338
19.4.1 Conditions for Sale 338
19.5 Packing and Labeling of Foods 340
19.5.1 Package of Food to Carry a Label 340
19.5.1.1 Oils 342
19.5.1.2 Milk and Infant Food Substitutes 342
19.5.1.3 Packaged Drinking Water 343
19.6 Indian Standards for Direct Food Contact 343
19.7 Methods of Analysis and Determination of Specific and Overall Migration Limits 343
19.8 Acceptability Criteria 343
19.9 Future 344

20 Southeast Asia Food Contact *345*
Caroline Li and Sumalee Tangpitayakul
20.1 Singapore 345
20.2 Malaysia 346
20.3 Thailand 348
20.4 Conclusions 354
References 354

21 Legislation on Food Contact Materials in the Republic of Korea *355*
Hae Jung Yoon and Young Ja Lee
21.1 Introduction 355
21.2 Food Sanitation Act 355
21.3 Food Code 357
21.3.1 General Provision 357
21.3.2 Common Standard and Specification for Food in General 358
21.3.3 Equipment, Containers, and Packaging for Food Products 358
21.4 Data and Information Required for Submission of Food Contact Material Prior to Use 364
21.5 Labeling Standard for Food and Food Additives 365
21.6 Requirements for Importing Food Contact Materials 366
21.7 The Future Direction for Packaging Material Regulations 366
References 367

22 Australia and New Zealand *369*
Robert J. Steele
22.1 Introduction 369
22.2 Australia 370
22.3 Australia's Regulatory Framework 370
22.4 Enforcement 372
22.5 The State Food Authority 373
22.6 New Zealand 373

22.7 New Zealand's Regulatory Framework 374
22.8 Relationship with Codex 375
22.9 Food Contact Legislation 375
22.10 Recycled Material 376
22.11 Food Recall Examples Under Section 1.4.3 of the FSANZ Code 377
22.12 Conclusions 377
References 378

Index 379

Preface

Food packaging has been used since ancient times to transport and store foods. While initially natural materials were used, and are still used, mankind has started to develop packaging materials that respond to their needs. The use of ceramics, paper, metal, glass, and natural resins is known for a long time. Only in the last century did the development of food packaging materials explode. Developments in polymerization of organic molecules have brought us large quantities of plastics with many properties. Modern community has increasing needs for preparation, transport, and storage of foods under safe conditions. Not only packaging materials but also equipment and utensils in contact with food are the major way to fulfill that need. Industrial food preparation, longer distribution lines, demand for long shelf-life, and increasing consumption of packaged small portions require specific properties of the packaging materials. Also, the appearance and advertising of the packaged food, which should persuade the consumer to buy the food, have an effect on the composition of packaging materials. In addition, environmental aspects have an increasing affect. Natural sources may not be infinite, and therefore recycling of materials is a growing market. To manage waste problems, an increase in biodegradable packaging is noticed. However, these new environment-driven changes in food contact materials may not jeopardize the food safety.

With the increasing use of packaging materials, a growing awareness of concerns became evident. Responsible authorities started to make inventories of risk of food contamination. Consumer protection is the driving force for any measure taken by authorities. Thus, migration of substances from packaging materials is restricted in one or another way. Many countries have developed their own system of regulation and as a consequence different approaches exist all over the world. Some countries recognize the regulations of other countries and accept the safety of a packaging material when it complies with such recognized regulations.

Safety of food, in general, and of packaging materials, in particular, rests on three pillars: toxicity of a substance, level of migration of the substance into a food, and level of exposure to that food. Toxicity is usually recognized by standard tests that establish an acceptable or tolerable level of daily exposure. However, establishing the level of exposure is complex and full of uncertainties. To determine exposure, it is

Global Legislation for Food Packaging Materials. Edited by Rinus Rijk and Rob Veraart

ISBN: 978-3-527-31912-1

necessary to know migration into food and consumption of that food. However, migration strongly depends on the type of food, while food consumption depends on individuals and food habits prevalent in a certain region. As a consequence, exposure can be established only as an average derived using some exposure models to cover most of the population.

Most regulatory systems apply conservative approaches to establish exposure and to determine maximum migration levels. Extraction, migration into food simulants, or in real foods may be prescribed. Some countries require restriction on packaging systems while others rely on migration of individual substances. Whatever the system there may be, authorities always have a mechanism of enforcement. Industry voluntarily or on comment put a lot of effort and resources to demonstrate compliance of their materials with the applicable rules.

In this book, we aim to provide information on rules established in a number of countries distributed all over the world. These rules may include legislation, recommendations, or voluntary guidelines. The authors have presented a structure and explanation of the rules applicable in their territory. These rules may include guidelines for new substances or to demonstrate compliance. In other cases, the process of getting an official "approval" may be a legal measure. In some cases, it appears that food packaging is fully incorporated into food law, and sometimes it is difficult to separate rules on food from rules on packaging. Many countries are frequently updating their regulations. It is not easy to keep updated with new developments. However, this book will provide you with sufficient background information, for example, web sites, to allow further search for latest developments. Some countries have just started to draft their regulations on food packaging. For such countries, it is difficult to get the latest information. Another serious problem is that many countries publish their laws only in their national language, and official translations into a more accessible language such as English are not available. We trust that in those situations this book provides the reader with sufficient information to understand the essential requirements of a particular regulation.

Rinus Rijk
Keller and Heckman LLP
Rob Veraart
Keller and Heckman LLP

List of Contributors

Haim H. Alcalay
Apa – Advanced Technologies
11 Tuval Street
Ramat Gan 52522
Israel

Deborah Attwood
Keller and Heckman LLP
1001 G Street
N.W. Suite 500 West
Washington, DC 20001
USA

Joan Sylvain Baughan
Keller and Heckman LLP
1001 G Street
N.W. Suite 500 West
Washington, DC 20001
USA

Sam Bian
Asia Pacific,
SABIC, Innovative Plastics
China

Vincent Dudler
Federal Office of Public Health
Food Safety Division
P.O. Box 3003
Bern
Switzerland

Bente Fabech
Ministry of Food
Agriculture and Fisheries
Danish Veterinary and
Food Administration
Mørkhøj Bygade 19,
2860 Søborg
Denmark

Alexandre Feigenbaum
UMR Emballages
Institut National de la
Recherche Agronomique
Campus du Moulin de la Housse
51086 Reims
France

Per Fjeldal
Norwegian Food Safety Authority,
Head Office
Department of Legislation
Ullevålsveien 76
0454 Oslo
Norway

Jean Gauducheau
Consultant
8 Avenue des Roitelets
44800 Saint Herblain
France

Global Legislation for Food Packaging Materials. Edited by Rinus Rijk and Rob Veraart

ISBN: 978-3-527-31912-1

Pirkko Kostamo
Finnish Food Safety Authority Evira
Mustialankatu 3
00790 Helsinki
Finland

Young Ja Lee
Korea Food and Drug Administration
Centre for Food Evaluation
#194, Tongil-no Eunpyung-gu
Seoul 122-704
Korea

Caroline Li
Asia Pacific Chemical Regulations
Competence Center
Environment Health & Safety
Asia Pacific
BASF
China

Sameer Mehendale
Associate Vice President
Cadbury India Limited
Packaging Development
19, Bhulabhai Desai Road
Mahalaxmi
Mumbai 400026
India

Roger Meuwly
Federal Office of Public Health
Food Safety Division
P.O. Box 3003
Bern
Switzerland

Maria Rosaria Milana
Istituto Superiore di Sanità
Viale Regina Elena 229
00161 Roma
Italy

Yasuji Mori
Toyo Seikan Kaisha, Ltd.
Technology and Packaging
Development Division
Customer Solution System Department
1-1-70 Yako, Tsurumi-ku
Yokohama 230-0001
Japan

Catherine R. Nielsen
Keller and Heckman LLP
1001 G Street
N.W. Suite 500 West
Washington, DC 20001
USA

Peter Oldring
The Valspar (UK) Corporation Ltd.
Packaging Coatings Group
Avenue One, Station Lane, Witney
Oxford OX8 6XZ
UK

Marisa Padula
Institute of Food Technology – ITAL
Packaging Technology Center – CETEA
Av. Brasil, 2880 – Jd. Brasil
!3-070-178 – Campinas SP
Brazil

Wichard Pump
Oberstr. 96
45468 Mülheim a.d. Ruhr
Germany

Rinus Rijk
Advisor
Bendienslaan 18
3431 RA Nieuwegein
The Netherlands

Luigi Rossi
Keller and Heckman LLP
Avenue Louise 523
1050 Brussels
Belgium

Anastase Rulibikiye
Bureau of Chemical Safety
Food Directorate, Health Canada
AL:22038
251 Sir Frederic Banting
Driveway
Ottawa
K1A OL2
Canada

Kristina Salmén
Normpack
Drottning Kristinas väg 61
Stockholm
Sweden

Annette Schäfer
European Commission
Directorate-General Health and
Consumer Protection
B232 04/25
1049 Brussels
Belgium

Robert J. Steele
CSIRO Food Science
11 Julius Avenue
Riverside Corporate Park
Delphi Road
North Ride, NSW 2113
Australia

Sumalee Tangpitayakul
Ministry of Science and Technology
Department of Science Service
Food Packaging Laboratory
Biological Science Program
Rama VI Road
Ratchathewi
Bangkok 10400
Thailand

Paul Tobback
Katholieke Universiteit Leuven
Kapucijnenvoer 35
3000 Leuven
Belgium

Rob Veraart
Keller and Heckman LLP
Avenue Louise 523
1050 Brussels
Belgium

Hae Jung Yoon
Korea Food and Drug Administration
Centre for Food Evaluation
#194, Tongil-no Eunpyung-gu
Seoul 122-704
Korea

1
EU Legislation

Annette Schäfer

1.1
Introduction

The European Union is joining until now 27 sovereign Member States in an alliance that aims, among others, at establishing an internal market and an economic union.

In the area of food safety, measures are being taken to remove trade barriers between the Member States while attaining a high level of health protection. In 2002 [1], the general food law established for the first time in the EU the general principles for food safety covering the whole food chain from farm to fork. Food contact materials play a major role in the whole food chain as they are used in the manufacture of food, such as food producing machinery, they are used to package and enable storage and transport of food, and they are used to consume food, such as tableware. Therefore, legislation is needed to ensure that the materials used to handle or protect food do not become a source of food contamination. The legislation on food contact materials has to be seen in the context of the general food law.

In the area of food contact materials, the first Community legislation was adopted in 1976 laying down the general principles in a Framework Directive [2]. At that time, the Member States had their national legislations on food contact materials and articles, but provisions were divergent and thus were posing a barrier to trade. The adoption of the Framework Directive was the first step in the harmonization of food contact materials legislation. In the meantime, specific Community legislation on certain food contact materials, such as plastics and ceramics, was adopted. As far as most of the specific materials, such as paper or rubber, are concerned, specific rules have not yet been adopted. At the EU level, two types of legislations exist in parallel for food contact materials: harmonized Community legislation adopted by the EU and nonharmonized national legislations adopted by individual Member States.

Therefore, national provisions on specific materials still exist in areas where Community legislation is not yet in place. The rule of mutual recognition applies to this national legislation. Any product lawfully produced and marketed in one Member State must, in principle, be admitted to the market of any other Member State. The only reason a Member State can reject a product is the protection of human health. Even under mutual recognition, a national legislation may foresee that the use of a

Global Legislation for Food Packaging Materials. Edited by Rinus Rijk and Rob Veraart

ISBN: 978-3-527-31912-1

substance lawfully manufactured and/or marketed in another Member State is subject to prior authorization provided certain requirements are fulfilled such as a simplified procedure for including the substance in a national list [3]. In nonharmonized areas, Member States may even adopt new national legislation. This has to be notified to the Commission and must not introduce a new unjustified trade barrier.

1.2 Community Legislation

The Community legislation comprises general rules applicable to all materials and articles laid down in the Framework Regulation (EC) No. 1935/2004 [4] and specific rules only applying to certain materials or certain substances. The two general principles on which legislation on food contact materials is based are the principles of inertness and safety. A general overview is presented in Figure 1.1.

Since 2005, Community legislation can be adopted in the form of a directive, a regulation, or a decision. While a regulation is directly applicable in each Member State, directives have to be transposed into national law with transposition times of up to 18 months. In the past, the 1976 and 1989 Framework Directives required a directive as the legal instrument to adopt the specific implementing measures, but with the new Framework Regulation, regulations have become the preferred implementing measure.

Directives and regulations can be changed and updated by the so-called amendments. These amending directives or regulations include only the changes to the original act but do not repeat the whole text. They follow the same numbering system as the directives or regulations. When a reference is made to a directive or regulation, it is always made to the legal act including its last amendment. Consolidated versions of the directives are usually made available at the Commission web site EURLEX [5]. They consolidate into one text the original directive and the amendments indicating the changes that have been introduced.

The Framework Regulation is adopted by the European Parliament and the Council, while specific directives and regulations are adopted by the Commission after consultation with the Member States in the Standing Committee on the Food Chain and Animal Health. The Commission can adopt only those proposals that gain a qualified majority support of Member States in the Standing Committee, which consists of administrators of the ministries concerned of the Member States.

1.2.1 Horizontal Legislation

1.2.1.1 Framework Regulation

The Framework Regulation (EC) No. 1935/2004 is the basic Community legislation that covers all food contact materials and articles. As a basic framework, it defines food contact materials and articles and then sets the basic requirements for them.

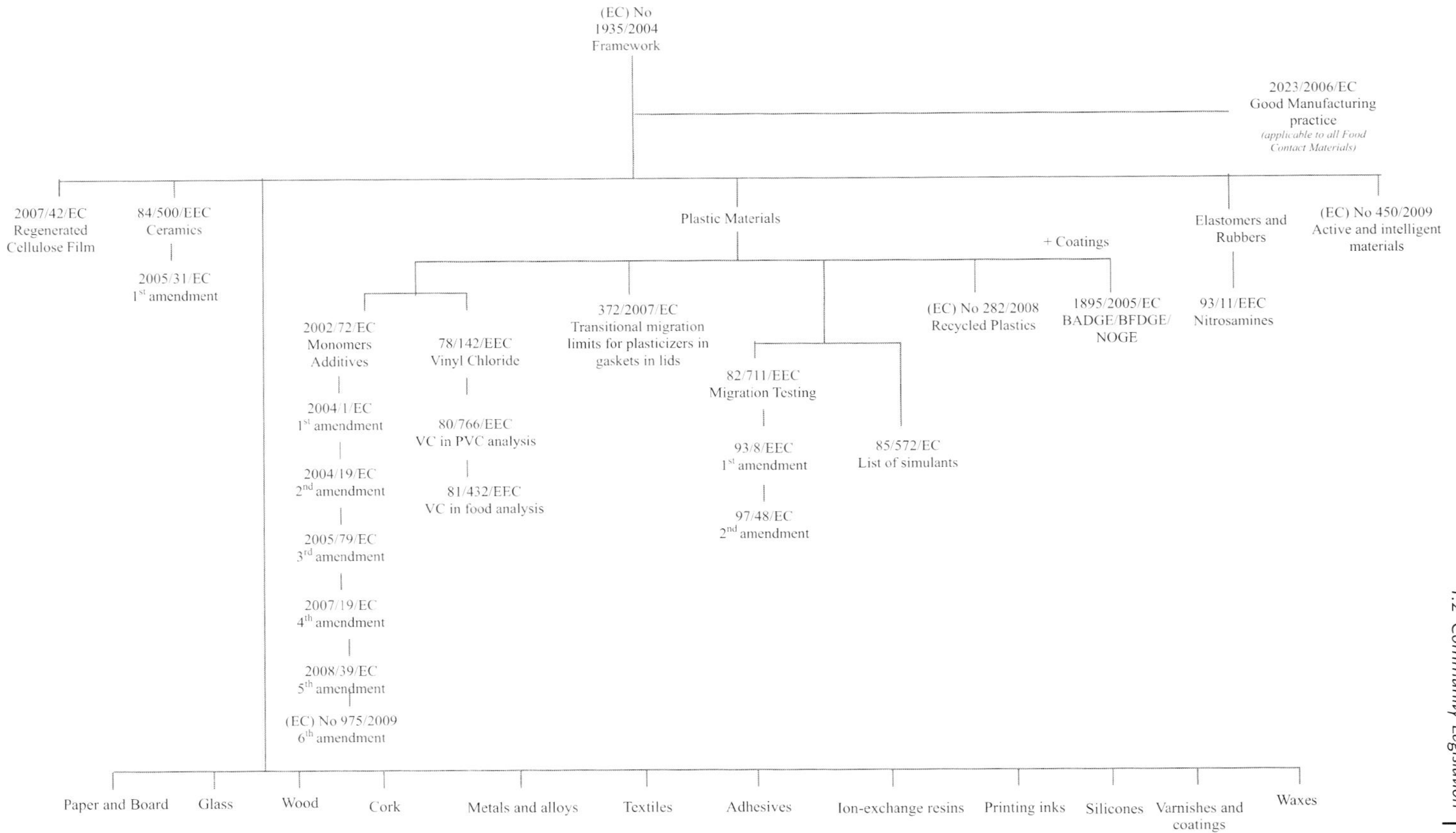

Figure 1.1 Overview of Community legislation (last update October 20, 2009).

Community legislation on food contact materials covers the following products: materials that are already in contact with food such as the packaging of prepackaged food; materials that are intended to come into contact with food, such as cups, dishes, cutlery, and food packaging not yet in use; materials that can be reasonably expected to be brought into contact with food such as table surfaces in food preparation areas or the inner walls and shelves of a refrigerator; and materials that can be reasonably expected to transfer their constituents to food such as a cardboard box around a plastic bag of cereals.

Basic requirements are set to ensure safe food and protect consumer interests. Four basic requirements are set to ensure safe food:

1) food contact materials shall not endanger human health;
2) food contact materials shall not change the composition of the food in an unacceptable way;
3) food contact materials shall not change taste, odor, or texture of the food;
4) food contact materials shall be manufactured according to good manufacturing practice (GMP).

Exemptions from the requirements 2 and 3 are made for active materials and articles (see Section 1.2.3). General rules for good manufacturing practice are laid down in a specific horizontal regulation (see Section 1.2.1.2). Labeling of food contact materials is required to ensure both safety and protection of consumer interests. The consumers, food packer, or converter should be informed on

1) the suitability of the product for food contact (for this purpose, a symbol presenting glass and fork, , can be used or the words "for food contact");
2) the person responsible for manufacturing or placing on the market of the product;
3) instructions for the safe use of the product;
4) means of identification of the product for traceability.

Traceability is a general obligation derived from the general food law to ensure, for example, retrieval of batches in case of need. Different ways of labeling are possible: on the product itself, on accompanying documents, or at the retailer on a sign near the product. The information that is provided on the label shall not mislead the consumer. If the food contact material is also covered by specific Community legislation, the producer has the obligation to declare that the food contact material used in his product conforms to these specific requirements (see Sections 1.2.2 and 1.2.3). The declaration of compliance of a packaging material, for example, has to contain all information on the product that is necessary for food industry to comply with limits set for certain substances in the food (see Section 1.2.2).

The Framework Regulation empowers the European Commission to set requirements for specific materials. These requirements are specifications of the general rules of the Framework Regulation. These specific requirements can be set for certain types of materials, such as plastic or ceramic, or they can cover only the use of certain substances. Specific requirements can comprise authorization of substances used in food contact materials, limits on substances used, authorization of manufacturing processes of certain materials, and rules on labeling and compliance testing. The authorization of substances is divided into a risk assessment procedure, performed by the European Food Safety Authority (EFSA), and a risk management decision by the European Commission. At this moment, this authorization procedure applies only to substances used in plastic food contact materials or in regenerated cellulose films that are regulated by a specific measure. A person interested in the authorization of a substance has to submit through a Member State an application including a dossier for safety evaluation by EFSA. EFSA will evaluate the substance following a conventional risk assessment procedure. After consultation with the Member States, the European Commission will, based on EFSA's safety evaluation and taking into account all other relevant factors, take a decision whether to authorize or not the substance. All authorizations granted are general authorizations. This means everybody can use an authorized substance. In principle, the authorization procedure could be adapted, in agreement with the Member States, to authorize the use of substances, materials, or processes only for the individual applicant. Authorized substances are listed in specific Community legislation.

The Framework Regulation contains a list of materials for which specific legislation may be adopted. This list contains 17 different materials. Only a few are yet covered by specific Community legislation (see Sections 1.2.2 and 1.2.3).

1.2.1.2 Good Manufacturing Practice

One of the four basic requirements of the Framework Regulation is the application of good manufacturing practice for the production of food contact materials. To ensure a harmonized application of GMP throughout the EU and across the different business sectors, the basic principles of good manufacturing practice are detailed in Regulation (EC) No. 2023/2006 [6]. These requirements are in force since August 1, 2008. GMP has to be applied at all stages of production of food contact materials and articles and in all sectors. Excluded are the stages of production of starting substances and raw materials. For example, for plastic production, the GMP requirements start with the plastic manufacturer followed by the converter including the printing process of the packaging material up to the production of the finished article. GMP starts with the selection of suitable raw materials for which specifications need to be set that ensure the production of a safe finished article. This would, for example, include the control of the purity of the chemicals used. The manufacturing operations have to be specified to ensure that the manufacturing process does not render the final material unsuitable for food contact, for example, by generating unsafe reaction or degradation products. The manufacturing process has to be accompanied by a quality assurance system ensuring that the pre-established criteria are adhered to. The quality assurance system also includes premises and equipment as well as the qualification of the personnel. The critical steps in the manufacturing process need to be controlled by an adequate quality control system including the specification of corrective actions in case of failure. All aspects of the GMP need to be adequately documented and the documentation should be available to control authorities.

Imports from third countries should also conform to GMP. For two applications, GMP requirements have been further detailed: for printing of nonfood contact surfaces and for recycling of plastics. For printing of nonfood contact surfaces, adequate care has to be taken to ensure that constituents of the printing ink are not transferred to the food contact side either via diffusion through the material or via set-off due to direct contact between printed and nonprinted surfaces. The responsibility lies with both the printing industry and the downstream users of the printed articles. For plastic recycling, the quality assurance system shall include quality plans, including those for input and recycled plastic characterization, suppliers' qualification, sorting processes, washing processes, deep cleansing processes, and heating processes.

1.2.2
Specific Measures

The Framework Regulation sets the general principles that apply to all materials coming into contact with food. Details for implementing these general rules taking into account the specific risks of the individual material are given in the specific legislative measures.

1.2.2.1 Plastics

Plastic materials and articles were the first materials to be covered by Community harmonization. The harmonization of the sector is not yet finalized; therefore, provisions applicable to plastic materials and articles exist both at Community level and at national level. Not all Member States have national legislation on plastics. An overview can be found in Section 1.3.

General Requirements in the Plastics Directive

The Directive 2002/72/EC [7] including its amendments covers plastic monolayer and multilayer structures that purely consist of plastic. It also covers plastic layers or coatings that form gaskets in lids. A monolayer structure may be a polyethylene (PE) bag, a multilayer structure may be a plastic tray for prepackaged food consisting of different plastic layers, for example, ethylene vinyl alcohol copolymer/polyethylene (EVOH/PE). Multilayers that consist of plastic and other materials such as plastic-covered paperboard, as in beverage cartons, do not fall under the specific Community legislation on plastics. In this case, national legislation applies. Usually, Member States require that each layer has to conform to the requirements set for the respective materials, while the finished article has to comply with the overall requirement of the Framework Regulation. For these multilayers, compliance with the Framework Regulation is interpreted by most Member States as complying with the migration limits laid down in the plastics Directive.

Plastic coatings, adhesives, and epoxy resins are covered only in part by the specific Community legislation on plastics. Usually, they are used on other substrates rather than plastic and thus do not fall under the plastics Directive. In addition, monomers and additives used only in plastic coatings, adhesives, or epoxy resins are not listed in the Community lists. Plastic coatings and adhesives are covered by national legislation in some Member States. Plastic coatings containing certain epoxy derivatives are regulated by Regulation (EC) No. 1895/2005 [8] (Section 1.2.3).

Biobased polymers and biodegradable polymers such as polylactic acid (PLA), polyhydroxybutyric acid (PHB), and polycaprolactone (PCL) or starch-based polymers are covered by the Community legislation on plastics.

The general principles laid down in the Framework Regulation are those of inertness and safety. These principles are interpreted in the specific legislation on plastic food contact materials as follows.

The principle of inertness is translated into an "overall migration limit" (OML). The overall migration comprises the total amount of all substances (except water) transferred from the plastic food contact material to the food. The OML is set at 60 milligrams per kilogram of food (mg/kg food). In addition, it has to be ensured that a substance migrating from the food contact material does not exhibit a technological function in the food (unless it is part of an active material, see Section 1.2.3). This may occur if the substance used in food contact materials is at the same time an authorized food additive, for example, antioxidant or preservative. In this case, the migration limit is defined by the amount of substance that does not exhibit a technological function in the food provided any limit on the amount of the food additive permitted in the food is not exceeded.

The principle of safety is translated into specific authorization of substances that are used for the manufacture of plastic materials after their favorable toxicological evaluation by the European Food Safety Authority. A general authorization for the use of the substances is given. Everybody may use the substances respecting the restrictions and specifications given in the authorization, not only the applicant who provided the data for the evaluation. Authorized substances and their restrictions and specifications are published in Community lists annexed to the plastics Directive 2002/72/EC. The list is regularly updated through amendments to the plastics Directive. If necessary for the safety of a material "specific migration limits" (SML) are laid down. The specific migration is the amount of a single substance that may be transferred from the plastic food contact material to the food. The SML is set individually and is based on the toxicological evaluation of the substance. The tolerable daily intake (TDI) of a substance expressed in mg/kg bodyweight is translated into an SML based on a conventional system. This system assumes that 1 kg of food is consumed daily by a 60 kg person. This 1 kg of food is packaged in a plastic material releasing the substance at the level of the TDI. The SML can vary from nondetectable (allowing analytical tolerances) to several milligrams per kilogram food. In cases where only a limited data set is provided for toxicological evaluation that does not allow TDI to be set, specific migration limits are established as follows. If the substance is shown not to be carcinogenic, mutagenic, or toxic to reproduction based on three *in vitro* tests, a specific migration limit of 50 μg/kg food is established. If, in addition, the substance is not toxic in a 90-day long study and is not likely to accumulate in the human body, a specific migration limit of 5 mg/kg food is established. The migration of a single substance or substances may not exceed the overall migration limit of 60 mg/kg food.

Harmonization of legislation on the substances used in food contact plastics started with monomers as these are reactive substances and thus of primary importance with regard to any potential health risk. Monomers and other starting substances are fully harmonized at Community level. This means that only the monomers listed in the plastics Directive can be used in food contact materials. An exemption exists for plastic coatings, adhesives, and epoxy resins. Monomers that are used only in their manufacture are not listed in the Community list.

In the second step, harmonization of additives used in plastic food contact materials was started. However, this step is not yet finalized. Therefore, additives listed both in the Community legislation and in national legislation can be used in food contact plastics (for national lists, see Section 1.3). Harmonization on additives will be almost finalized by 2010. Until December 31, 2006, all parties interested in Community authorization of additives permitted at national level had to supply EFSA with a valid application for the evaluation of those additives. A provisional list of additives for which a valid application has been received is available on the Commission web site [9]. Additives will be removed from the provisional list if they are being authorized or if a decision is taken not to authorize them or if the applicant did not respect the time limit set by EFSA for submission of additional information. By January 2010, the list of additives annexed to Directive 2002/72/EC will become a positive list. From that date onward, only those additives on that list can be used in the manufacture of food contact plastics. In addition, a substance that is still on the provisional list at that time may continue to be used according to national law until its safety evaluation is finalized by EFSA and a decision on its authorization is taken by the Commission. This Community list on additives contains those additives that are used solely in plastics and those that are used both in plastics and in coatings. However, it does not contain additives used only in plastic coatings, adhesives, and epoxy resins. It also does not include substances used only as polymer production aids. The list does not contain solvents and aids to polymerization, which are not intended to remain in the finished product and colorants.

Impurities, reaction, and degradation products of the authorized substances are usually not evaluated unless listed in restrictions and specifications for the authorized substance. They remain under the responsibility of the producer of the material and article who has to ensure that they do not migrate in quantities that pose a health risk.

Declaration of Compliance

The purchaser of a food contact material should receive an assurance from the manufacturer that the food contact material complies with the applicable legislation. The finished article can be compliant only if throughout the production process requirements of the plastics Directive have been adhered to. Therefore, a declaration of compliance is necessary from the moment a substance, mixture, or plastic is intended for food contact. Each manufacturer has to declare compliance for the manufacturing steps under his responsibility. For example, a producer of a monomer has to ensure that the monomer is authorized and conforms to the specifications relevant to it. The producer of a plastic has to ensure that monomers and additives are authorized and as far as under his responsibility indicate the conditions of use under which migration limits can be complied with. The manufacturer of the final article has to indicate conditions of use under which restrictions and migration limits can be complied with. The information is, in particular, relevant for the so-called dual-use additives, additives that are used both in plastic manufacture and as food additives. The addition of food additives to food is strictly regulated and it has to be ensured that migration from the food contact materials do not violate those rules of food additives legislation. The manufacturer has to maintain documentation substantiating the declaration of compliance. This documentation has to be available to control authorities on demand.

The Functional Barrier Concept

In multilayer materials, a layer can function as a barrier to migration of substances into food. Since June 2008, when such a functional barrier layer is applied to ensure no migration into food takes place, it will no longer be necessary to authorize the substances used in that layer if those substances are not carcinogenic, genotoxic, or toxic for reproduction. However, the use of the functional barrier concept needs to be declared in the declaration of compliance. Adequate information on the nonauthorized substances used and the demonstration of effectiveness of the functional barrier have to be provided to control authorities on request.

Verification of Compliance with Migration Limits

Analysis of migration from food contact materials can be performed according to different protocols. Verification of the migration limit can be performed on the food itself in case it is already in contact with the packaging material. Verification of migration limits may also be performed on food simulants, usually in case of packaging that is not yet in contact with food. The legislation foresees five food simulants representing different possible extraction properties of food (Directive 82/711/EEC [10] and amendments). These five simulants are water for aqueous food, 3% acetic acid for acidic food, 10% ethanol for alcoholic food, olive oil for fatty food excluding dairy products, and 50% ethanol for dairy products. A correlation list is laid down in legislation that indicates which food is represented by which food simulant (Directive 85/572/EEC [11]). Verification can also be performed by extracting the residual amount of a substance from the food contact material. The residual amount can then either be directly compared with the SML or be subject to mathematical migration modeling giving the migration potential of the application. For proof of noncompliance with SML values, only the migration testing into food and food simulants can be accepted. In cases where the substance is not stable in food or food simulant, the value is expressed as residual content per square decimeter of contact surface (QMA).

Migration testing has to be performed under worst foreseeable contact time and temperature for the envisaged application. A long-term storage at room temperature is, for example, represented by storage for 10 days at 40 °C. A correlation table with migration test conditions is laid down in the legislation (Directive 82/711/EEC). Analytical methods for migration testing of overall migration and specific migration have been standardized at European level by the European standardization body CEN. Information on migration testing can be found at the web site of the Community Reference Laboratory (CRL) for food contact materials [12].

Reduction Factors Applicable in Migration Testing

Two types of correction factors have been established to correct the overestimation of real exposure to or real migration into fatty foods. These are, respectively, the fat consumption reduction factor (FRF) and the simulant D reduction factor (DRF).

When migration limits for substances are set, a conventional system is applied to calculate exposure. It is assumed that a 60 kg person will consume 1 kg of packaged food per day. However, a different convention is sometimes necessary under certain circumstances. One such convention arises in the case of lipophilic substances, which readily migrate into fatty foods. The consumption of fat is usually only 200 g or less per day. For these substances, a reduction factor is therefore established for use in compliance testing, taking into account the lower consumption of fat. A list of substances for which this FRF is applicable is annexed to the plastics Directive.

As migration testing into food is not always feasible, migration testing with food simulants is an alternative to test compliance with migration limits. In particular for fat simulants, the extraction power of the fat simulant (simulant D) is often greater than the expected migration into food that it is representing. Therefore, correlation between foods and fat simulant has been established and on this basis a list of correction factors (D reduction factor) has been created and provided in Directive 85/572/EEC. In case migration testing is carried out in simulant D, both FRF and DRF can be combined to a maximum reduction factor of 5.

1.2.2.2 Recycled Plastics

Recycling of plastic materials has come into focus as the sustainability of production and environmental issues have become more important. As plastic is using up oil resources, targets for recycling plastic packaging waste have been set within the EU. Recycled plastics could qualify as a source for the manufacture of food contact materials provided the strict safety requirements for food contact materials are respected. As the plastics Directive regulates only substances used in the manufacture of plastics such as monomers and additives, the rules laid down therein were not regarded as sufficient to ensure a safe use of recycled plastics in food contact materials. Recycling of used PET beverage bottles into new beverage bottles is increasingly becoming common in Member States. Requirements for the use of recycled plastic in contact with food vary between EU Member States from a ban on authorization schemes to no requirements at all. Some Member States, such as the United Kingdom, apply the rules for virgin plastics to recycled plastics. EU harmonization of the rules is necessary to ensure an equal treatment to recycled plastics in all Member States.

Therefore, a new approach was developed for using recycled plastics in plastic food contact materials. Two different processes of plastic recycling can be distinguished: in chemical recycling, plastics are depolymerized into monomers or oligomers that are purified and isolated and then used again as starting substances. Monomers and oligomers derived from such chemical recycling need to respect the same safety and purity criteria as identical authorized monomers or oligomers derived from chemical synthesis. For this type of recycling, the requirements of the plastics Directive are regarded as sufficient to ensure product safety. The second recycling process is the mechanical recycling of plastics in which the plastic is simply melted and subjected to certain purification steps. This type of recycling process is not sufficiently covered by the current rules of the plastics Directive and therefore a specific Regulation (EC) No. 282/2008 [13] has been adopted to ensure that products derived from this process can be safely used in food contact plastics. The regulation foresees the individual authorization of the recycling process at Community level based on the safety evaluation of the recycling process performed by EFSA. Critical points in recycling process are the sourcing of the material that is being recycled and the capacity of the process to reduce contamination. Only those plastics that respect the compositional requirements of the plastics Directive can be used as a source for mechanical recycling. As the recycling processes are unique based on the technology used, individual authorization dedicated to the applicant will be issued. All recycling processes shall be accompanied by an adequate quality assurance system that should be audited by Member States. Both recycled plastic and the materials and articles containing recycled plastics need to be accompanied by a declaration of compliance. A transitional phase of 2 years is foreseen during which applications for already existing or new processes can be submitted. All safe processes from this transitional phase will be authorized at the same time at Community level once they all have been evaluated. From the date of this first authorization, only recycled plastics from authorized processes can be used in plastic materials and articles. Subsequent authorizations will follow the adapted general authorization procedure. Until the date of the first Community authorization, national legislation remains in force. The regulation also covers recycled plastics from third countries. Also, these can be used only if the recycling process is authorized. Requests for authorization have to be addressed to a Member State's contact point. Premises in third countries that use the authorized recycling processes have to be notified to the Commission. A level playing field is established for products from third countries and those originating from within the EU.

1.2.3 Other Materials

Specific Community legislation exists not only for plastics but also for some other materials, namely, ceramics and regenerated cellulose film (RCF, cellophane). For rubber teats and soothers, migration of nitrosamines is regulated. For coated materials, plastics, and adhesives, the substances BADGE, BFDGE, and NOGE are regulated. Specific Community legislation exists for active and intelligent materials, and some general rules for these materials have been laid down in the Framework Regulation.

1.2.3.1 Ceramic Articles

Ceramic articles may pose a risk to the consumer through heavy metals used in glazing and coloring. Substances of major concern in the past have been lead and cadmium. Community legislation (Directive 84/500/EEC [14] amended by Directive 2005/31/EC [15]), therefore, imposes limits for lead and cadmium leaching from ceramic articles into a 4% (v/v) acetic acid solution. Ceramic articles have to be accompanied by a declaration of compliance indicating the manufacturer and importer, if any, as well as the conformity to the limits for lead and cadmium. Rules for migration testing and performance criteria of the analytical method are set in the legislation. For other heavy metals, the general rules of Article 3 of the Framework Regulation apply. Some Member States have national restrictions for some of the other heavy metals and separate limits for migration from the mouth rim of cups and beakers (see Section 1.3).

1.2.3.2 Regenerated Cellulose Film (Cellophane)

At Community level, specific rules for materials and articles made of cellophane exist (Directive 2007/42/EC [16]). Exempted are synthetic casing such as those used for sausages. In these exempted cases, national legislation applies. The legislation contains a positive list of substances that can be used in manufacturing cellophane. The restrictions in the positive list are usually expressed as residual content in the film because migration testing with pure cellophane film into liquid simulant is in general not feasible due to the absorption of water by the film. The positive list does not include dyes, pigments, and adhesives. Substances used for these purposes shall not migrate into food in detectable amounts. From July 29, 2005, the legislation also covers plastic-coated cellophane. For plastic coating, only substances in the list of authorized substances in the plastics Directive (Directive 2002/72/EC as amended) shall be used. The whole film has to comply with overall migration and specific migration limits in the plastics Directive. Analytical methods for compliance testing are published on the web site of the Community Reference Laboratory (http://crl-fcm.jrc.it).

1.2.3.3 Rubber Teats and Soothers

In the 1980s, it became evident that rubber teats and soothers may release carcinogenic nitrosamines, which are reaction and degradation products from accelerators and stabilizers used in the rubber. Legislation contained in Directive 93/11/EEC [17] mandates that nitrosamines and nitrosatable substances that can be transformed into nitrosamines in the stomach shall not be released from teats and soothers in detectable quantities. Methods for analysis are proposed with the detection limit set at 0.01 mg/kg rubber for nitrosamines and 0.1 mg/kg rubber for nitrosatable substances.

1.2.3.4 BADGE, BFDGE, and NOGE in Coated Materials, Plastics, and Adhesives

In the 1990s, high amounts of BADGE (bisphenol A diglycidyl ether) were discovered in fish in oil in tins. The source for the contamination was the coating where BADGE was added as an additive. As the substance contains epoxy groups, it was a suspected carcinogen although it was not considered to be genotoxic, measures were taken to reduce the migration of BADGE from the coating, the plastic, and any adhesive. The measure also covered the replacement products BFDGE (bisphenol F diglycidyl ether) and NOGE (novolac glycidyl ether) that are similar in structure to BADGE. The toxicity of BADGE has now been more thoroughly investigated and studies have clarified that BADGE is not carcinogenic in humans. Toxicity of BFDGE and NOGE is, however, still not clear. The Community legislation takes account of the new toxicological results (Regulation (EC) 1895/2005) and sets a new, higher migration limit for BADGE and its hydrolysis products at 9 mg/kg/food, but for BADGE chlorohydrins it maintains a limit of 1 mg/kg/food. The use of BFDGE and NOGE has been prohibited from January 1, 2005 and March 1, 2003, respectively. Exempted from this ban are heavy-duty coatings in tanks with a capacity greater than 10 000 l and attached tubing. Analytical methods have been developed by CEN.

Although specific to these substances, this legislation has been the first to explicitly set any rules for coatings and adhesives and those plastics that are not within the scope of the rules on food contact plastics. This last point arises from the fact that the legislation covers all plastic materials and articles, not just those within the scope of Directive 2002/72/EC, as amended. A declaration of compliance needs to be issued also for coatings and adhesives with regard to BADGE, BFDGE, and NOGE.

1.2.3.5 Active and Intelligent Materials and Articles

The main function of packaging as regarded in the past was to protect the food from contamination and spoilage and enable the transport of the food. Derived from this concept are the basic principles of food contact materials legislation: packaging should be inert; it should not release substances into food that pose a risk to human health; and it should not release substances into food that change the taste, odor, and composition of the food.

Recent technological developments have made it possible to assign new functions to packaging: it can inform the consumer about the condition of its content and may even interact with the food by releasing or absorbing substances. In view of these additional functions, food contact material legislation was revised in 2004. Two new concepts, apart from inert packaging, have therefore been introduced in the legislation: intelligent food contact materials and active food contact materials. The basic principles of food contact materials have been adjusted in the Framework Regulation to take account of these new features.

Intelligent food contact materials are those that provide the consumer with information on the condition of the packaged food or the atmosphere in the packaging. This information may, for example, indicate storage conditions the food has undergone using time/temperature indicators that turn from green to red when the food has been stored for a certain time at elevated temperatures. Other examples are indicators for oxygen level in the food or for the presence of microorganisms that spoil the food. The general principle of inertness and the requirements that control substance migration continue to apply to this type of food contact material. However, given the extra function of packaging, it has to be ensured that the information provided to the consumer is not misleading. A freshness indicator, for example, should not misleadingly indicate freshness when the food is already spoiled.

Active food contact materials are those that actively change the composition of the food or its surrounding atmosphere. Two functions have been distinguished: those of absorbers and releasers. Absorbers are constructed such that they absorb substances released by the food or from the atmosphere around the packaged food, for example, oxygen scavengers that reduce the oxygen level around and in the food and thus prevent microbiological growth and reduce oxidation of the food. Releasers are the opposite, they release substances into the food to improve the food or its condition, for example, packaging that releases preservatives into the food. The new now permitted characteristic of the packaging allows to add the active substance to the material to be intentionally released into the food. However, traditional packaging that releases its natural constituent into the food such as wooden barrels used in wine and whiskey production are not covered by the definition of active food contact material, neither are materials to which an antimicrobial substance is added to keep the surface of the material free of microbiological growth. The function in this case is exhibited on the material itself and not on the food. Examples are antimicrobials in chopping boards or conveyer belts.

Thus, in contrast to the traditional concept that food contact materials are inert and perform no intended function on the food, active materials may change the composition of the food, for example, by releasing preservatives, and may change the environment around the food by the absorption of oxygen; they may also change the taste of the food, for example, by releasing flavors; and they may change the color of the food by releasing colorants. To take account of this and to ensure the safe application of the material, the principle of inertness was modified in the Framework Regulation. Active materials may release substances into food but only under certain specified conditions. The substance released has to be a substance that is authorized in the context of food legislation [18], for example, an authorized food additive or an authorized flavoring. The substance may be released only into foods in which it is authorized for release by food legislation; for example, sorbic acid may be added to prepacked sliced bread but not to whole bread. The substance may be released only in quantities authorized in food legislation, for example, sorbic acid 2000 mg/kg prepacked sliced bread. The change in the composition, odor, or taste of the food shall not mislead the consumer about the quality of the food; for example, an absorber may not mask food spoilage and a colorant may not mask poor food quality. Information has to be provided to all operators in the food chain and to the consumer to ensure the correct application of and compliance with food legislation. Therefore, strict labeling rules have been established. The producer of the material has to provide to the food packer information on the identity of the substance used and levels released. The food packer has to list the released substance in the list of ingredients. Labeling also has to clearly show when active or intelligent materials are used. Nonedible parts of the packaging, for example, absorbing sachets in food packaging, have to be clearly labeled as nonedible.

In addition to these general requirements laid down in the Framework Regulation, additional rules are laid down in a specific measure; Regulation (EC) No, 450/2009 [19] adopted in 2009.

Specific Measure on Active and Intelligent Materials and Articles

Basic requirements for active and intelligent materials have been set in the Framework Regulation that includes provisions for released active substances that have to comply with the food legislation and labeling rules. However, some issues need to be regulated in more detail. These cover in particular the following:

(A) Safety of substances used in active and intelligent materials
(B) Relation to material specific requirements, for example, on plastic food contact materials
(C) Labeling of parts that can be mistaken for food
(D) Declaration of compliance

(A) Safety of substances used in active and intelligent materials

Substances intended to be released into food with an intentional function in the food According to Regulation (EC) No. 1935/2004, substances released into the food need to be authorized and used in accordance with the applicable food legislation. The specific measure confirms and applies this principle. Specific measure clarifies that the same rules and legislation apply if a substance is added directly to the food or via packaging. A duplication of authorization should be avoided; therefore, no authorization scheme would be necessary for these substances in the context of active packaging. Regulation would remain within food legislation.

The following aspect is covered in the specific measure. If legislation on food provides for a limit in food for the "released active substance," the total quantity of this substance in food should not exceed this limit independent of the source from which it derives (released via packaging or added directly to the food). The released substances should be listed in the declaration of compliance (see point D) and adequate information should be given to the consumer or food packer to be able to comply with food legislation. The released substance needs to be listed in the ingredients list.

Substances that contribute to the active or intelligent function but that are not intended to be released into food and that do not have a function in the food These substances have not yet undergone a safety assessment and they might migrate into the food. Therefore, the specific measure on active and intelligent materials applies the same approach as for plastic materials. These substances should undergo a safety assessment by the European Food Safety Authority and a Community authorization. An authorized substance should then be listed in a Community list (positive list) specifying its identity, function, and, if necessary, conditions or restrictions of use. In certain cases, a combination of substances may be inserted, for example, when the safety assessment is linked to this combination of substances due to their interaction. Once authorized, the substance could be used by all operators provided they comply with the conditions of authorization. Exempted from the Community authorization should be substances that are separated from the food by a functional barrier that reduces migration of the substance to a nondetectable level if the substance is not classified as proved or suspected to be "carcinogenic," "mutagenic," or "toxic to reproduction." As active and intelligent materials are already on the market, it should be provided for a transitional period to set up the Community positive list.

(B) Relation to other material-specific requirements, for example, on plastic food contact materials

The specific measure covers only the component responsible for the active or intelligent function and does not regulate the basic material into which the component is incorporated. This applies not only to ceramics, regenerated cellulose films, and plastics for which specific Community measures exist but also to paper, rubber, metals, and so on that are regulated at the national level. For example, in the case of an active plastic absorber, the plastic material has to be manufactured in accordance with the plastics Directive and the active absorber component would need to be manufactured in accordance with the rules set out in the specific measure. In the particular case of a "releasing active material," if the material-specific measure, for example, the plastic Directive, foresees an overall migration limit, the measured overall migration value should not include the amount of the intentionally released substance.

(C) Labeling of parts that can be mistaken for food

For nonedible parts of active and intelligent materials, in particular, sachets containing substances that can be mistaken for food, the consumer should be informed that they are not for human consumption. The specific measure foresees, for example, labeling with the words "do not eat" and a symbol such as .

(D) Declaration of compliance

All specific measures should require a declaration of compliance. For active and intelligent materials and articles, the declaration of compliance covers the following aspects. The active and intelligent materials shall not mislead the consumer. Therefore, they need to be effective and suitable. Information with regard to their effectiveness and suitability should be included in the declaration of compliance and demonstrated in the supporting documentation. The declaration of compliance should contain adequate information related to the substances for which restrictions are in place. This information shall allow the user of the material to ensure compliance with those restrictions. The declaration of compliance should contain adequate information on the released active substances to allow the user to ensure compliance with the restrictions in the relevant food legislation including the labeling requirements of Directive 2000/13/EC.

1.2.4
Control of Food Contact Materials in the EU

The basic rule in the Community food legislation specifies that only safe food shall be placed on the market (Article 14 General Food Law). Consequently, food contact materials should not transfer their substance into the food in concentrations that can endanger human health (Article 3 Framework Regulation). The main players to ensure the safety with regard to food contact materials are the packaging industry, food industry, competent authorities in the Member States, and the European Commission.

1.2.4.1 Role of the Business Operators: Food Industry and Packaging/Contact Material Industry

Both the food industry and the food contact material industry have a shared responsibility for the material in contact with the food and, as a consequence, for the food itself. In the case of food packaging, the food packer has to ensure that only packaging that is suitable for the food is used and that it conforms to the Community and/or national legislation on food contact materials. The packaging industry has to supply packaging that is suitable for food contact. This means that they have to make sure that substances they use in the food contact material are authorized (if positive lists exists) and/or are not transferred into food in concentrations that pose a danger to human health. They have to confirm this in a declaration of compliance. An intensive dialogue between the two parties is, therefore, essential for the compliance of the legislation to be achieved.

The food business operator has the obligation to withdraw unsafe food from the market and to collaborate with the national control authorities on that (Regulation (EC) No. 178/2002 – General Food Law). The European Commission has published guidelines to help business operators to comply with this obligation [20].

1.2.4.2 Role of the Member States

Member States have the responsibility of enforcing the Community and their own national legislations and must ensure that the legislation requirements are fulfilled by business operators (General Food Law). Inspection and control measures on food contact materials shall be carried out according to the Regulation (EC) No. 882/2004 on official feed and food control (OFFC) [21]. In the OFFC, it is specified that control of the application of the rules on materials and articles in contact with food is within its scope. Member States are required to carry out official controls regularly and with appropriate frequency that should be based on the level of assessed risk. The controls shall include those on materials and substances including those covering food contact materials. They shall equally treat products for EU and the local market, as well as imports and exports. The official control can cover the following actions: monitoring, surveillance, verification, audit, inspection, sampling, and analysis. The text explicitly mentions inspections of materials and articles in contact with food. Member States have to lay down a catalog of sanctions and measures including dissuasive penalties for nonconformity with the food legislation. The measures taken by Member States may include prohibition of placing on the market, order and monitor withdrawal of goods from the market, and recall and destruction. Furthermore, they have the right to detain consignments from third countries.

When Member States take measures that affect other Member States, such as withdrawal from the market when the article originates from, or is distributed to, another Member State, they should inform the Commission and the other Member States via the electronic Rapid Alert System for Feed and Food (RASFF). Other Member States affected by their action are then also able to act.

Member States have to lay down their control activities in multiannual control plans from 2007.

1.2.4.3 Role of the European Commission

The European Commission's Food and Veterinary Office carries out Community controls in Member States and in third countries in order to check their national control systems. On food contact materials, desk studies on the systems in place have been performed in all Member States. An inspection of the national control system on food contact materials has been undertaken in several Member States and in China.

1.2.4.4 Methods for Sampling and Analysis in the Official Control

The OFFC Regulation establishes a hierarchy of methods used for sampling and analysis in applying official controls. First priority is given to methods laid down in Community legislation. If these do not exist, methods according to internationally established rules such as those of CEN or those in national legislation should be applied. In the absence of these methods, other methods fit for the purpose or developed in accordance with scientific protocols shall be used. In the area of food contact materials, analytical methods are laid down in Community legislation for vinyl chloride in PVC and food, lead and cadmium leaching from ceramic ware, nitrosamines and nitrosatable substances in rubber and food, as well as rules for migration testing. The majority of methods in the area are standardized CEN methods covering migration testing procedures (series EN1186) and the analysis of specific migrating substances (series EN13130) [22]. Examples of methods laid down in national legislation are the methods according to §64 of the German Lebensmittel-, Bedarfsgegenstände- und Futtermittelgesetzbuch [23].

Other methods not used in the context of official controls such as those used for in-house control purposes may be single laboratory validated according to internationally accepted protocols (e.g., IUPAC harmonized guidelines). General criteria for the characterization of methods of analysis exist.

In order to achieve uniformity in the application and the performance of laboratories in the official control, a system of Community Reference Laboratory (CRL) and National Reference Laboratories (NRLs) was established. This system became fully functional in 2006 with the laboratory on contact materials at the Joint Research Centre IHCP in Ispra acting as CRL. The CRL provides national control laboratories with harmonized analytical methods and coordinates the Network of NRLs to favor exchange of information, capacity building, and training.

1.3 Specific National Legislation

Community legislation was introduced to harmonize national legislation and to remove trade barriers. A Member State's legislation is based on different principles and this can still be observed in those areas that are not yet covered by Community legislation.

Four main legal systems can be distinguished.

1) **Premarket approval system**: This system was applied by some new member states. All materials and articles had to be approved by a central authority before they could be placed on the market. This system no longer exists.
2) **System of authorized substances and migration limits comparable to the Community system**: This system was applied in the Netherlands (warenwet) and to some extent in France and Italy. It still exists for those specific areas where no Community legislation is yet in place.
3) **System of recommendations and quantities of substances recommended to be used in the finished material or article**: This system is applied in Germany (BfR recommendations).
4) **System of no specific legislation but industry code of practice defining due diligence of the business operators**: This system is applied in the United Kingdom.

Most of the 27 Member States do not have specific national rules, newer EU Member States replaced their national legislation with Community legislation before accession to the EU.

Member States that do not have specific national legislation on food contact materials will sometime refer to other Member State's legislation, such as Dutch warenwet or the German recommendations, when testing for safety compliance. Some Member States may also refer the application of the general safety clause included in the Framework Regulation to the resolutions and policy statements of the Council of Europe. In the area of food contact materials, the Council of Europe can take initiatives in those sectors that are not yet harmonized at Community level.

The Council of Europe is an intergovernmental organization but not an institution forming part of the European Union. The Council of Europe currently has 47 member countries, including all 27 European Union Member States.

Within the Council of Europe, the activities on materials coming into contact with food are delegated to the European Directorate for the quality of medicines and health care. The work of the subcommittee commonly results in resolutions, which act as recommendations to the 18 members of the Partial Agreement, 17 of which are European Union Member States.

Table 1.1 gives an overview of the Member States in which national legislation exists and in which sectors national legislation is applicable. Norway, Iceland, and Liechtenstein, as part of the European Economic Area (EEA), also apply the Community legislation while keeping additional national legislation. Switzerland has adopted regulations corresponding to the Community legislation.

1.4 Future Trends

1.4.1 Plastic

The completion of harmonization of rules for plastic food contact materials and articles is within sight. The near future will bring a codification of the measures on plastics and it is envisaged at the same time to separate explanatory rules on compliance testing into guidelines.

1.4.2 Nanomaterials

In food contact materials, as in other areas, substances may be used in manufacturing materials and articles and added in the form of nanoparticles to increase the functionality of the material. The use of this technology is developing. The challenge for the industry and European authorities is to assess if the migration behavior from the nanomaterial is different from that of traditional materials and whether substances migrating are more reactive and have a different toxicological profile from regular substances. The answers to these questions will determine if specific requirements are necessary for nanomaterials and if there is a need for a specific implementing measure. The EU is developing general strategies on policies on nanomaterials, nanoparticles, and nanotechnology on horizontal level. In the meantime, it should be kept in mind that the safety evaluation of an authorized substance to be used in plastics has not taken into account its toxicological profile in nanoform unless specifically mentioned in the evaluation. According to the Framework Regulation, a business operator using an authorized substance has the obligation to inform the Commission of any new scientific or technical information that might affect the safety assessment of the authorized substance.

Table 1.1 Summary of the national legislation. +: National legislation apply –: No national specific legislation.

Last update	Member States	Other	Adhesives	Ceramics	Glass	Enamel	Metals alloys	Cork	Wood	Textile
07/11/2005	Austria	–	–	+	–	+	–	–	–	–
27/07/2009	Belgium	–	–	–	+	–	+	–	–	–
23/07/2009	Bulgaria	–	–	–	–	–	–	–	–	–
20/08/2009	Cyprus	–	–	–	–	–	–	–	–	–
07/05/2007	Czech Republic	–	–	+	+	+	+	+	+	–
23/07/2009	Denmark	Mandatory registration[a]	–	+ [b]	+	–	–	–	–	–
07/11/2005	Estonia	–	–	–	–	–	–	–	–	–
07/11/2005	Finland	–	–	–	–	–	+	–	–	–
02/09/2009	France	–	–	+	+	+	+	–	+	–
07/11/2005	Germany	–	+ [c]	+ [d]	+ [d]	+ [d]	–	–	–	–
07/11/2005	Greece	–	–	–	–	–	+	–	–	–
30/07/2009	Hungary	–	–	+	–	+	+ [e]	–	–	–
23/07/2009	Ireland	–	–	–	–	–	–	–	–	–
15/09/2009	Italy	–	–	+	+	+ [f]	+ [g]	–	–	–
23/07/2009	Latvia	–	–	–	–	–	–	–	–	–
14/08/2009	Lithuania	–	–	–	–	–	+	–	–	–
07/11/2005	Luxembourg	–	–	–	–	–	–	–	–	–
07/11/2005	Malta	–	–	–	–	–	–	–	–	–
23/07/2009	Netherlands	–	–	+	+	+	+	+	+	+
07/11/2005	Poland	–	–	+ [h]	–	+ [h]	–	–	–	–
30/07/2009	Portugal	–	–	–	+ [h]	+ [h]	–	–	–	–
15/09/2009	Romania	Yes: colours[e]	–	–	–	–	–	–	–	–
23/07/2009	Slovakia	–	–	–	+	–	+	+	+	+
04/08/2009	Slovenia	Yes: colours[i]	+	–	+	+	+	–	+	+
20/08/2009	Spain	Yes[j]	–	–	–	–	–	–	–	–

(Continued)

Table 1.1 (*Continued*)

Last update	Member States	Other	Adhesives	Ceramics	Glass	Enamel	Metals alloys	Cork	Wood	Textile
11/08/2009	Sweden	–	–	–	–	–	+	–	–	–
23/07/2009	UK	–	–	–	–	–	+ [k)]	–	–	–
07/08/2009	Norway	Mandatory registration [n)]	–	+	+	+	+	–	–	–

Last update	Member States	Paper board	RCF	Plastics	Varnish coating	Printing inks	Silicone	Wax	Rubber	Ion-exchange resin
07/11/2005	Austria	–	–	–	–	–	–	–	–	–
23/07/2009	Belgium	+ [o)]	–	–	+ [o)]	–	–	–	–	–
02/10/2009	Bulgaria	–	–	–	–	–	–	–	–	–
20/08/2009	Cyprus	–	–	–	–	–	–	–	–	–
07/05/2007	Czech Republic	+	–	–	+	+	+	–	+	–
23/07/2009	Denmark	–	–	–	–	–	–	–	–	–
07/11/2005	Estonia	–	–	–	–	–	–	–	–	–
07/11/2005	Finland	+	–	–	–	–	–	–	–	–
02/09/2009	France	+	–	+	+	–	–	–	+	–
07/11/2005	Germany	+ [c)]	–	+ [c)]	–	–	+ [c)]	+ [c)]	+ [c)]	–
07/11/2005	Greece	+	–	+	+	–	–	–	–	–
30/07/2009	Hungary	–	–	+	–	–	+ [l)]	–	+ [l)]	–
23/07/2009	Ireland	–	–	–	–	–	–	–	–	–
15/09/2009	Italy	+	+	+	+	–	+	–	+	–
23/07/2009	Latvia	+ [m)]	–	–	–	–	–	–	–	–
14/08/2009	Lithuania	+	–	–	–	–	–	–	–	–
07/11/2005	Luxembourg	–	–	–	–	–	–	–	–	–
07/11/2005	Malta	–	–	–	–	–	–	–	–	–
23/07/2009	Netherlands	+	+	+	+	+	+	+	+	–
07/11/2005	Poland	+ [h)]	–	–	–	–	–	–	–	–
30/07/2009	Portugal	–	–	–	–	–	–	–	–	–

15/09/2009	Romania	–	–	–	–	+ [e)]	–	–	+ [e)]	–
23/07/2009	Slovakia	+	–	–	+	–	–	–	+	–
04/08/2009	Slovenia	+	–	–	+	–	–	–	+	–
20/08/2009	Spain	–	–	+	–	–	–	–	–	–
11/08/2009	Sweden	–	–	–	–	–	–	–	–	–
23/07/2009	UK	–	–	–	–	–	–	–	–	–
07/08/2009	Norway	–	–	–	–	–	–	–	–	–

a) Mandatory registration for producers and importers of: all materials covered by the GMP regulation and the reference to Regulation (EC) No 1935/2004, Annex I.
b) Also for glass and ceramic products.
c) BfR recommendation.
d) DIN standard.
e) National legislation and standards.
f) Limitation on lead.
g) Specific measures for: stainless steel, tin free steel, tin containers.
h) National standards.
i) Rules on the requirements concerning the hygiene suitability of consumer goods.
j) Polymeric materials legislation, Prohibition of regenerated polymeric material and plastic and sanitary Register for Industries of substances and food contact material Industries.
k) Old legislation from 1972 on cooking utensils.
l) For teats.
m) Paper and cardboard materials and articles can not release more than 0,5 mg cadmium from a kg of paper and not more than 3 mg of lead from a kg of paper.
n) Mandatory registration for all producers, importers and wholesalers of food contact materials.
o) Resolutions (adoption pending).

1.4.3
Risk Assessment

For the majority of materials and articles, specific Community legislation is not yet in place. In the area of plastic materials and articles, monomers and additives are toxicologically evaluated, but possible impurities, and reaction and degradation products are not taken into consideration in the authorization unless they have been evaluated in the risk assessment. Therefore, it is the manufacturer's responsibility to assess and ensure the safety of such substances that migrate from their products. To ensure the safety of the product, the manufacturer should apply scientifically based risk analysis including exposure assessment in those instances where an established migrant into the food is not specifically regulated in law. At this moment, an EU-wide research project is generating data on exposure to food contact materials and is exploring the feasibility of refined exposure assessment in the legislation of food contact materials.

1.4.4
Other Materials

For materials not yet harmonized at Community level, the Council of Europe resolutions could be taken as a basis for discussion on new rules. Paper and board, and coatings and adhesives are the sectors most likely to follow.

References

1 General Food Law: Regulation (EC) No. 178/2002 of the European Parliament and of the Council of 28 January 2002, laying down the general principles and requirements of food law, establishing the European Food Safety Authority, and laying down procedures in matters of food safety. OJ L 31, 1.2.2002, p. 1.

2 Council Directive 76/893/EEC of 23 November 1976 on the approximation of laws of the member states relating to materials and articles in contact with foodstuffs. OJ L 340, 9.12.1976, p. 19.

3 Judgment of the European Court, 5 February 2004, in Case C-24/00.

4 Regulation (EC) No. 1935/2004 of the European Parliament and of the Council of 27.10.2004 on materials and articles intended to come into contact with food and repealing Directives 80/590/EEC and 89/109/EEC. OJ L338, 13.11.2004, p. 4.

5 http://eur-lex.europa.eu.

6 Commission Regulation (EC) 2023/2006 of 22 December 2006 on good manufacturing practice for materials and articles intended to come into contact with food. OJ L 384, 29.12.2006, p. 75.

7 Commission Directive 2002/72/EC of 6 August 2002 related to plastic materials and articles intended to come into contact with foodstuffs. Corrigendum OJ L39, 13.02.2003, p. 2.

8 Commission Regulation (EC) No. 1895/2005 of 18 November 2005 on the restriction of the use of certain epoxy derivatives in materials and articles intended to come into contact with food. OJ L 302, 19.11.2005, p. 28.

9 Provisional list of additives available at the following web site: http://ec.europa.eu/food/food/chemicalsafety/foodcontact/documents_en.htm.

10 Council Directive 82/711/EEC of 18 October 1982 laying down the basic rules for testing migration of the constituents of plastic materials and articles intended to come into contact with foodstuffs. OJ L297, 23.10.1982, p. 26.
11 Council Directive 85/572/EEC of 19 December 1985 framing the list of simulants to be used for testing migration of the constituents of plastic materials and articles intended to come into contact with foodstuffs. OJ L372, 31.12.1985.
12 CRL FCM web site: http://crl-fcm.jrc.it/.
13 Commission Regulation (EC) 282/2008 of 27 March 2008 on recycled plastic materials and articles intended to come into contact with food and amending Regulation (EC) No. 2023/2006. OJ L 86, 28.3.2008, p. 9.
14 Council Directive 84/500/EEC of 15 October 1984 on the approximation of the laws of the member states related to ceramic articles intended to come into contact with foodstuffs. OJ L 277, 20.10.1984, p. 12.
15 Commission Directive 2005/31/EC of 29 April 2005 amending Council Directive 84/500/EEC with regard to a declaration of compliance and performance criteria of the analytical method for ceramic articles intended to come into contact with foodstuffs. OJ L 110, 30.4.2005, p. 36.
16 Commission Directive 2007/42/EC of 29 June 2007 concerning the materials and articles made of regenerated cellulose film intended to come into contact with foodstuffs. OJ L 172, 30.6.2007, p. 71. (This text codifies Directives 93/10/EEC, 93/111/EC, and 2004/14/EC.)
17 Commission Directive 93/11/EEC of 15 March 1993 concerning the release of the *N*-nitrosamines and *N*-nitrosatable substances from elastomer and rubber teats and soothers. OJ L93, 17.4.1993, p. 37.
18 Council Directive 89/107/EEC of 21 December 1988 on the approximation of the laws of the member states concerning food additives authorized for use in foodstuffs intended for human consumption. OJ L40, 11.21989, p. 27, Flavouring Directive, nutritional fortification, ...
19 Commission Regulation (EC) No 450/2009 of 29 May 2009 on active and intelligent materials and articles intended to come into contact with food of L135, 30.5.2009, p. 3–11.
20 http://europa.eu.int/comm/food/food/foodlaw/guidance/index_en.htm.
21 Regulation (EC) No. 882/2004 of the European Parliament and of the Council of 24 April 2004 on official controls exercised to ensure compliance with the feed and food law, animal health, and animal welfare. Corrigendum OJ L191, 28.5.2004 p. 1.
22 www.cen.eu/CENNORM/Sectors/SECTORS/food.
23 Lebensmittel-, Bedarfsgegenstände- und Futtermittelgesetzbuch. Ausfertigungsdatum: 1 September 2005. Verkündungsfundstelle: BGBl I 2005, 2618 (3007), http://bundesrecht.juris.de/bundesrecht/lfgb_2005/index.html, http://www.methodensammlung-lmbg.de/. European web sites: http://ec.europa.eu/food/food/chemicalsafety/foodcontact/index.en.htm.

2
Petitioning Requirements and Safety Assessment in Europe

Paul Tobback and Rinus Rijk

2.1
Introduction

In Europe, the safety assessment of food contact materials lies within the remit of the European Food Safety Authority (EFSA).

Following a series of food crises (e.g., BSE, dioxins) in Europe in the late 1990s, the EFSA was created in January 2002, as an independent source of scientific advice and communication on risks associated with the food chain.

The EFSA was legally established by the European Parliament and Council Regulation (EC) No. 178/2002, adopted on January 28, 2002. The regulation laid down the basic principles and requirements of food law. It also stipulated that EFSA should be an independent scientific source of advice, information, and risk communication in the areas of food and feed safety.

In the European food safety system, risk assessment is done independent of risk management. As a risk assessor, the EFSA produces scientific opinions and advice to provide a sound foundation for European policies and legislation and to support the European Commission, European Parliament, and EU Member States in taking effective risk management decisions.

The EFSA has its own legal personality and is independent of the Community institutions. It is managed by an *Executive Director*, who is answerable to a *Management Board*.

An *Advisory Forum* connects the EFSA with the national food safety authorities of all 27 EU Member States. The members of the Forum represent the respective national bodies responsible for risk assessment in the EU. The Forum is at the heart of EFSA's collaborative approach to working with the EU Member States. Through the Forum, the EFSA and the Member States join forces in addressing European risk assessment and risk communication issues. The Forum is used to advise the EFSA on scientific matters, its work program and priorities, and to address emerging risk issues.

To keep the EFSA and the Advisory Forum members informed of developments with regard to risk assessments and science in their countries as well as on communications within the areas of the Advisory Forum's responsibility, a network

Global Legislation for Food Packaging Materials. Edited by Rinus Rijk and Rob Veraart

ISBN: 978-3-527-31912-1

of "National Focal Points" has been established. These concern the national networks of risk managers, national authorities, research institutes, consumers, and other stakeholders in the field of risk assessment within the remit of the EFSA.

The network of Focal Points is to be responsible for the organization and coordination of risk assessment institutes of the Member States and are closely involved in the preparation and implementation of EFSA and national authorities' work programs.

The EFSA's remit covers food and feed safety, nutrition, animal health and welfare (AHAW), plant protection, and plant health (PLH). In all these fields, the Authority is committed to provide objective and independent science-based advice and communication on the basis of up-to-date scientific information and knowledge.

Requests for scientific assessments are received from the European Commission, the European Parliament, and EU Member States. The EFSA also undertakes scientific work on its own initiative (self-tasking).

The EFSA provides independent scientific advice through its *Scientific Committee* (SC), 10 *Scientific Panels,* and 6 *Supporting Units*. An overview of these bodies and their remit is summarized in Tables 2.1 and 2.2.

A key element in the EFSA's mandate is to communicate on risks associated with the food chain. This communication activity aims at providing appropriate, consistent, accurate, and timely communications on food safety issues to all stakeholders and to the public at large.

The EFSA also has an important role in collecting and analyzing scientific data to ensure the European risk assessment is supported by the most complete scientific information available. In accordance with its Founding Regulation, the EFSA is legally obliged to publish on its web site (http://www.efsa.europa.eu/EFSA/efsa_locale-1178620753812_home.htm) outcomes of its scientific work. All of the EFSA's activities are guided by a set of core values: excellence in science, independence, transparency, and responsiveness.

2.2
EFSA and its Role in Safety Evaluation of Food Contact Materials

The EFSA CEF Panel [9] has in its remit the safety evaluation of substances used in the production of materials and articles to come into contact with food.

As stated in the "European Parliament and Council Regulation (EC) No. 1935/2004 of 27 October 2004 on materials and articles intended to come into contact with food" a substance shall be authorized only for use in food contact materials if it is sufficiently demonstrated that it does not present a risk to human health. Therefore, a favorable opinion of the EFSA is needed.

To obtain an EFSA opinion on a new substance, which later may be inserted into an EU directive, regulation, or decision, a petition has to be submitted to the competent authority of a Member State [5]. For substances originating from countries other than the Member States, a petition has to be submitted to the competent authority of any Member State of choice.

Table 2.1 Scientific bodies of the EFSA and their remit.

- Scientific Committee (SC)
  The SC deals with scientific advice in the area of new and harmonized approaches for risk assessment of food and feed. It also provides strategic advice to the executive director
- Panel on food additives and nutrient sources added to food (ANS)
  The ANS Panel deals with questions of safety in the use of food additives, nutrient sources, and other substances deliberately added to food, excluding flavorings and enzymes
- Panel on food contact materials, enzymes flavorings and processing aids (CEF)
  The CEF Panel deals with questions on the safety of the use of materials in contact with food, enzymes, flavorings and processing aids, and also with questions related to the safety of processes
- Panel on animal health and welfare (AHAW)
  The AHAW Panel deals with animal health and welfare issues
- Panel on biological hazards (BIOHAZ)
  The BIOHAZ Panel deals with biological hazards in relations to food safety and food-borne diseases
- Panel on contaminants in the food chain (CONTAM)
  The CONTAM Panel deals with contaminants in the food chain
- Panel on additives and products or substances used in animal feed (FEEDAP)
  The FEEDAP Panel deals with additives and products or substances used in animal feed
- Panel on genetically modified organisms (GMO)
  The GMO Panel deals with genetically modified organisms and genetically modified food and feed
- Panel on dietetic products, nutrition, and allergies (NDA)
  The NDA Panel deals with questions related to dietetic products, nutrition and food allergies as well as associated subjects such as novel foods
- Panel on plant protection products and their residues (PPR)
  The PPR Panel deals with plant protection products (commonly known as pesticides) and their residues
- Panel on plant health (PLH)
  The PLH Panel deals with organisms posing a risk to plant health. These include both plant pests that threaten crop production and species, and thus biodiversity

An overview of substances evaluated in the past, by the predecessor of the EFSA, the SCF (Scientific Committee on Food) and their SCF listing may be found in the Synoptic Document. Substances evaluated by the EFSA are found on the EFSA website.[1]

Within the EFSA, the Working Group on Food Contact Materials (FCM WG) carries out the initial safety assessment of FCM substances. This working group, created by the CEF Panel, is composed of independent experts (chemists, toxicologists, experts in dietary exposure, microbiologists, and food technologists), selected by the EFSA on the basis of their expertise, integrity, and independence.

A representative of the European Commission (DG SANCO) functions as "observer" in the FCM WG meetings.

1) Register of opinions and statements in the area of food contact materials: http://www.efsa.europa.eu/EFSA/ScientificPanels/efsa_locale-1178620753812_CEF.htm.

Table 2.2 Supporting units of the EFSA and their remit.

• Pesticide Risk Assessment Peer Review Unit (PRAPeR) PRAPeR is responsible for the EU peer review of active substances used in plant protection products
• Animal Disease Transmissible to Humans (Zoonoses) Unit The unit analyzes and reports data of zoonoses, antimicrobial resistance, microbiological contaminants and food-borne outbreaks
• Scientific Cooperation Unit The unit fosters scientific collaboration, projects, and the exchange of scientific information between EFSA and national food safety agencies of EU Member States
• Data Collection and Exposure Unit (DATEX) DATEX collects, collates, and analyzes data on food consumption and chemical occurrence in food and feed for exposure assessments at European level
• Emerging Risks Unit (EMRISK) EMRISK establishes procedures to monitor, collect, and analyze information and data to identify emerging risks in the field of food and feed safety with a view to their prevention
• Assessment Methodology Unit (AMU) AMU provides technical support in the field of statistics, modeling, data management and risk assessment

To exclude any possible conflict of interest of the FCM WG members with their work in the panel, the EFSA requires that the selected experts submit a yearly "Declaration of Interest" (DoI) in which they list their involvement in activities related to FCM matters outside the EFSA. These declarations are published on the EFSA web site

2.3 Data Requirement on a Substance for its Safety Assessment by the EFSA

2.3.1 Introduction

The general problem arising from the use of food contact materials derives from the capability of chemical substances to migrate into the food in contact. Therefore, to protect the consumer, an assessment of the potential hazards from oral exposure to those migrants into the food must be made.

To establish the safety from ingestion of migrating substances, both toxicological data indicating their potential hazard and data on the likely human exposure to these substances need to be combined. However, for most substances used in food contact materials, human exposure data are not readily available. Therefore, data from studies on migration into food or food simulants may be used.

When considering human exposure, in the safety assessment by the EFSA the conventional assumption is made that a person with a body weight of 60 kg may consume daily up to 1 kg of food in contact with 6 dm^2 of the relevant food contact material.

Guidelines on the submission of a dossier on a substance to be used in food contact materials were developed to guide the petitioner on the scope of the data required for the safety assessment of the substances. The "Note for Guidance" includes a number of chapters. Chapter I contains administrative guidance including, for example, model letters to be used when a dossier is sent through the national authority to the EFSA. A list of national authorities is included. In Chapter II, the guidelines established in 2001 by the SCF are given. The SCF Guidelines are very general and are considered to cover any situation. Chapter III has been drafted by the EFSA Working Group on food contact materials and contains detailed requirements and clarifications, including examples, on the intended information required in various items. This chapter may be changed depending on evolving scientific developments. To assure the availability of the latest version, the document should be downloaded at the moment of use. Chapter IV is drafted by the Commission services and guidelines are given on general issues related to migration protocols as they are established in various EU directives (see Chapter II).

The quantity of toxicity data to be supplied by a petitioner depends on the extent of the likely migration into food.

As a general principle in the safety assessment, the greater the exposure through migration, the more toxicological information will be required in the submission for acceptance of a food contact substance. The rules are as follows:

1) In case of high migration (i.e., 5–60 mg/kg food), an extensive data set is needed to establish the safety.
2) In case of migration between 0.05 and 5 mg/kg food, a reduced data set may suffice.
3) In case of low migration (i.e., <0.05 mg/kg food), only a limited data set is needed.

2.3.2 Data to be Supplied within a Submission

The submission of a petition for a substance to be used in materials and articles intended to come in contact with food need to contain, in view of its safety evaluation, three sets of data:

1) Nontoxicity data, including
   (a) general data on the identity of the substance,
   (b) specific data related to the chemical/physicochemical properties of the substance,
   (c) information on the intended use of the substance,
   (d) information on the authorization in various regulations,
   (e) data on the migration of the substance, including overall migration (OM) and quantification and identification of migrating substances,
   (f) data on the residual content of a substance in the finished article.
2) Microbiological properties (only in case the substance has antimicrobial properties, i.e., biocides)
3) Toxicological data

2.3.2.1 General Information on the Substance

To identify the substance under evaluation, the petitioner is requested to provide general information. At first the petitioner should classify the substances for which he would like to obtain authorization. There are four classes of substances:

- **Individual substances**: This group comprises relatively simple substances with a unique chemical structure, for example, styrene or 2,6-di-*tert*-butyl-*p*-cresol.
- **Defined mixtures**: Defined mixtures are related only to process mixtures. Synthetic mixtures obtained by mixing two individual substances are excluded from this group because they are evaluated as individual substances. Typical examples are mixtures of isomers manufactured using a repeatable process and which are always present in the same ratio.
- **Nondefined mixtures**: This group comprises substances or mixtures of substances that may vary in composition from batch to batch but within a certain range. Substances of natural source such as vegetable oils are typical examples of this group of substances, but substances obtained by epoxidation or ethoxilation may also be found in this category.
- **Polymeric additives**: A substance is considered a *polymeric additive* when it contains repeating units obtained through polymerization. It may include a polymer and/or prepolymer and/or oligomers. A polymeric additive may be added to plastics in order to achieve a technical effect but it is not suitable for the manufacture of finished materials and articles. It also comprises polymeric substances that are added to the medium in which polymerization occurs. For the evaluation of the polymeric additives, the fraction of constituents with molecular weight below 1000 g/mol is decisive for the migration and toxicity tests to be performed. In addition, the appearance of the monomers, used for the manufacture of the polymeric additive, in a list of evaluated substances is essential. If the monomers are not evaluated before, then data on the monomers are also most likely required. Specific guidelines on polymeric additives are provided in Annex 2 of Chapter III of the Note for Guidance [10].
- Substances with a polymeric structure, composed of authorized monomers, and which are used as monomers, are not included in this group as they are, in principle, not subject to the evaluation process.

After making a decision on the group of substances, the petitioner is requested to provide the following general information:

- The identity of the substance, its chemical name, synonym(s), trade name(s), CAS Nr, molecular and structural formula, molecular weight, spectroscopic data (e.g., FTIR, UV, NMR, and/or MS).
- Manufacturing details, that is, the production process, including starting substances, production control, and reproducibility of the process.
- The percent purity of the substance and how the purity was established.
- The impurities (percentage), their identity and typical percentage range, their origin (e.g., derived from the starting substance, side reaction products and

degradation products), the individual levels of impurity, the analytical methods to determine the impurities and supporting documentation (e.g., chromatograms). If a potential safety concern about the impurities exists, migration and/or toxicity data on these impurities might be requested.
- The specifications of the substance (e.g., level of purity or impurity, type of polymer to be used, molecular weight range). The specifications may be included in a future directive. Therefore, the specifications should be substance specific, and usually they do not refer to production control parameters. Providing the specifications may avoid the need for establishing specific migration (SM) restrictions. Typical examples are petroleum hydrocarbon resins (Ref. no. 72081/10), carbon black (Ref. no. 42080), or petroleum-based waxes (Ref. no. 95859). In other cases, a maximum use limit may be established, for example, activated charcoal (Ref. no. 43480).

The same information as listed above should be submitted for all groups of substances (individual, defined, nondefined or polymeric substances).

In case the substance is considered a *polymeric additive*, a number of additional data are required, namely, the structure of the polymeric additive, the weight averaged molecular mass (Mw) and the number average molecular mass (Mn), the molecular mass range and its distribution curve, the percent level of the constituents with molecular weight below 1000 g/mol, the viscosity, melt flow index, density (mg/kg) and the level of residual monomers. In case the monomer(s) are not evaluated in a risk assessment, additional information, most likely, will be needed on the monomer itself.

2.3.2.2 Information on Physical and Chemical Properties of the Substance

The petitioner is requested to provide information on the following:

- **The physical properties of the substance**: The melting point (°C) and the boiling point (°C), the decomposition temperature (°C), the solubility (g/l) in organic solvents or in food simulants, the octanol/water partition (log Po/w), and any other relevant information related to lipophilicity of the substance.
- Information on the solubility is considered useful to understand the behavior of the substance. However, in case the migration in olive oil has been replaced with the substitute simulants *iso*-octane and/or 95% ethanol, the solubility in these simulants is essential to establish the suitability of the substitute simulants and the reliability of the migration values. Solubility in olive oil should match with at least one of the substitute simulants. If the solubility in 95% ethanol is significantly less than in olive oil, then 95% ethanol should be considered not appropriate for the determination of migration. On the other hand, if the solubility in 95% ethanol is significantly higher than in olive oil, then this simulant may be used, but it should be considered that excessive high migration values are obtained that may have further consequences in the request for toxicity data.
- The log Po/w is requested only when migration exceeds 0.05 mg/kg food. The log Po/w, when higher than 3, is used as an indication for the potential accumulation

of the substance in the human body. In that case, additional toxicity data may be required. The log Po/w may, however, also be used to claim the so-called FRF (food consumption reduction factor), which is a correction for the maximum amount of fat consumed daily and which could result in a reduction factor of 5 for the migration values obtained with olive oil. Special conditions are outlined in the Directive 2002/72/EEC.

The petitioner is requested to provide chemical information on the following:

- The nature of the substance, that is, whether it is "acidic," "basic," or "neutral" and its reactivity.
- The stability of the substance in the polymer toward light, heat, moisture, air, ionizing radiation, oxidative treatment and so on. For additives, a TGA (thermographic analysis) curve should be provided. Decomposition temperature should be less than the maximum use temperature as indicated in Item 3.3. Otherwise, an explanation should be given why the additive can still be used. In some cases, where the maximum use temperature exceeds the decomposition temperature, additional attention should be given to any decomposition product.
- Hydrolysis data, since they may simplify the petition in case chemicals that would already have been evaluated are formed in high yield in body fluid simulants. Experiments may be performed according to the Note for Guidance Annex 1 to Chapter III. Details and results of hydrolysis tests should be provided.
- Intentional decomposition or transformation of the substance during the manufacturing of a food contact material or article. If there is concern about decomposition products, migration and/or toxicity data on these products may be requested and specifications or restrictions for these products may be set. Typical examples are antioxidants that decompose intentionally to protect the polymer or an HCl scavenger that scavenges the HCl set free during heat treatment of PVC. In principle, in Item 5.3 of the Petitioner Summary Data Sheet (P-SDS),[2)] the migration of such substance should be identified and quantified.
- Where relevant, information should be provided on unintentional decomposition or transformation products of the pure substance or on products formed in the material during the manufacture of a final article, for example, oxidation, or during various treatments likely to be applied to the finished material or article (e.g., ionizing treatments).
- Any reaction of the substance with food substances. This item is important for making decisions on the type of restriction to be established. A typical example is the reaction of primary amines with oil constituents. This will result in low recovery values of substances added to food simulant and stored under the same conditions as the migration experiments are performed. The EFSA will advise the EU Commission to set an SML (specific migration limit) in food or food simulant (mg/kg food), a QM (maximum permitted quantity) of the "residual" substance in the material or article (mg/6 dm^2) or a QMA (maximum permitted quantity) of

2) A model of a P-SDS is found in Annex 6 to Chapter III on the following web site: http://www.efsa.europa.eu/EFSA/Scientific_Document/afc_noteforguidancefcm_en1,0.pdf.

substance in the finished material or article (mg/6 dm^2). A QMA restriction assumes that all residual substance will migrate into the food. It is considered a worst-case situation, but unless the migration of the substance can be determined by other means, this is the only way to efficiently protect the consumer.

2.3.2.3 Information on the Intended Application of the Substance

The petitioner is requested to provide information on the following:

- What type of polymers the substance is intended to be used in and/or in what type of food contact material (e.g., all kinds of polyolefins, ABS used for the manufacture of household machines, only in PET beverage bottles). This information may be important for estimating the real exposure since the indication of a very restricted or a very broad field of application will influence the final authorization and the restrictions on migration that will be set for the substance. If a very restricted use is claimed and if the reason for this is explained, the substances for which a petition is submitted may be authorized only for a very restricted use or for very restricted conditions of use. In addition, this information will be used to establish a worst-case material used in the migration experiments.
- The technological function of the substance in the production process or in the finished product (e.g., whether the substance is used as a monomer or a comonomer in the production of a particular polymer or whether it is used as an antioxidant, antistatic agent or preservative, etc.). In addition, relevant information to demonstrate the functionality of the substance in the final product needs to be given.
- The maximum process temperature (°C) in the manufacturing process of the polymer and of the final food contact material. If one of these temperatures is higher than the decomposition product, special attention should be given to the migration of possible decomposition products.
- The maximum percentage (expressed on a dry weight basis) of the substance used in the formulation and/or the final food contact material must be given together with the maximum percentage to achieve its technological function in practice. This information is necessary because, for example, typically in the case of additives, the maximum percentage will influence the migration of the substance. Selection of a relevant sample for migration testing will be checked with information provided in this section. It should be assured that always a worst-case sample is selected.
- The conditions of contact of the finished product or article with the food, in practice (i.e., whether it is used only in a particular foodstuff or in all types of foods) need to be given.
- In addition, information on the approximate time and temperature of contact in practice is requested together with any possible restriction of time and temperature.
- If no restrictions are indicated, according to Directive 82/711/EEC, the food contact material should be able to withstand test conditions at 175 °C with olive oil for 2 h.

- Also, information on the treatment of the food contact material prior to contact with food, for example, sterilization, cleaning with pressurized steam, rinsing, irradiation, e-beam or UV light treatment and so on, is requested.
- Information on the surface to volume approximate ratio of dm^2 food contact materials to kg food in practice is requested knowing that in the EU approach for materials intended for general application, this ratio is conventionally set at $6\,dm^2/kg$ food. In specific applications, other ratios may be applicable and accepted in the evaluation process. However, this may result in a restriction related to the S/V (surface/volume) ratio applied for.

2.3.2.4 **Information on Authorization of the Substance**

Information is requested on the authorization of the substance in individual Member States, the United States of America, Japan or any other country. In addition, it is requested to indicate whether the substance should be considered, according to Directive 67/548/EEC, as a new substance [6].

2.3.2.5 **Information on the Migration of the Substance**

For a safety evaluation, information on the migration of the relevant substance from a representative sample is necessary. The request for migration data serves two major purposes. In the first place, the level of migration determines the extent of the set of toxicity data that should be provided (see Section 2.3.2.8). In the second place, the migration data set should include a method of determination suitable for enforcement purposes. In addition, results of the migration experiment should demonstrate that the extent of migration can meet the restriction that may be established on the basis of the toxicity data set.

Although data on the specific migration of a substance should be provided, there are a number of exceptions that do not allow or require the determination of the specific migration, for example, when the substance is reactive with the simulant(s), when the specific migration can be determined through the overall migration or when the residual content is low enough to demonstrate that an SML cannot be exceeded.

As a rule, the migration of the substance that is the subject of the petition should be determined. However, if there are indications that decomposition or reaction products may be present, then migration of these decomposition or reaction products may be required. A typical example is the oxidation of a phosphite antioxidant into phosphate. For such substances, the sum of the phosphite and the phosphate compound has to be determined. Other more complex situations may occur and the need for the determination of decomposition products should be considered on a case-by-case situation. It should be taken into account that if the migration of such substances is relevant from a toxicological point of view, then they have to be determined.

If migration experiments are performed, always a worst-case sample should be used in the test protocol. Such a worst-case sample contains the highest foreseeable amount of the substances, the thickest layer and the polymer with the highest diffusion coefficient. These parameters will be responsible for the highest migration. Proper information on the composition and physical properties of the test sample should be provided. To allow the EFSA evaluator to understand the migration experiments, information on the dimensions of the test sample and the test specimen should be provided including information on the thickness of the sample. Only samples with a thickness above 0.5 mm should be brought into contact with the simulant by total immersion to allow the calculation of both sides of the sample in the final calculation of the migration.

The basic rules for specific and overall migration testing are laid down in Directive 82/711/EEC [7].

Various types of foodstuffs with complex compositions are possible (e.g., fatty foods, aqueous, acidic or alcoholic foods, etc.), and it is therefore not always possible to use real foodstuffs for testing food contact materials. To overcome this problem, alternative means of migration testing have been introduced.

Migration tests for the determination of specific and overall migration are usually carried out using the so-called *food simulants* under *conventional migration test conditions* as laid down in the above-mentioned directive. If for technical reasons the migration in olive oil is not feasible, then a substitute simulant can be used. It should be noticed that acceptability of the substitute simulants is strongly related to their similarity in migration behavior, that is, the solubility of the substance in the substitute simulants (*iso*-octane and 95% ethanol) should be comparable (but not less) to the solubility of the substance in olive oil. If substitute simulants are used, solubility both in the oil and in the substitute should always be provided. Also, details on the testing conditions (i.e., time and temperature conditions for migration testing) are provided in Directive 82/711/EEC.

As it is assumed that the S/V ratio is 6 dm^2/kg food, this ratio should preferably be maintained in migration experiments. However, in some cases it may be necessary to increase the S/V ratio. This can be allowed, provided the substance is sufficiently soluble in the simulant. In case saturation of the simulant occurs, the ratio of 6 dm^2/kg should be maintained, and enrichment of the substance in the simulant should be achieved using a proper analytical method.

The analytical method used to determine the migration of the substance should, in principle, be described according to the CEN (European Standardization Committee) template as it was drafted in the Note for Guidance (Annex 1 of Chapter 4). The use of the template is highly recommended, although other formats are often accepted provided they contain all necessary information for a proper evaluation. The method shall be suitable for enforcement purposes, which excludes methods based on radioactive labeled substances.

The detection limit and how it was established is of great importance when the migration is reported to be "not detectable." Detection limits may be calculated using the lowest standard solution or it may be calculated from the calibration curve.

Precision data should be provided. They may be calculated from triplicate migration experiments or from the triplicate recovery experiment. Other ways of calculation of the precision may also be acceptable.

Recovery experiments should be performed by spiking the substance at a relevant concentration to a blank simulant. This means spiking should be close to the level of migration, but in no case should it exceed the restriction that will be covered by the set of toxicity data; for example, when three negative mutagenicity tests are available, then the spiking level should not exceed the level of 50 μg/kg. The spiked simulant should be submitted to the same storage conditions as applied in the migration experiment and subsequent determination of the recovery of the spiked substance. By following this procedure, the EFSA will obtain information both on the analytical properties of the method used and on the stability of the substance in the simulant(s). There is no guideline on the level of recovery. In principle, any recovery level may be acceptable. Low values will indicate that the substance is reactive, has been lost (e.g., by evaporation) during the storage period, or that the analytical method is not capable of extracting the substance from the medium. Low values should always be clarified with valid arguments to be accepted by the EFSA.

It should be noted that this type of recovery experiments differ from the protocol described by the US FDA, where only analytical recovery is tested.

The results of migration and recovery test should be accompanied by sufficient detailed information and, if relevant, with some chromatograms or other instrumental evidence to allow the EFSA to check the method and the results.

Overall migration is usually not required; however, in some cases the overall migration may represent the migration of the petitioned substance. In these cases, the CEN methods (EN 1186) should be followed as close as possible. Requirements for a test sample and test conditions are similar to the requirements for the specific migration determination.

Composition of polymers and therefore the migrating substances may be influenced by the use of a new monomer, polymer production aid, or additive. For a proper risk assessment, there is a need to be informed on the identity and quantity of migrating substances. In this respect it should be emphasized that particularly substances with a Mw < 1000 g/mol are of toxicological relevance. For most polymers, the migration in oil is higher than in aqueous simulants; however, identification of migrant in olive oil is an analytical challenge that may not be feasible. Although many protocols may lead to an acceptable result in the identification and quantification of migrants, the following sequence is usually successful: (i) determine the total migration in *iso*-octane and 95% ethanol applying relevant test conditions while using a larger surface area than in the CEN protocols (EN 1186); (ii) if the migration is significant above the detection limit (e.g., >2 mg/6 dm^2), then the residue should be further analyzed to identify the substances in the residue. Substances with Mw > 1000 g/mol may be removed first and the substances with Mw < 1000 g/mol may be identified with more advanced analytical methods, for example, NMR, IR, and MS techniques.

Test sample and test conditions should be similar to the conditions used for the specific migration determination. Quantification may be problematic as there may be no reference material available. However, the main intention is to obtain a reliable impression about the identity of the substances and only then the quantity is of interest.

2.3.2.6 Information on the Residual Content

Information on the residual content from a sample should be determined in case the petition is for an additive. This requirement is mainly intended to demonstrate that the test sample under investigation indeed contained the substance at the intended level. For monomers, the determination of the residual content is not required. The residual amount of a monomer or additive may be determined to demonstrate that under worst-case conditions of use (assuming 100% migration), the restriction to be set and covered by the set of toxicity data cannot be exceeded. In these situations there is usually no method available for determining the specific migration and thus a QMA may be proposed as restriction. Similar to the sample requirements mentioned for migration testing, the sample should be a representative worst-case sample. The analytical method should show that extraction of the substance is exhaustive. Therefore, it may be necessary to have relative long extraction times or repeated extraction periods with fresh extractant. In other situations the sample may be cut to small pieces. If the residual content determination is intended to serve a QMA or QM restriction, then the method should be validated in-house and described properly. In addition, the method should be suitable for enforcement purposes. Recovery experiments are not easily performed because it is not feasible to make any inclusion in the polymer structure by adding the substance to the polymer. Therefore, the standard addition may be added to the extractant that is then submitted to the same conditions as is done with the test sample. If the substance is not reactive with the polymer, then a standard addition may be made to the polymer that is then extracted. The level of the standard addition should be related to the residual level of the substance in the polymer.

2.3.2.7 Antimicrobial Substances

To facilitate hygiene conditions at food production locations, the practice of insertion of antimicrobial substances into plastic materials is expanding. Such substances are intended to have an effect on the surface of the food contact material. They should, however, not show any effect on the food itself. Antimicrobial substances are also used to protect an aqueous polymer emulsion before making a final food contact material out of it. In these applications there should be any effect neither on the surface nor on the food. Besides the nontoxicity and toxicity data, as required for any substance, there are additional requirements about the antimicrobial activity of the substance.

For antimicrobials that protect the aqueous emulsion from spoilage it is required to demonstrate by proper means that there is no effect on the surface or on the food. This can be achieved using certain microbial tests or using the MIC (minimum inhibitory concentration) and the migration values. For "surface active" antimicrobials more information is required. The level of activity, the potential risk for overgrowth, the efficacy under the actual conditions of use or upon repeated use and demonstration of the absence of activity on the food are items that should be covered in the petition.

2.3.2.8 Toxicological Information

The general requirements for toxicological studies to be supplied for substances in food contact materials follow a tiered approach depending on the level of migration. The rationale for the approach has been described by Barlow [1] and the requirements are summarized below.

However, not all chemicals used in the manufacture of a food contact material will migrate into food. While many substances migrate in the same chemical form in which they were incorporated into food contact materials, others will migrate only partially or will migrate totally but in another chemical form. In such a case the toxicological requirements may also apply to the transformation or reaction products.

Other substances might disappear during the production process, while yet others will decompose completely to yield either no residues or vanishingly small residues.

The following core set of toxicological tests is required for any substance migrating in excess of 5 mg/kg, up to the maximum 60 mg/kg of food or food simulant, which is the overall limit for all substances migrating out of any food packaging. If it is assumed as a "worst case" that 1 kg of food wrapped in a particular type of packaging may be consumed by an individual in any 1 day, the maximum possible intake of a single substance by an adult consumer (with a standard body weight of 60 kg) is 1 mg/kg bodyweight/day. The core set of toxicological tests has been drawn up bearing in mind (i) this potential maximum exposure, (ii) the need to have adequate knowledge of potential toxicity, if any, and, for those substances that are toxic, (iii) the need to establish the size of safety margin. Only then can a decision be made on whether a substance remains acceptable for use.

Core Set of Toxicological Tests [11]

1) Mutagenicity studies *in vitro*:
   - A test for induction of gene mutations in bacteria according to the EC Method B.13/14 and the OECD Guideline 471.
   - A test for induction of gene mutations in mammalian cells *in vitro*, (preferably the mouse lymphoma to assay), according to the EC Method B.17 and the OECD Guideline 476.
   - A test for the induction of chromosomal aberrations in mammalian cells *in vitro* according to the EC Method B.10 and the OECD Guideline 473.
2) General toxicity studies
   - A subchronic (90-day) oral toxicity study according to the EC Method B.26 and the OECD guideline 408.
   - A chronic toxicity/carcinogenicity study according to the EC Method B.33 and the OECD guideline 453.
   - A reproduction/teratogenicity study according to the EC Methods B.34–B.35 and the OECD guidelines 421–422.
3) Metabolism studies
   - Studies on absorption, distribution, metabolism, and excretion (ADME)
   - Study on accumulation in man

In circumstances where migration is below 5 mg/kg of food or food simulant, not all core tests may be required. As a general guide, if migration is between 0.05 and 5 mg/kg, then only the three types of studies listed below are required.

The rationale for this approach is that, for this migration range, intakes from food will not exceed 0.1 mg/kg bodyweight/day and that at this low level of exposure, long-term, reproductive or teratogenic effects are extremely unlikely to occur. Indeed data in literature have shown [2, 3] that there are very few effects other than carcinogenicity that are not detected in a thorough short-term, repeat-dosing study. Thus, provided the mutagenicity tests are all clearly negative, which rules out the possibility of the substance being a genotoxic carcinogen, a reduced set of testing is acceptable since nongenotoxic carcinogens are generally active only at relatively high, sustained exposures [4].

Reduced Set of Toxicological Tests

1) The mutagenicity tests mentioned above
2) A 90-day oral toxicity study
3) Data to demonstrate the absence of potential for accumulation in man

The EFSA requires that these studies be carried out according to prevailing EU or OECD guidelines, including "Good Laboratory Practice" [12]. The substances tested should also be of the same (technical) specification as described under Sections 3.3.2.1 and 3.3.2.2 (general, chemical, and/or physicochemical data).

Special Investigations/Additional Studies

If from the above-mentioned studies or from prior knowledge or from structural considerations of the molecule (e.g., structure–activity considerations) there would be an indication that other biological effects such as neurotoxicity, immunotoxicity or endocrinological events may occur, additional studies may be required.

At present, no validated methods are available for studies in laboratory animals that would allow assessment of a substance's potential to cause intolerance and/or allergic reactions in susceptible individuals following oral exposure. However, studies on dermal or inhalation sensitization may give information relevant for possible hazards from occupational exposure and could be helpful in assessing consumer safety.

Under certain circumstances, particularly those related to the chemical nature of the substance, the tests normally to be provided for the safety evaluations and risk assessments may be modified as outlined below.

If the chemical structure of the substance would suggest that rapid hydrolysis occurs in the food and/or the gastrointestinal tract into components that already have been toxicologically evaluated, the rate of hydrolysis and its degree of completeness will determine the extent of toxicological testing necessary for an evaluation. In particular, it will depend on these parameters. Whether the unhydrolyzed substance also needs to be included in the testing program depends on the outcome of the hydrolysis studies.

Because in toxicological assessments only the fraction with molecular weight below 1000 g/mol is regarded as toxicologically relevant, a distinction has been made between polymeric additives with a *weight averaged molecular mass* (Mw) below 1000 g/mol and those with Mw above 1000 g/mol. For polymeric additives with Mw > 1000 g/mol, only the reduced set of toxicological data may be required. In deciding which data are needed, the data available on the monomers involved, the size of the fraction with molecular masses below 1000 g/mol, and the proportion of the additive in the plastic will be taken into account.

2.4
Evaluation Process of a Food Contact Substance

As already outlined in Section 2.2, formal submissions by petitioners applying for the use of substances as FCMs are submitted to the European Food Safety Authority.

After a petition is received by the EFSA and the submitted data are analyzed by the EFSA's scientific secretariat or by an FCM WG expert, the petitioner receives a letter acknowledging its submission. In this letter, the reference number allocated to the substance and the document reference number are mentioned as well as the official name as allocated by the EU Commission services.

The letter will confirm whether or not the request is in compliance with the instructions set out in the *Note for Guidance*. This letter is called the *Administrative Acceptability of the Petition* (AAP).

If the request does not comply with the instructions in the Note for Guidance, the applicant will be asked to appropriately modify the request. In such a case, the petitioner will receive an *AAP negative*. It should be stressed that the acceptance of the petition (AAP positive) does not imply that the documentation provided necessarily fully complies with the requirements set out in the Note for Guidance. The EFSA reserves the right to request additional information as necessary for a complete assessment of the substance. It has also to be stressed that any deviation must be justified both in the technical dossier and in the *Petitioner Summary Data Sheet* (P-SDS).

For facilitating the petitioner in compiling a valid dossier for evaluation by the CEF Panel, a checklist of the documents to be submitted is provided in Table 2.3.

The evaluation of a submission is performed by the FCM WG on the basis of the so-called *summary data sheet* (SDS). This SDS, drafted by a rapporteur (usually an expert, member of the FCM WG), contains the essential data submitted by the petitioner in its P-SDS and in the technical annexes.

The evaluation consists in a detailed and critical analysis of all the data submitted by the petitioner in the P-SDS (e.g., the identity of the substance and its purity; its physical and chemical properties; the intended use; the authorization of the substance in countries outside the EU; data on migration of the substance into the prescribed food simulants; data on the quantification and identification of possible migrating oligomers and reaction products, starting substances, and additives; data on residual content of the substance in the FCM, microbiological properties of the substance (for biocides); and data on the core set of toxicological tests).

Table 2.3 Checklist of the documents to be provided for an evaluation an FCM by EFSA.

1	Model letter[a]	No. 1 for evaluation and No. 2 for re-evaluation
2	Letter explaining the background of the request for evaluation (*only for re-evaluation of a substance*)	Alternatively the reasons for asking the re-evaluation can be given in the model letter
3	P-SDS	Document summarizing all data with appropriately marked confidential information and, in the case of re-evaluation, the new data
		Reference to the technical annexes attached has to be made in every section of the P-SDS
		Verifiable justification should be provided as to why the disclosure of information marked as confidential would significantly harm the petitioner's competitive position
4	Technical annexes	The necessary technical information, for example, scientific reasoning, full reports of experiments, and bibliographic references cited
5	Table of contents for the annexes	A table giving the contents of each annex and the relevant point on the P-SDS
6	CD-ROM with complete information	All the information in hard copy should also be on the CD
		The P-SDS should be provided in Word format. The other files may be either in Word format or in Adobe Acrobat Reader
		Appropriate labels should be attached on the CD jewel case, including the following information: name of the substance, REF No. (when it is known), company, date of submission, and CD-ROM number (if more than one per dossier, e.g., disk # of #)
		Each CD-ROM should contain a file detailing the name of the files contained in the disk and their contents. A printout of this file should accompany the CD-ROM, clearly indicating the different files and where they can be found
7		Only the information that is not considered as confidential by the petitioner should be on this CD-ROM. This information will be readily available to anyone who might so request, according to Regulation (EC) No. 1935/2004, Art. 19

a) Copies of the model letters are found on the web site: http://www.efsa.europa.eu/EFSA/Scientific_Document/afc_noteforguidancefcm_en1,0.pdf, pp. 15, 16.

The data presented in the P-SDS by the petitioner are summarized, evaluated on their merit, and commented on by the FCM rapporteur in its SDS and afterward discussed by the members of the FCM WG.

After this process of safety assessment, the FCM WG formulates its conclusions and a *draft opinion* is drawn up. This "draft opinion" presents the essential elements of the nontoxicity and toxicity studies on the substance under consideration together with a discussion on possible safety issues in relation to the proposed use of the substance and a conclusion on the acceptability of the substance as food contact material. This acceptance is expressed on the basis of either a *Tolerable Daily Intake* (TDI, in mg/kg body weight/day) or a *Restriction in migration* ("*R*" in mg/kg food) of the substance.

As outlined in Section 2.3.1, the greater the exposure through migration, the more toxicological information is required to prove the safety of the FCM. Conversely, in the safety evaluation by the CEF Panel, the extent of the toxicological data set submitted by the petitioner will determine the acceptable extent of migration into the food in contact that will be allocated.

If the full core of toxicological studies is supplied, a TDI may be allocated. In case only the three required mutagenicity tests are supplied together with a 90-day oral toxicity study and a demonstration of the absence of potential for accumulation in man, a restriction in migration of up to $R = 5$ mg/kg food is allocated. If only mutagenicity tests are provided, a restriction in migration of $R = 0.05$ mg/kg food is allocated.

After evaluating the petition, the FCM WG submits the draft opinion to the CEF Panel for further discussion and its final acceptance (or refusal). On acceptance, the substance is then put on the *SCF Listing* and the opinion is published on the EFSA web site. Different SCF listings are given in Table 2.4.

Regarding this SCF listing it should be stressed that the classification of a substance into the SCF List is only a tool used for tackling authorization dossiers and do not prejudice the management decisions that will be taken by the European Commission on the basis of the scientific opinions of the CEF Panel and in the framework of the applicable legislation.

2.4.1 Re-Evaluation of Substances

It should be noted that the re-evaluation of substances can be requested if, during the evaluation of a petition, the EFSA considers it necessary to require more information including additional studies.

Also, if a substance was classified in the SCF lists 0–5 and the petitioner would have obtained more information on the substance and he believes that the additional data might permit a different classification or restriction for that substance, a request for re-evaluation can be introduced.

The third case for requesting a re-evaluation would be if a new petitioner would have obtained more information on a substance currently in the Synoptic Document [17] and believes that the additional data might permit a different classification. In that case, Article 21 on data sharing of Regulation (EC) No.

Table 2.4 Definition of the SCF lists.

List 0	Substances, for example, foods, which may be used in the production of plastic materials and articles, for example, food ingredients and certain substances known from the intermediate metabolism in man and for which an ADI need not be established for this purpose
List 1	Substances, for example, food additives, for which an ADI (acceptable daily intake), a t-ADI (temporary ADI), an MTDI (maximum tolerable daily intake), a PMTDI (provisional MTDI), a PTWI (provisional tolerable weekly intake), or the classification "acceptable" has been established by this Committee or by JECFA
List 2	Substances for which a TDI or a t-TDI has been established by this Committee
List 3	Substances for which an ADI or a TDI could not be established, but where the present use could be accepted
	Some of these substances are self-limiting because of their organoleptic properties or are volatile and therefore unlikely to be present in the finished product. For other substances with very low migration, a TDI has not been set, but the maximum level to be used in any packaging material or a specific limit of migration is stated. This is because the available toxicological data would give a TDI that allows to fix a specific limit of migration or a composition limit at levels very much higher than the maximum likely intakes arising from present uses of the additive
List 4 (for monomers)	4A: Substances for which an ADI or TDI could not be established, but which could be used if the substance migrating into foods or in food simulants is not detectable by an agreed sensitive method
	4B: Substances for which an ADI or TDI could not be established, but which could be used if the levels of monomer residues in materials and articles intended to come into contact with foodstuffs are reduced as much as possible
List 4 (for additives)	Substances for which an ADI or TDI could not be established, but which could be used if the substance migrating into foods or in food simulants cannot be detected by an agreed sensitive method
List 5	Substances that should not be used
List 6	Substances for which there exist suspicions about their toxicity and for which data are lacking or are insufficient. The allocation of substances to this list is mainly based upon similarity of structure with that of chemical substances already evaluated or known to have functional groups that indicate carcinogenic or other severe toxic properties
	List 6A: Substances suspected to have carcinogenic properties. These substances should not be detectable in foods or in food simulants by an appropriate sensitive method for each substance

(Continued)

Table 2.4 *(Continued)*

	List 6B: Substances suspected to have toxic properties (other than carcinogenic). Restrictions may be indicated
List 7	Substances for which some toxicological data exist, but for which an ADI or a TDI could not be established. The required additional information should be furnished
List 8	Substances for which no or only scanty and inadequate data was available
List 9	Substances and groups of substances that could not be evaluated either due to lack of specifications (substances) or due to lack of adequate description (groups of substances). Groups of substances should be replaced, when possible, with individual substances actually in use. Polymers for which the data on identity specified in "SCF Guidelines" are not available
List W	"Waiting list." Substances not yet included in the EU lists, as they should be considered "new" substances, that is, substances never approved at national level. These substances cannot be included in the EU lists, lacking the data requested by the Committee
	List W7: Substances for which some toxicological data exists, but for which an ADI or a TDI could not be established. The required additional information should be furnished
	List W8: Substances for which no or only scanty and inadequate data are available
	List W9: Substances and groups of substances that could not be evaluated either due to lack of specifications (substances) or due to lack of an adequate description (groups of substances)

1935/2004 [16] applies. The new applicant should enquire with the Commission and the European professional organizations about an agreement on data sharing with the original applicant. If such an agreement is reached, the petitioner should include the written agreement signed by all parties involved in the application and supply only the new data. If the original petitioner and the new petitioner have not agreed on data sharing, the latter has to submit a new petition including all available data.

2.5 Public Access to Data

In accordance with Articles 38, 39, and 41 of Regulation (EC) No. 178/2002 [15] and Articles 2, 4, 7, 8, and 10 of Regulation (EC) No. 1049/2001 [14], the EFSA has the obligation to make available to the public its scientific opinions, statements, and minutes of panel meetings. Also, the minutes of WG meeting will be made available to the public.

Petitions for authorization of food contact materials, supplementary information from applicants, with the exclusion of any confidential information, are made accessible to the public.

The complete information will also be made available to the European Commission and Member States that, however, have to respect the confidentiality of any commercial and industrial information provided.

The European Commission will determine, after consultation with the applicant, which information should be kept confidential, as stated in Article 20 of the Regulation (EC) No. 1935/2004. However, the following information cannot be considered confidential: (i) the name and the address of the applicant and the chemical name of the substance, (ii) information of direct relevance to the assessment of the safety of the substance, and (iii) the analytical method or methods.

Verifiable justification has to be provided by the petitioner as to why the disclosure of information that is claimed to be confidential would harm the competitive position of the petitioner.

References

1 Barlow, S.M. (1994) The role of the Scientific Committee for Food in evaluating plastics for packaging. *Food Additives and Contaminants*, **11** (2), 249–259.

2 Lumley, C.E. and Walker, S.R. (1985) The value of chronic animal toxicology studies of pharmaceutical compounds: a retrospective analysis. *Fundamental and Applied Toxicology*, **5**, 1007–1024.

3 Lumley, C.E. and Walker, S.R. (1986) A critical appraisal of the duration of chronic animal toxicity studies. *Regulatory Toxicology and Pharmacology*, **6**, 66–72.

4 Weisburger, J.H. and Williams, G.M. (1981) Carcinogen testing: current problems and new approaches. *Science*, **214**, 401–407.

5 Competent Authorities of the EU member states. http://ec.europa.eu/food/food/chemicalsafety/foodcontact/nat_contact_points_en.pdf.

6 Council Directive 67/548/EEC on the approximation of laws, regulations, and administrative provisions for the classification, packaging, and labeling of dangerous substances, and subsequent adaptations to technical progress.

7 Council Directive 82/711/EEC of 18 October 1982 laying down the basic rules necessary for testing migration of constituents of plastic materials and articles intended to come into contact with foodstuffs. *Official Journal*, 297, 1982, 26–30.

8 Council Directive 2002/72/EC. Corrigendum to Commission Directive 2002/72/EC of 6 August 2002 related to plastic materials and articles intended to come into contact with foodstuffs (*Official Journal*, L220, 2002, 0018–0058.), *Official Journal*, L0392, 2003, 01–42. Last amended by Commission Directive 2008/39/EC, *Official Journal of the European Union*, L63, 2008, 6–13.

9 EFSA, European Food Safety Authority. Web site: http://www.efsa.europa.eu/EFSA/ScientificPanels/efsa_locale-1178620753812_CEF.htm.

10 Note for Guidance for petitioners presenting an application for the safety assessment of a substance to be used in food contact materials prior to its authorization. http://www.efsa.europa.eu/EFSA/Scientific_Document/afc_noteforguidancefcm_en1,0.pdf.

11 OECD *Guidelines for Testing of Chemicals*, Organisation for Economic Co-operation and Development, Paris.
12 OECD (1983) *Principles of Good Laboratory Practice*, Organisation for Economic Co-operation and Development, Paris.
13 Register of opinions and statements in the area of food contact materials evaluated by the EFSA. http://www.efsa.eu.int/science/afc/afc_opinions/catindex_en.html.
14 Regulation (EC) No. 1049/2001 of the European Parliament and the Council of 30 May 2001 regarding public access to European Parliament, Council, and Commission documents. *Official Journal of the European Communities*, L145/43, 31-5-2001.
15 Regulation (EC) No. 178/2002 of the European Parliament and of the Council of 28 January 2002 laying down the general principles and requirements of food law, establishing the European Food Safety Authority and laying down procedures in matters of food safety. *Official Journal of the European Union*, L31, 12-02-2002, pp. 1–24. Last amended by Commission Regulation (EC) No. 575/2006, *Official Journal of the European Union*, L100, 08-04-2006, p. 3.
16 Regulation (EC) No. 1935/2004 of the European Parliament and the Council of 27 October 2004 on materials and articles intended to come into contact with food and repealing Directives 80/590/EEC and 89/109/EEC. *Official Journal of the European Union*, L338, 13-11-2004, pp. 4–17.
17 Synoptic Document. Provisional list of monomers and additives notified to European Commission as substances that may be used in the manufacture of plastics or coatings intended to come into contact with foodstuffs (updated on June 2005) http://ec.europa.eu/food/food/chemicalsafety/foodcontact/synoptic_doc_en.pdf.

3
Council of Europe Resolutions

Luigi Rossi

3.1
Introduction

The Council of Europe (CoE) is a political organization that was founded on May 5, 1949 by 10 European countries in order to promote greater unity between its members. It has now 48 Member States, and a multinational European Secretariat helps the various bodies and committees in their activities. When a lesser number of Member States of the CoE wish to engage in some action which not all their European partners desire to join, they can conclude a "Partial Agreement" that is binding on themselves alone. The Partial Agreement in the social and public health fields was concluded on this basis in 1959 and the area of activity was the "protection of public health." At present, the Partial Agreement in the public health field has 16 countries, namely, Austria, Belgium, Cyprus, Finland, France, Germany, Ireland, Italy, Luxembourg, The Netherlands, Norway, Portugal, Slovenia, Spain, Sweden, and Switzerland.

The work of the Partial Agreement committees may result in the elaboration of conventions or agreements. In the field of the materials and articles intended to come into contact with foodstuffs (FCMs), the more usual outcome of the Committee of Experts is the drawing-up of Resolutions of the Member States or guidelines or Technical Documents (TDs). The Resolutions, prepared by the Committee of Experts, after an approval of the Public Health Committee, are adopted by the Committee of Ministers. The TDs are not submitted to the Committee of Ministers and they are adopted by the Committee of Experts. In principle, the CoE adopts guidelines when the subject is not relevant from the health point of view or when there are no specific restrictions laid down in the document but only a description of the "state of the art" on the subject.

The resolutions and the TDs are not binding documents. Only when a country transposes, totally or partially, such documents into a national law, do they become binding. However, because the governments have actively participated in their formulation and the Committee of Ministers and the Committee of Experts adopted respectively the Resolutions and the TDs, these documents may be considered reference documents for the interpretation of the European law and mainly for the so-called "safety clause," that is, the materials should not release substances that endanger human health.

Global Legislation for Food Packaging Materials. Edited by Rinus Rijk and Rob Veraart

ISBN: 978-3-527-31912-1

The European Commission (EC) has a particular status of participation within the Committee of Experts and may take part in the preparation of the documents even if it cannot participate in the vote and in the Council of Ministers.

The European professional associations are not entitled to send representatives to the meetings of the Committee of Experts. But they may be represented at the level of the *ad hoc* groups, which are advisory bodies to the Committee of experts, without any decision-making power. Hearings are regularly organized between the Committee of Experts and the European professional organizations.

The aim of the CoE Partial Agreement on public health activities is to raise the level of health protection of consumers and food safety in its widest sense. In particular, the aim of the Committee of Experts is to protect the consumers from the risks related to the release of chemicals present in FCM into the food.

3.2 Activity of CoE in the 1960s–1970s

The CoE is the first European institution that started in the early 1960s to prepare a European law in the field of FCM but not binding on the member states. A group of chemists and toxicologists of the six initial member states of the EU, that is, Belgium, France, Italy, Luxembourg, The Netherlands, and Germany, started to meet in Strasbourg to prepare a resolution on plastics.

3.2.1 Resolution on Plastics

The CoE draft was based mainly on the Italian and FDA system, on which France and Benelux quickly agreed. The system proposed was based on the principle of a single positive list for all the plastics, the overall (OML) and specific migration limits (SML) or % use of the substances and the use of simulants and standardized testing conditions. Germany was quite reluctant to accept the principle of the overall migration and also the system based on a single general list and SML. In fact, the first German recommendations were based on the specific list for each type of plastics and on the restriction based on the % use of the substances in each polymer. Other three main parts of the draft resolution were (1) the guidelines for the evaluation of substances, (2) the evaluation of the exposure, and (3) the conventional classification of the foodstuffs.

3.2.2 Guidelines for the Evaluation of Substances

These guidelines, proposed by the French delegation, contained all the data to be submitted to the CoE Committee to obtain the insertion of a substance in the CoE plastic list. The data used were quite similar to the current SCF-EFSA guidelines, that is (a) identity, (b) physical and chemical properties, (c) intended use, (d) authorization in other countries, (e) migration data, and (f) toxicological data. The toxicological data requested were quite different from the present EFSA guidelines but quite similar to 1976 SCF guidelines, when the EU started to evaluate the substances in plastics. For instance, no mutagenicity tests were requested, as they did not yet exist in the 1960s. The system was mainly based on the evaluation of a 90-day oral study, which permitted the establishment of a tolerable daily intake (TDI). Many evaluations were based on the personal experience of the toxicologists participating in the CoE meetings and not on the data available, often missing. To maintain a trace of the basis of CoE evaluation, the CoE Committee charged the Dutch delegation to insert the data available in toxicological sheets. These sheets were used by the EU to evaluate substances in the 1980s.

3.2.3
Estimation of Exposure

Since the beginning, the CoE experts recognized the impossibility of basing the system of the evaluation of the substances on the exposure due the lack of the needed data (consumption of food packaged in various materials and in various countries). Therefore, it was decided that the safety assessment of chemicals used in food contact materials would be based on the assumption that a person of 60 kg bodyweight will eat daily, during his lifetime, 1 kg of foodstuffs in contact with a surface of 6 dm^2 of the same type of plastic packaging, containing the migrant substance at the maximum value compatible with the established specific migration limit. Hence, for a substance having a TDI, the SML was established as follows:

$$\text{migration} \leq \text{SML (mg/kg food)} = \text{TDI(mg/kg/bw)} \times 60\text{(kg/bw)} \times 1\text{ kg food.}$$

Another assumption was that this 1 kg of foodstuff could have an "aqueous," "acidic," "alcoholic," or "fatty" character and could be simulated by 1 kg of food simulant (respectively, "water," "3% acetic acid," "10% ethanol," and "heptane/5"; the last one was replaced in the 1970s with olive oil or other fat simulants). In this system, the total dose of the migrant absorbed per day can be derived only from 1 kg of food in contact with plastic and that this 1 kg is composed either of 1 kg of fatty food or of 1 kg of one of the other types of food, but never of the sum of the various types of food.

The assumptions of the present system can be summarized as the following:

- No material use factor (plastic, paper and board, glass, metal, etc.)
- The same plastic packaging material type (= no plastic use factor)
- 100% market share for the migrants under review
- Lifetime exposure day by day
- All packaging materials release the migrants at the maximum value

A material is considered suitable for packaging any type of food if the migration into each of the four simulants is below the SML. If the migration into a given food exceeds the SML, the material is considered unsuitable for the corresponding class of food.

3.2.4
Conventional Classification of Food

Some countries, in particular France and Italy, were very active at that time in the CoE Committee of Experts and proposed a scheme in accordance of which for each food the simulant to be used was indicated. However, these recommendations were not harmonized.

Finally, the differences between the approaches suggested in the draft resolution on plastics by Germany and other CoE delegations were so vast that it was impossible to reach a compromise and the CoE activity came to cease in the late 1960s.

3.3
EU Activity and Relationship Between EU and CoE

At the end of the 1960s, the European Commission started the process of harmonization of the laws based on the German approach. A German expert assisted the EC in preparing the working documents. However, the majority of the other five EU countries and the professional organizations, with the exception of the German industry, were against the German approach. The major criticism was that the German recommendations reflected the German technology and not the toxicological results. Therefore, after various meetings, the EC decided to stop attempting the harmonization of national laws.

Some years later, first the Dutch delegation in Noordwijk (1970) and then the Italian delegation in Rome (1971) organized two international conferences to relaunch the process of harmonization. Thereafter, an Italian expert was engaged, and in 1976 a framework directive was adopted and, later, directives related to specific groups of materials were also adopted.

When the EC started to study the positive list of substances for plastics, it was clear that the activity of EC might overlap the CoE activity in the same field. Therefore, in a meeting held in Strasbourg between the representatives of the EC and those of the CoE, an informal agreement was reached to avoid overlapping between the two institutions. It was agreed that the CoE could continue and increase its activities in those fields where there was no action from the EC. Moreover, the EC actively participate in the CoE activities and both institutions agreed to coordinate their activities as much as possible.

3.4
Resolutions of the CoE

3.4.1
Procedure for the Adoption of a Resolution and Guidelines and Technical Documents

The documents (Resolutions or Guidelines and TDs) are drafted within the Committee of Experts by a panel selected as rapporteur from among the delegates having a certain amount of expertise on the subject. Upon the proposal of the rapporteur, the committee may decide to institute an *ad hoc* working group (WG) in which, together with some representatives offering their support to the selected rapporteur, experts from the European professional organizations may also be requested to participate. The rapporteur prepares a draft that is transmitted to the Committee of Experts. When an agreement is reached within the Committee of the Experts, two different procedures are followed.

If the document is a draft Resolution, it is sent first to the Public Health Committee for approval and then to the Council of Ministers for formal adoption.

If the document is a Guideline or a TD accompanying the resolution/guidelines, it is formally adopted by the Committee of Experts. Both documents may be amended, when necessary, following the same procedure as their adoption. After their adoption, the texts are published in English and in French on the web site of the CoE: http://www.coe.int/T/E/Social_Cohesion/soc-sp/Public_Health/Food_contact/.

The Resolution/Guidelines and their TDs constitute the so-called "Package" and are contained in a CoE document called "Policy Statement."

The CoE countries may transpose into national law totally or partially any document of the Package.

3.4.2
CoE Resolutions, Guidelines, and TDs

The following Resolutions and Guidelines have been adopted until 2007. Some have been updated.

- Resolution AP (1989) 1 on the use of colorants in plastic materials coming into contact with food
- Resolution AP (1992) 2 on the control of aids to polymerization for plastic materials and articles
- Resolution AP (2002) 1 on paper and board materials and articles intended to come into contact with foodstuffs
- Framework Resolution AP (2004) 1 on coatings intended to come into contact with foodstuffs (superseding Resolution AP (96) 5)
- Resolution AP (2004) 2 on cork stoppers and other cork materials and articles intended to come into contact with foodstuffs
- Resolution AP (2004) 3 on ion exchange and adsorbant resins used in the processing of foodstuffs (superseding Resolution AP (97) 1)
- Resolution AP (2004) 4 on rubber products intended to come into contact with foodstuffs
- Resolution AP (2004) 5 on silicones used for food contact applications (superseding Resolution AP (99) 3)
- Resolution AP (2005) 2 on packaging inks applied to the nonfood contact surface of food packaging materials and articles intended to come into contact with foodstuffs
- Guidelines on metals and alloys (13 February 2002)

- Guidelines on lead leaching from glass tableware into foodstuffs (2 December 2003)
- Guidelines on tissue paper kitchen towels (9 November 2005)

Many other TDs have been prepared, some of them have been adopted, and others are only drafts that need to be completed before their adoption. In the following paragraphs is briefly summarized the current situation on different groups of materials or substances in accordance with their temporal adoption.

3.5 Status of the Packages (Resolutions, Guidelines, and Technical Documents)

3.5.1 Resolutions on Colorants in Plastics

3.5.1.1 Inventory of the Documents

- Resolution AP (1989) 1 on the use of colorants in plastic materials coming into contact with foodstuffs

The document is available on the web site of the Partial Agreement Division in the Social and Public Health Field.

3.5.1.2 Chronological Development

The Resolution was adopted in 1989 (Resolution AP (99) 3). No further action was taken.

3.5.1.3 Content of the Resolution

The Resolution contains the following requirements:

- Safety clause (migrated substances should not endanger human health)
- No visible colorants' migration from finished articles and description of the method
- Purity criteria expressed in % (w/w) related to some metals (As, Ba, Cd, Cr^{VI}, Hg, Pb, Sb, and Se)
- Specifications on some substances such as aromatic amines, carbon black, PCBs, and so on

3.5.2 Resolutions on Control of Aids to Polymerization for Plastic

3.5.2.1 Inventory of the Documents

- Resolution AP (1992) 2 on control of aids to polymerization for plastic materials and articles intended to come into contact with foodstuffs

The document is available on the web site of the Partial Agreement Division in the Social and Public Health Field: www.coe.int/soc-sp or the more specific http://www.coe.int/t/e/social_cohesion/soc-sp/public_health/food_contact/COE%27s%20policy%20statements%20food%20contact.asp#TopOfPage.

3.5.2.2 Chronological Development

The Resolution was adopted in 1992 (Resolution AP (92) 2). No further action was carried out. The existing restrictions take into account the SCF evaluations or, in the absence of an SCF opinion, the national limitations existing at that time.

3.5.2.3 Content of the Resolution

- Definition
- Safety clause
- Inventory lists of substances banned
- Specific inventory list of classes of substances to be used as aids to polymerization under certain restrictions that take into account SCF evaluations and national restrictions existing at that time. Also, authorized plastic substances may be used.
- OML and SML in accordance with EU plastics rules
- Migration testing according to EU plastic rules (82/711/EEC and amendments)

3.5.3 Resolution on Silicones

3.5.3.1 Inventory of the Documents

Policy Statement No. 1 contains

- Resolution AP (2004) 5 on silicones to be used for food contact applications
- Technical Document No. 1 – List of substances to be used in the manufacture of silicones used for food contact applications

The documents are available on the web site of the Partial Agreement Division in the Social and Public Health Field: www.coe.int/soc-sp.

3.5.3.2 Chronological Development

The first Resolution was adopted in 1999 (Resolution AP (99) 3) and the second in 2004 (Resolution AP (2004) 5) as it was necessary to update the list.

3.5.3.3 Content of the Resolution

- Definition
- Description of the silicone group

- Safety clause
- GMP (ISO 9002 or other)
- Inventory lists (in accordance with TD N.1). The list is subdivided into two lists. List 1, substances approved and List 2, substances not approved in accordance with TD 1
- List of banned substances
- OML and SML
- Compliance testing in accordance with the EU plastics rules

3.5.4
Resolution on Paper

3.5.4.1 Inventory of the Documents

All the updated documents are included in Policy Statement Version 3 dated September 11, 2007.

- Resolution AP (2002) 1 on paper and board materials and articles intended to come into contact with foodstuffs
- Technical Document No. 1 – List of substances to be used in the manufacture of paper and board materials and articles intended to come into contact with foodstuffs (Version No. 2)
- Technical Document No. 2 – Guidelines on test conditions and methods of analysis for paper and board materials and articles intended to come into contact with foodstuffs (Version No. 2)
- Technical Document No. 3 – Guidelines on paper and board materials and articles, made from recycled fibers, intended to come into contact with foodstuffs (Version No. 2)
- Technical Document No. 4 – CEPI Guide for good manufacturing practice for paper and board for food contact, prepared by CEP December 19, 2004
- Technical Document No. 5 – Practical Guide for users of Resolution AP (2002) 1 on paper and board materials intended to come into contact with foodstuffs (Version No. 1)

All these documents are available on the web site of the Partial Agreement Division in the Social and Public Health Field: www.coe.int/soc-sp.

3.5.4.2 Chronological Development

No remark. Rapporteurs: United Kingdom and Germany.

3.5.4.3 Content of Package

The Resolutions Contains:

- Field of application

- Definition
- Safety clause
- GMP (in accordance with TD No. 4)
- Inventory lists (in accordance with TD No. 1, RD 6.1/1-44 of October 11, 2005) that contains
  - List 1: additives approved by the CoE
    - Temporary appendix to List 1 of additives
  - List 2: list of additives not approved by CoE
    - Appendix A: monomers assessed
    - Appendix B: monomers approved by CoE
    - Appendix C: monomers not yet assessed
- Recycled fibers in accordance with TD No. 3
- Suitable microbiological quality
- Released substances should not have an antimicrobial effect
- SML and other restrictions
- Criteria purity for Cd, Pb, Hg, and pentachlorophenol, dioxins
- Compliance in accordance with TD No. 2

The content of the TDs is described in their titles.

3.5.5 Resolution on Coatings

3.5.5.1 Inventory of the Documents

Policy Statement No. 1 contains

- Resolution AP (2004) 1 concerning coatings intended to come into contact with foodstuffs
- RD 11/1-44 – Consolidated list on coatings
- RD 11/2-44 – CEPE – Update of Technical Document No. 1 on inventory list for coatings

The documents are available on the web site of the Partial Agreement Division in the Social and Public Health Field: www.coe.int/soc-sp.

3.5.5.2 Chronological Development

No remark. Rapporteur: Belgium

3.5.5.3 Content of the Resolution

- Field of application
- Definition
- Safety clause
- GMP

- Inventory lists of monomers and additives subdivided in accordance with CoE recent criteria (in accordance with RD 11.1 and 2 containing List 1: substances approved divided into five categories and List 2: substances not approved).
- OML and SML

3.5.6 Resolution on Cork Stoppers and Other Cork Materials

3.5.6.1 Inventory of the Documents

- Resolution AP (2004) 2 on cork stoppers and other cork materials and articles intended to come into contact with foodstuffs
- Technical Document No. 1 – List of substances to be used in the manufacture of cork stoppers and other cork materials and articles intended to come into contact with foodstuffs (under preparation)
- Technical Document No. 2 – Test conditions and methods of analysis for cork stoppers and other cork materials and articles intended to come into contact with foodstuffs

The documents are available on the web site of the Partial Agreement Division in the Social and Public Health Field: www.coe.int/soc-sp.

3.5.6.2 Chronological Development

No remark. Rapporteur: Spain.

3.5.6.3 Content of the Resolution

- Field of application (at least 51% of cork)
- Safety clause
- GMP in accordance with International Code of Cork Stoppers Manufacturing Practice
- Specific Inventory List and lists in EU Rules
- Restrictions in accordance with EU Rules
- Criteria purity for pentachlorophenol and trichlorophenols
- Migration testing in accordance with TD No. 2 (reference to EU Rules)
- Other minor rules

3.5.7 Resolution on Ion Exchange and Adsorbant Resins Used in the Processing of Foodstuffs

3.5.7.1 Inventory of the Documents

- Resolution AP (2004) 3 on ion exchange and adsorbent resins used in the processing of foodstuffs

- Technical Document No. 1 – List of substances to be used in the manufacture of ion exchange and adsorbent resins used in the processing of foodstuffs

These documents are available on the web site of the Partial Agreement Division in the Social and Public Health Field: www.coe.int/soc-sp.

3.5.7.2 Chronological Development

The first resolution was adopted in 1997 (Resolution AP (1997) 1) and the second in 2004 (Resolution AP (2004) 3) as it was necessary to update the list.

3.5.7.3 Content of the Resolution

- Definition
- GMP and safety clause
- Specific Inventory in TD No. 1 containing the list of classes of substances to be used as monomers, chemical modifiers, or polymerization aids. The list, in accordance with CoE criteria, is subdivided into
  - List 1: substances approved
  - List 2: substances not approved
- No OML
- SML fixed in accordance with EU plastic rules
- Migration testing with 3% acetic acid or 15% (v/v) ethanol in accordance with AFNOR test T 90-601

3.5.8 Resolution on Rubber Products

3.5.8.1 Inventory of the Documents

Policy Statement Version 1 contains

- Resolution AP (2004) 4 on rubber products
- Practical Guide containing the inventory list

The documents are available on the web site of the Partial Agreement Division in the Social and Public Health Field: www.coe.int/soc-sp.

3.5.8.2 Chronological Development

No remark. Rapporteur: The Netherlands

3.5.8.3 Content of the Resolution

- Field of application
- Definitions
- Safety clause

- GMP
- Inventory lists of monomers and list of additives, as well as aids to polymerization and vulcanizing agents are contained in Practical Guide. They are subdivided in accordance with CoE recent criteria (List 1: substances approved divided into five lists and List 2 substances not approved)
- Criteria purity on *N*-nitrosamines, *N*-nitrosable substances, aromatic amines
- OML and SML
- Compliance testing in accordance with the scheme in Practical Guide

3.5.9 Resolution on Packaging Inks

3.5.9.1 Inventory of the Documents

- Policy statement version is not yet prepared.
- Resolution AP (2005) 2 on packaging inks
- Technical Document No. 1: Inventory list of the following substances and requirement for the selection of inventory list (Document RD 8/1-43#5, approved by the Committee in its 45 M)
  - additives
  - binders
  - dyes
  - pigments
- Technical Document No. 2: Good Manufacturing Practices derived one by CEPE and the other one by flexible and fiber-based packaging for food (not approved)
- Technical Document No. 3: Migration testing conditions (CD-P-SP (2005) 14 Appendix 3) (approved by the Committee in its 45 M)

The documents are available on the web site of the Partial Agreement Division in the Social and Public Health Field: www.coe.int/soc-sp.

3.5.9.2 Chronological Development

No remark. Rapporteur: Switzerland

3.5.9.3 Content of the Resolution

- Field of application
- Definitions
- Safety clause
- No migration (DL = 10 ppb) for substances not listed
- Not in direct contact with food
- GMPs for the inks and for converters contained in TD No. 2

- Requirements for the finished printed materials: OML, SML, no visible set-off
- Responsibility of the ink suppliers
- Traceability
- Migration testing in accordance with TD. No. 2

The content of the technical documents is indicated by their titles.

3.5.10 Guidelines on Metals and Alloys

3.5.10.1 Inventory of the Documents

- "Guidelines on metals and alloys used as food contact materials" (09.03.2001)

The documents are available on the web site of the Partial Agreement Division in the Social and Public Health Field: www.coe.int/soc-sp.

3.5.10.2 Chronological Development

No remark. Rapporteur: Denmark.

3.5.10.3 Content of the Guidelines

The following metallic materials are covered by the guidelines:

- Aluminum
- Chromium
- Copper
- Iron
- Lead
- Manganese
- Nickel
- Silver
- Tin
- Titanium
- Zinc
- Stainless
- Other alloys

The guidelines also cover the metals listed as follows that may be present as impurities in some metallic materials and then migrate:

- Cadmium
- Cobalt
- Mercury

3.5.11 Guidelines on Glass

3.5.11.1 Inventory of the Documents

Policy statement version 1 (22.09.2004) contains:

- Guidelines for lead leaching from glass tableware into foodstuffs

The document is available on the web site of the Partial Agreement Division in the Social and Public Health Field: www.coe.int/soc-sp.

3.5.11.2 Chronological Development

No remark. Rapporteur: Italy

3.5.11.3 Content of the Resolution

The guidelines contain the following requirements:

- Introduction
- Field of application (crystalline glass and crystal glass tablewares)
- Lead leaching from glass hollow and flatwares
- Safety aspects
- Test methods
- Conclusions
- Appendix 1: Parameters influencing lead leaching
- Appendix 2: Limit values in certain foodstuffs

3.5.12 Guidelines on Tissue Paper Kitchen Towels

3.5.12.1 Inventory of the Documents

Policy Statement No. 1 contains

- Guidelines on tissue paper kitchen towels and napkins

The documents are available on the web site of the Partial Agreement Division in the Social and Public Health Field: www.coe.int/soc-sp.

3.5.12.2 Chronological Development

No remarks.

3.5.12.3 Content of the Guidelines

- Field of application

- Definitions
- Raw materials
  - Nonfibrous components (functional additives – processing aids – exclusion list)
  - Substances typically used in printing inks (binders, colorants, additives banned substances)
- Test conditions and methods of analysis
- Recycled fibers
- GMP

3.6 Transposition of the Resolutions, Guidelines, and Technical Documents

In 2006, an inquiry was carried out by a consultant with the aim:

1) to know the level of the transposition into the national laws;
2) to know whether, in the absence of a transposition, the CoE documents are used at national level as reference document for the enforcement of the safety clause;
3) to investigate the major criticism on CoE activity;
4) to know the groups of materials for which an action at CoE level is necessary to amend or complete the current documents;
5) to prepare a working plan for the future.

The results are summarized in the Annex and here briefly described:

- Three countries (Belgium, Switzerland, and Greece) have transposed or are transposing some of the CoE Resolutions into national law.
- Ten out of the fourteen countries that answered the questionnaire declared that they use the CoE Resolutions/Guidelines as reference documents for the verification of the compliance with the safety clause of the EU Framework Regulation 1935/2004.
- The majority of the countries were favorable to continue the CoE activity and they indicate in the evaluation of substances included in the inventory lists (to be completed) the first priority for an action.
- Paper and packaging inks are the materials for which an immediate action was requested, even if some other materials were mentioned.

3.7 The Future of the CoE

During 2007, the Committee of Experts discussed in two meetings a possible working plan. On the basis of the results of the inquiry, it was proposed to concentrate the activity in the following fields:

- The update of the Resolutions, Guidelines, and Technical Documents with the priority attributed to paper, inks, and metals and try to use the same format of the EU measures.

- The completion of the inventory lists transformed into positive lists.
- The evaluation of the new substances in accordance with a new adopted procedure described below.

3.7.1 Procedure of Authorization of a New Substance at National/CoE Level

The Committee of Experts decides that any petitioner, whether belonging to CoE Partial Agreement or not, should transmit the request for adding a new substance the technical dossier and the petitioner Summary Data Sheet (SDS) to the national authority of one of the countries that have a positive list in their law or recommendations ("selected country"). The country should inform the petitioner and, at the same time, all other Partial Agreement countries (PA countries) on the Administrative Acceptability of the Petition (issue of an AAP, identical to EFSA AAP). The national authority should prepare an SDS in accordance with the model that will appear on the CoE web site and with the EFSA Note for Guidance. It will also contain the assessment and the restrictions, if any. The country should transmit the SDS to all PA countries. If no objection is transmitted within *a month* from the receipt of the SDS, the country may insert the substance in its national list. When the substance is inserted in the national list, the country should inform the Secretariat of the CoE, which should automatically insert the substance in the CoE list and provide the information to all PA countries. There is no obligation for the other PA countries having a positive list to introduce the substance in their national list. In any case, the principle of mutual recognition applies.

3.8 Conclusions

In December 2007, the Council of Ministers decided to continue the activity of the CoE in the field of FCM. Sixteen CoE countries voted in favor of this continuation. In 2008, the Committee of Experts was to have prepared a detailed program and to have confirmed the priorities already set in 2007. The success of the CoE action will depend both on the collaboration of different stakeholders (governments and professional organizations) and on the adoption of new procedures for elaboration and adoption of the documents.

Annex: Answers to the Questionnaire Used During the Enquiry

	The CoE texts have been transposed?	Are the CoE Res/TD used as guidance?	Description of the major stakeholders' remarks	Answers of the stakeholder to the proposal of consultant regarding the materials to be regulated, that is, coatings → paper → inks[a]
Austria[b]	No	Yes	Inventory lists not evaluated Need for an evaluation	Yes but in the following order: 1. paper; 2: coatings; 3. inks
Belgium	Some are proposed to be adopted in the near future	Yes	More collaboration between EC and EFSA to CoE activity	Yes
Cyprus	n.a.[c]	n.a.	n.a.	n.a.
Finland	No	No	Participation of all MS/meeting in city easily to be reached by flight/better organization of the meeting	1. Paper; 2. coatings
France (Feigenbaum)[d]	No	Yes	Inventory lists not evaluated Need for an evaluation	Yes. To be added also: adhesives; active and intelligent packaging, ion exchange resins, silicones
France (Gaquerel)[d]	No	No	Inventory lists not evaluated	Yes. Inks in priority
Germany[d]	No	Yes	Inventory lists not evaluated Need for an evaluation	Yes
Ireland[d]	No	Yes	Inventory lists not evaluated Need for an evaluation	Yes but in the following order: 1. coatings; 2. inks; 3. paper
Italy	No	No	—	Yes but in the following order: 1. inks; 2. paper; 3. coatings
Luxembourg	n.a.	n.a.	n.a.	n.a.

(Continued)

	The CoE texts have been transposed?	Are the CoE Res/TD used as guidance?	Description of the major stakeholders' remarks	Answers of the stakeholder to the proposal of consultant regarding the materials to be regulated, that is, coatings → paper → inks[a]
The Netherlands	No	Yes	Inventory lists not evaluated Need of an EFSA networking	Yes. To be added ion exchange resins
Norway	No	Yes	Need for EFSA evaluation of substances	Yes but in the following order: 1. inks; 2. coatings; 3. silicones
Portugal	No	Yes	Inventory lists not evaluated Need of an EFSA networking	Yes. To be added: 4. cork and 5. adhesives
Slovenia	No	No	Better collaboration between CoE and EFSA and EC	Yes. To be added: 4. rubber; 5. metals and alloys; 6. wood
Spain	n.a.	n.a.	n.a.	n.a.
Sweden[d]	No	Yes	Inventory lists not evaluated Need of an EFSA networking	Yes but in the following order: 1. paper; 2. coatings; 3. inks
Switzerland	Yes. Resolutions on colorants, silicones and inks	Yes	Inventory lists not evaluated.	Yes. To be added 4. Biodegradable materials
United Kingdom	No	No. But the courts can take account of their existence	Need for an evaluation If possible, evaluation of inventory lists by EFSA or national agencies	Yes

a) Subjects indicate in priority in the questionnaire: 1. Coatings. 2. Paper and board. 3. Inks.
b) The member states labeled by an asterisk gave an informal focal point opinion that does not necessarily reflect the official position of its government.
c) n.a. = no answer.
d) Personal opinion of the representatives of the country in the Committee of Experts.

4
National Legislation in Germany

Wichard Pump

4.1
Introduction

The fundamental requirements for all food contact materials are laid down in Regulation (EC) No. 1935/2004, the so-called Framework Regulation, which was adopted by the European Parliament and the Council of the EU on October 27, 2004 [1], replacing two predecessors. This regulation is referred to by Article 31 of the German Food and Feed Act (LFGB: "Lebensmittel- und Futtermittelgesetzbuch"), enforced on September 7, 2005, modified in the version of the *Bulletin* of April 26, 2006 (BGBl: "Bundesgesetzblatt" No. I, p. 945), and recently modified by Article 12 of the law on February 26, 2008 (BGBl: No I, p. 215) [2].

Also, this code had predecessors, one being the amended Food Law of 1958 and the last one being the Food and Commodities Law of 1974.

4.1.1
Commodities Defined in LFGB

According to §2 of the Food and Feed Act (LFGB) the term "commodities" encompasses a wide range of products that the consumer comes into contact with. There are the following categories:

1) Materials and articles in the sense of Article 1, paragraph 2 of the Regulation (EC) No. 1935/2004. This comprises, for example, all kinds of materials and articles, including polymers, both plastics and rubbers, regenerated cellulose, glass, metal and alloys, paper and board, as well as adhesives, printing inks, ceramics, ion exchange resins, silicones, and varnishes and coatings, waxes, cork, textiles, wood, and so on. It includes not only packaging materials but also utensils, equipment, and containers used in industrial and domestic food processing, storage, and transport.
2) Packagings, containers, and other wrappings intended to come into contact with cosmetic preparations.

Global Legislation for Food Packaging Materials. Edited by Rinus Rijk and Rob Veraart

ISBN: 978-3-527-31912-1

3) Articles intended to come into contact with the mucous membranes of the mouth (e.g., pacifiers, teethers, babies' bottle teats, tooth brushes, etc.) – if these articles are in contact with food, they are also subject to the requirements for food contact materials.
4) Articles intended for personal hygiene.
5) Toys and play articles – the requirements for food contact materials are also imposed on toys.
6) Articles intended to not only momentarily come into contact with the human body, like clothing articles, bed linen, masks, perukes, hair parts, artificial eyelashes, and bracelets.
7) Cleaning agents and care products intended for domestic use and for commodities according to no. 1.
8) Impregnating and other finishing agents for commodities according to no. 6 intended for domestic use.
9) Agents and articles for odor improvement in rooms intended for human habitation.
10) Items 2–9 will not be described further in the following paragraphs as they are out of the scope of this chapter.

4.1.2 Basic Requirements on Commodities

The main requirements to be fulfilled by all commodities are spelt out in §§30 and 31 of the Food and Feed Act.

Section 30 states that it is prohibited to manufacture or use commodities in such a way that they are, in conventional or foreseeable use, capable of affecting human health by their substantial composition, particularly by toxicologically active substances, or by impurities.

It further states that it is prohibited to put on the market commodities as described before.

Finally, it states that it is prohibited to use commodities during the professional manufacturing or treating of foodstuffs in such a way that the commodities are capable of affecting human health when these foodstuffs are consumed.

Section 31, called "Transfer of Substances onto Food," is equivalent to Article 3, paragraph 1 of the Regulation (EC) No. 1935/2004 and lays down the following rules:

1) It is prohibited to use or to put on the market materials and articles, described as commodities in §2, which do not fulfill the requirements on their manufacturing of Article 3, paragraph 1 of the Regulation (EC) No. 1935/2004.
2) By statutory regulation, it can be ruled that materials and articles, described as food contact materials in §2, must be manufactured only in such a way that they, under the usual and foreseeable conditions of their use, do not release substances onto foodstuffs or their surfaces in quantities that are capable
   (a) to endanger the human health,
   (b) to compromise the composition or odor, taste, or appearance of the foodstuffs.

For certain substances in the commodities, it is ruled, whether and in which defined quantities, can be transferred to foodstuffs.

To put the rules of the law in §31 in other words, one could define a *principle of inertness* for all food contact materials. The only deviation from this principle, but under well-defined conditions, is allowed for "active" materials and articles according to Article 4 of the Regulation (EC) No. 1935/2004.

This general requirement of inertness, which in comparable words was already part of former provisions, had to be transformed into precise instructions. Therefore, the competent national authorities, often together with the affected industry branches, have elaborated recommendations, standards, and ordinances since 1957. From around 1975, the former EEC Commission and later on the Council of Europe started working on a harmonized Europe-wide legislation [3].

The material-specific rules and requirements, nowadays effective in Germany, are always to be consulted for the interpretation of Article 3, paragraph 1 of the Regulation (EC) No. 1935/2004 whenever migration limits or guide values have to be observed. The material-specific requirements can come both from the specific directives that have been transformed into the German Commodities Ordinance [4] and from the German or European recommendations or standards. They are outlined in Section 4.2 and can also be retrieved from the web site of the German Federal Institute for Risk Assessment (BfR) [5].

The Commodities Ordinance will be discussed in more details later.

4.1.3
BfR Recommendations

The first detailed German rules for food contact materials and articles were the well-known "BGA Plastics Recommendations," their full name being "Recommendations on the health assessment of plastics and other high polymers." They were first issued in 1958 by the Federal Health Office (Bundesgesundheitsamt, BGA) and consecutively amended, from 1994, by the BGA successor, the Federal Institute for Consumer's Health Protection and Veterinary Medicine (Bundesinstitut für gesundheitlichen Verbraucherschutz und Veterinärmedizin, BgVV), and from 2002, by the Federal Institute for Risk Assessment (Bundesinstitut für Risikobewertung, BfR).

From the beginning, the recommendations structured were material specific, that is, one for PVC, one for PE, one for PP, and further ones for all plastics used in food contact, as well as for rubber and for paper and board. All recommendations had their own positive list of admissible monomers, auxiliaries, and additives with substance input limits in the material. Additives, admissible for one polymer, for example, for PS, were not automatically allowed for another polymer class, for example, for PE, but had to be separately evaluated and then authorized, often restricted by a different input limit. In this way, different material properties could be taken into account. Substances listed in the recommendations have been introduced on the basis of a petition. The authorization procedure including migration and toxicological evaluation had always been accurately prescribed according to the state of the art in the past and is now laid down in the periodically updated EU Note for Guidance [7]. The toxicological evaluation of migration is based on worst-case transfer of the substance in question, that is, on its migration from a material manufactured with the highest intended concentration of the substance and its use under worst-case conditions in terms of time, temperature, surface to volume ratio, and food characteristics.

Through the structure of recommendations, a definite allocation of substances to specific types of materials is given, which is of advantage for the official control in the field of food contact materials. This applies to the specification of compositional limits, too. Compared to the Specific Migration Limits (SML) of Directive 2002/72/EC on plastics for food contact, this type of restriction is easier to comply with by the manufacturer and to be controlled by the authorities. On the other hand, a substance that is to be used in several materials has to be requested for every purpose because its migration behavior is strongly influenced by the type of material.

The Plastics Recommendations are not legal norms. If materials are produced in a manner that deviates from the provisions in the recommendations, responsibility for any complaints based on food law provisions lies solely with the manufacturer and commercial user.

Such material-specific rules had been established in many other European countries, such as France, Austria, Great Britain, the German Democratic Republic, Greece, Czech republic, and Slovakia (as well as in the United States, Japan, and Australia), whereas in Italy, Belgium, the Netherlands, and Spain in addition to the specific positive lists an overall migration limit (OML) was imposed as a measure of material inertness toward the foodstuff in contact.

During the years 1990–1992, that is, with the transformation of the EC Directives into national laws of the EC member states, the material-specific positive lists have been replaced with the system of universal positive lists of monomers and of additives, admissible for any polymer class and, where necessary, restricted by substance-specific migration limits, combined with the contact area or mass-related overall migration limit.

Since this structure has also been transformed into the German Commodities Ordinance, the recommendations on plastics had to be rearranged accordingly. For this purpose, substances subject to European regulations are continuously removed from the relevant recommendations in order to be in line with the legal provisions. This process is described in detail in the chapters on the recommendations in the BfR web site [5, 6].

Besides the Commodities Ordinance, the recommendations still today have their role in the safety assessment of plastics and other materials, for example, paper and board, rubber, silicones, coatings, and so on used in food contact in the fields not yet covered by harmonized European legislation. The materials covered by the recommendations are discussed in some detail in Section 4.2. Some recommendations deal with specified applications, for example, absorber pads, artificial sausage casings, conveyor belts, and so on.

When a food contact material complies with the relevant recommendation, it is justified to assume that the material is sufficiently inert according to §31 of the law. In the case of noncompliance, the enforcement authorities have to prove the breach by migration testing and a toxicological evaluation of the transfer into foodstuffs including an exposure assessment in order to demonstrate that there is a danger for human health.

4.1.4 Further Requirements

In the EU framework regulation, the protection against misleading the consumer has been newly introduced in Article 3, paragraph 2. Accordingly, §33 of the Food and Feed Act contains a comparable determination.

Also, the labeling duties for food contact articles have been broadened; the rules of Articles 15 and 17 of the framework regulation are laid down correspondingly in §35 of the Food and Feed Act. The main purpose is to provide adequate information to the consumer on the article and its safe use (especially, in the case of "active" commodities) and to ensure a reliable traceability of the article, its producers, its raw materials, and their quality.

4.2 Legal Assessment of Different Materials Classes in Food Contact

4.2.1 Plastic Materials

The Commodities Ordinance (Bedarfsgegenständeverordnung, last amendment on September 23, 2009, consolidated version: http://www.gesetze-im-internet.de/bundesrecht/bedggstv/gesamt.pdf) encompasses all kinds of commodity materials that have been enumerated under "Commodities" as the content of Article 1, paragraph 2 of Regulation (EC) No. 1935/2004. Comprehensive rules are laid down only for regenerated cellulose films and for plastics, whereas for other commodities only isolated limits or bans are described. Rubber and elastomers, paper and board, and wax casings, for example, for cheese, ion exchange resins, and silicones are not covered by specific rules.

In other words, the ordinance largely lays down the adequacy criteria for all thermoplastic materials and for thermosets except those laminates in which one layer does not consist of plastic (but, for example, of paper or metal). Only for certain substances, for example, catalysts, initiators, nonharmonized additives, and polymer production aids, it is necessary to additionally consult the relevant BfR recommendation.

Of course, the ordinance is based on the regularly updated EU Directives [11].

The main elements are

- Positive lists of authorized monomers and for the time being still incomplete lists of additives.
- The "overall migration limit."

- Limits for listed monomers and additives, when toxicologically justified, either in the form of "specific migration limits" related to food or in the form of maximal residual quantities (QM and QMA) in the material.
- Conventional exposure assumptions for the determination of SML levels (these are not spelt out in the wording of the ordinance, but implicitly used): the consumer with an average body weight of 60 kg has a total lifetime intake of 1 kg foodstuffs per day packed in plastics; this quantity is in contact with 6 dm^2 plastic surface. (This ratio of surface to food quantity of 6 $dm^2/1$ kg is derived from the packaging situation: 1 kg or 1 l food with the density 1 is enclosed by 6 surfaces × 10 cm × 10 cm packaging material). Furthermore, the food contact plastic contains the monomer or additive in question in its maximal concentration exhausting the relevant SML.
- Not described in the ordinance itself, but as background rules necessary for setting SMLs: graded requirements for toxicological information on listed substances or for the application to approve new monomers and additives, as laid down in the Note for Guidance.

In order to allow the reader an overview on the structure of the ordinance, its directory is shown and explained as far as food contact materials are concerned.

§2 Definitions (Begriffsbestimmungen): definitions of commodities covered by the ordinance and of functional barrier

§3 Prohibited substances (Verbotene Stoffe): azodicarbonamide in food contact plastics

§4 Authorized substances (Zugelassene Stoffe): refers to positive lists for regenerated cellulose and for plastics

§5 Prohibited manufacturing processes (Verbotene Verfahren): refers to rubber pacifiers or feeding bottle teats, manufactured in a way that they release detectable quantities of *N*-nitrosamines or nitrosamine precursors

§6 Maximum permissible concentrations (Höchstmengen): refers to substances with a limit of the residual quantity in the material, for example, vinyl chloride

§7 Prohibition of use (Verwendungsverbote): refers to the prohibition of professional use of materials and articles exceeding SMLs, QMs, or QMAs

§8 Transfer of substances to food (Übergang von Stoffen auf Lebensmittel): refers to detailed rules for compliance with SMLs, restrictions on the use of certain epoxy derivatives, introduction of the functional barrier concept, particular obligations in case of dual-use additives, to the applicability of modeling migration, and to details regarding the OML

§10 Labeling, obligation of proof (Kennzeichnung, Nachweispflichten): refers to obligatory written declarations of compliance with the requirements of this ordinance and with those of Regulation (EC) No. 1935/2004, as well as about analytical data of any dual-use additives

§10a Labeling of shoe articles (Kennzeichnung von Schuherzeugnissen): not relevant for food contact

§11 Analytical methods (Untersuchungsverfahren): refers to the analytical methods enumerated in Annex 10

§12 Criminal and administrative offences (Straftaten und Ordnungswidrigkeiten): refers to criminal and administrative offences regarding commodities
§13 Not-affected regulations (Unberührtheitsklausel): states that chemicals law and toy regulation are not affected

Annex 1 referring to §3
Substances not to be used in the manufacture or treatment of specific commodities (Stoffe, die bei dem Herstellen oder Behandeln von bestimmten Bedarfsgegenständen nicht verwendet werden dürfen): ban of azodicarbonamide for plastics.
Annex 2 referring to §4 para 1 and 1a and §6 sentence 1 No. 1
Substances authorized for the manufacture of regenerated cellulose films (Stoffe, die für die Herstellung von Zellglasfolien zugelassen sind): positive lists for regenerated cellulose films.
Annex 3 referring to §4 para 2–4, §6 No. 2 and §8 para 1, 1a, and 1b
Substances and products for manufacturing of food contact materials (Stoffe und Erzeugnisse für die Herstellung von Lebensmittelbedarfsgegenständen): positive lists for plastics and thermoplastic coatings on regenerated cellulose films.
Annex 4 referring to §5
Processes banned in the manufacture of specific commodities (Verfahren, die beim Herstellen bestimmter Bedarfsgegenstände nicht angewendet werden dürfen): refers to rubber pacifiers or feeding bottle teats, manufactured in a way that they release detectable quantities of *N*-nitrosamines or nitrosamine precursors.
Annex 5 referring to §6 No. 3
Commodities that may contain specific substances up to a specified amount (Bedarfsgegenstände, die bestimmte Stoffe nur bis zu einer festgelegten Höchstmenge enthalten dürfen): refers to the limit for residual vinyl chloride in PVC articles.
Annex 6 referring to §8 para 3
Commodities that may release substances into food up to a specified amount (Bedarfsgegenstände von denen bestimmte Stoffe nur bis zu einer festgelegten Höchstmenge auf Lebensmittel übergehen dürfen): refers to the limit for vinyl chloride migration from PVC articles.
Annex 10 referring to §11
Processes for analysis of specific commodities (Verfahren zur Untersuchung bestimmter Bedarfsgegenstände): cross-reference to the Official Compilation of Tests Procedures Pursuant to Section 64 of the Food and Feed Act regarding the basic rules and analytical methods for the determination of overall and specific migrations, for the release of lead and cadmium from ceramics, for the residual content and migration of vinyl chloride, and for *N*-nitrosamines and nitrosamine precursors [9].

Plastic materials and articles are regulated by the Commodities Ordinance (monomers, regulated additives), which is based on the relevant European directives. However, not all substances used in the manufacture of plastics are subject to the directives. Therefore, additional requirements for specified polymers and copolymers are listed in several BfR recommendations. The additional requirements include listing of catalysts, initiators, aids to polymerization, and not yet legally regulated additives for the specified polymer. The authorization of various substances is mainly through their maximum concentration in the polymer [6]. Despite the frequent update of the recommendations, it is necessary to examine any deviation from EU harmonized regulations as laid down in the Commodities Ordinance.

4.2.2
Colorants

In BfR Recommendation IX, colorants are defined as all substances that have coloring properties, including those possibly used as vehicles or production and processing aids, as well as any technically unavoidable contaminants.

Therefore, a list of production and processing aids is included with maximum use levels. For the colorants themselves, purity requirements on metal release and primary aromatic amines are established. In addition, any release of colored substances is not allowed.

4.2.3
Plastics Dispersions (BfR Recommendation XIV)

Coatings are defined as the layers of coating substances at large that are applied on a substrate. Coatings can typically consist of binding agents (resins and, as the case may be, hardeners), solvents, additives, pigments, and fillers.

The substrates onto which coatings are applied typically are metals, glass, and mineral substrates, whereas dispersions according to Recommendation XIV typically are used for paper and board and in some cases for plastic substrates.

Two types of dispersion coatings are discussed in the Recommendation. One concerns the coating on a support while the second type is a coating on cheese and not intended to be eaten. For coatings on a substrate, a rather extensive list of monomers, additives, and production aids is included because positive lists for coatings are not yet harmonized. Among the production aids are catalysts, polymerization regulators, emulsifiers, stabilizers, antimicrobials to protect the emulsion, slip agents, defoaming agents, and antioxidants. Many substances are restricted by the input limits. The use of antimicrobials should not have a preservative effect on the food.

For the cheese coatings, a relative short list of substances is included and it includes about the same categories as listed for coatings on substrates. Dyes are listed and with a reference to natural colored foodstuffs and allowed food additives.

For coating systems in contact with food outside the scope of Recommendation XIV, the general requirements of Article 3 of Regulation (EC) No. 1935/2004 are valid as a basic principle. For the interpretation of these requirements, the positive lists of monomers and of additives in plastics of the Commodities Ordinance or otherwise of Directive 2002/72/EU and its amendments can be used.

One has to bear in mind though that these substance lists in the case of coatings do not have the legal status of (exclusive) positive lists. That means these starting substances are admitted for the production of coatings and also more substances may be used.

Furthermore, migration limits for BADGE (bisphenol A-diglycidylether), BFDGE (bisphenol F-diglycidylether), and NOGE (novolak-glycidylether) according to the so-called Epoxy-Directive 2002/16/EU and its amendments [8] have been incorporated into the Commodities Ordinance, cf. Anlage 3, Chapter 3.

The assessment of coatings with respect to the migration of substances is more complex than with plastics. This relates to the fact that the raw materials for coatings mostly are not the listed monomers but the already partly polymerized or even cross-linked oligomeric resins. (So what constitutes the starting substance: the monomer or the resin?). Also, the intermediate products of the resin can be toxicologically relevant, the more so, as molecules up to 1000 Da can be seen as capable of migrating.

For coatings in contact with drinking water, the Federal Environment Agency has issued a "guideline for the hygienic assessment of epoxy coatings and organic coatings, respectively, in contact with drinking water" [10], which is described as recommendation based on §17, Chapter 1 of the Drinking Water Ordinance 2001 [11]. Annex 1 of the guideline contains a positive list of starting substances admissible for the production of their corresponding resins and hardeners, and of solvents, additives, and further assisting agents. The authorization of substances requires an application procedure including toxicological assessment according to the Note for Guidance.

4.2.4 Silicones (BfR-Recommendation XV)

The recommendation sets requirements for silicon oil, silicon resins, and silicon elastomers. Monomers and production aids and additives are listed with their restrictions when relevant. For each type of silicon, a product-specific positive list is established. For silicon elastomers used in paper coating, some additional substances are authorized. Specific requirements are established for silicon teats and the like. The list of authorized production aids is very limited. General restrictions on volatile and extractable substances are set, while peroxides should not be detectable.

4.2.5 Rubber and Elastomers (BfR Recommendation XXI)

This term comprises a large variety of materials with elastomer properties, including natural and synthetic rubbers, but not silicone rubbers, which are covered by the BfR Recommendation XV (cf. Section 4.2.4). All of them are characterized by their rubber-elastic state, due to a wide meshed cross-linking of the macromolecules, achieved by vulcanization (curing), by bifunctional agents, including elemental sulfur and vulcanization accelerators or organic peroxides.

Rubber recipes are multisubstance mixtures comprising polymers, vulcanization agents, fillers, plasticizers, and so on.

The rubber ingredients admissible for food contact are listed in the BfR Recommendation XXI (commodities based on natural and synthetic rubber) [6] that is applicable for the evaluation of compliance with Article 3 of Regulation (EC) No. 1935/2004. This recommendation contains positive lists for starting substances, additives, and production aids that can be used for manufacturing solid rubbers and latices, as well as requirements for finished articles with respect to their components subject to potential migration such as *N*-nitrosamines, secondary aliphatic amines, primary and secondary aromatic amines, and allergenic proteins.

The majority of rubber articles for food contact are neither with their total surface nor for longer time in contact with food. These specific-use conditions should be taken into account in the assessment.

According to Recommendation XXI, food contact commodities made from rubber are divided into four categories and one special category, cf. Table 4.1. Each category is characterized by its own positive list, such as starting substances, polymerization agents, vulcanization agents, accelerators, based on the principle of input limits for all admissible ingredients. The categories also contain their specific conditions (time, temperature) for the migration determination and their specific overall migration

Table 4.1 Examples of food contact articles made from elastomers.

Use	Category	Contact conditions
Preserve rings, juice caps	1	Long-term contact
Sealings for bottle closures	1	Long-term contact
Sealings for pressure cooking jars	2	Mid-term contact
Tubes for coffee machines	2	Mid-term contact
Tubes for transport of foodstuffs	2	Mid-term contact
Dough scraper	3	Short-term contact
Teat rubbers	3	Short-term contact
Tubes for milking machines	3	Short-term contact
Conveyor belts	3	Short-term contact for fatty food, insignificant contact for others
Gloves, aprons worn during food processing	3	Short-term contact
Sucking and pressure tubings	4	Insignificant contact
Sealings for conduits, pumps, plug valves for liquid foodstuffs	4	Insignificant contact
Pacifiers and teats for feeding bottles	Special	Specific contact conditions

Long-term contact: contact time longer than 24 h up to several months; mid-term contact: contact time up to maximal 24 h; short-term contact: contact time maximal 10 min; insignificant contact: no release to be expected.

Table 4.2 Test conditions and restrictions set for the total migration.

Category	Test conditions	Overall migration limit (mg/dm^2)	
		Water/10% ethanol	3% acetic acid[a]
Category 1	10 d at 40 °C	50	150
Category 2	24 h at 40 °C	20	100
Category 3	2 h at 40 °C	10	50
Category 4	—	—	—
Special	24 h at 40 °C	50[b]	—
		20[c]	—

a) Migration of organic substances shall not exceed the values set for water.

b) Overall migration in water toys and toy balloons.

c) Overall migration in all other examples.

limits. Some groups of articles cannot be attributed to this pattern; for them, the migration conditions have to be orientated at their practical use conditions (Table 4.2).

For the categories 1–3 and the special category, specific migration limits of, for example, amines are established and requirements for cleaning are set.

4.2.6 Hard Paraffins, Microcrystalline Waxes and Mixtures of these with Waxes, Resins, and Plastics (BfR Recommendation XXV)

Mixtures of hard paraffins, microcrystalline waxes with resins and plastics intended for the manufacture of impregnations, coatings, and contact adhesives for foodstuff packaging and other commodities are regulated. In an additional part, the use of waxes as cheese coating not meant to be eaten are regulated. The recommendation mainly contains specifications of the various components that may be mixed to manufacture the coating. The coatings should not be in direct contact with fatty foodstuffs.

4.2.7 Conveyor Belts Made from Gutta-Percha and Balata (BfR Recommendation XXX)

Raw materials allowed are listed while contact with fatty foods is not allowed. An extraction restriction and a maximum contact time are defined as well.

4.2.8 Paper, Carton, and Board

Food contact materials made from paper, carton, and board consist from fibers mostly of plant origin, which by dewatering of a fiber suspension on a sieve form a felt-like area-measured material. (In Germany, the raw materials for fibers are pulp and secondary fibers obtained by paper recycling.) To achieve certain properties, fillers, production aids, and paper finishing agents are added.

The large diversity of additives and production agents can only be mentioned by keywords: fillers and pigments, gluing agents, retention agents and dewatering accelerators, precipitating and fixing agents, foam prevention agents, preserving substances, wet-hardening agents, water repellent-making agents, and oleophobic-rendering agents.

Paper and cartons are mostly used in contact not only with dry foodstuffs but also with wet, liquid, or fatty food. These conditions have, of course, to be observed in the safety assessment.

As concrete interpretation support for the general requirements of Article 3 of the Framework Regulation (EC) No. 1935/2004 with respect to the release of substances into foodstuffs, the BfR Recommendations XXXVI are used [6].

In these recommendations, the raw materials, the production aids, and paper finishing agents corresponding to their intended use conditions are listed following technological and toxicological criteria and their input limits. In addition, requirements on the finished article are laid down; some examples are listed here:

- Limits for heavy metal ions in the paper extract
- Limitation for the release of formaldehyde, glyoxal, and the epichlorohydrin hydrolysis products due to the use of wet-hardening agents
- Prohibition of the release of colorants and optical brighteners
- Prohibition of a preserving effect on the contacted foodstuff due to the use of slime-preventing agents or preserving substances
- Prohibition of the release of primary aromatic amines, for example, due to the use of polyurethane-containing gluing agents
- Minimization dictate for diisopropylnaphthalene (DIPN), which can be present in the food contact paper due to the use of recycled fibers as raw input and which is then transferred to the food via the gas phase.

In the following sections, an overview of the paper recommendations is given.

4.2.8.1 XXXVI: Paper, Board for Food Contact

Recommendations for paper and board are one of the most consulted and respected areas of food contact materials that are not harmonized at European level. The scope comprises paper and cartons for the contact with dry, moist, or greasing foodstuffs, for example, sugar, cocoa, flour or cereal (dry food), gateau lace, snack dish, pizza cartons, and "baker's silk" or whipped cream covering paper (in contact with moist or greasing foodstuffs).

4.2.8.2 XXXVI/1: Cooking and Hot-filter Paper and Filter Layers

The materials and articles referred to in this recommendation concern contact with aqueous foods only, such as tea and coffee filter papers. A reduced list of raw materials (recycled materials not allowed) and production aids is given. For specified applications, some additional components are listed and some restrictions are established.

4.2.8.3 XXXVI/2: Papers, Cartons, and Boards for Baking Purposes

Such paper should withstand a temperature of 220 °C without decomposing. Raw materials including some polymer fibers and some recycled paper are defined. Compared to the Recommendation XXXVI, a reduced list of production aids is given. Restrictions are mainly based on the quantity in the finished material or in a hot water extract. The papers must not be used above 220 °C, but it is recognized that in case of contact with moist foodstuffs the temperature will not exceed 100 °C. Baking papers usually are treated with a sizing agent on the food contact side. The authorized size coatings are listed including silicone coating in order to enhance the release of the baked good.

4.2.8.4 XXXVI/3: Absorber Pads based on Cellulosic Fibers for Food Packaging

The absorber pads serve to absorb water released from packaged food such as meat, fish, poultry, and so on. The absorber pads may be manufactured from listed substances, for example, raw materials and production aids. The positive list is rather limited. The finished product should not show a preservative effect on the food while transfer of some metal ions is restricted.

4.2.9 Artificial Sausage Casings (BfR Recommendation XLIV)

European legislation on regenerated cellulose excludes artificial sausage casings; therefore, a recommendation has been prepared to cover artificial sausage casings that are not intended to be eaten. In this recommendation, a long series of artificial casings are inserted:

1) Artificial casings of cellulose hydrate (cellophane)
2) Artificial casings made of real parchment
3) Artificial casings made of protein-coated woven fabric
4) Artificial casings made of hardened protein
5) Artificial casings made from polyterephthalic acid diol esters
6) Artificial casings made of polyamides
7) Artificial casings made from copolymers of vinylidene chloride
8) Artificial casings made from polypropylene
9) Artificial casings made from polyethylene
10) Artificial casings made from plastic-coated woven fabric
11) Artificial casings made from plastic-coated knit fabric of polyamide or polyterephthalic acid diol esters
12) Artificial casings made from protein-coated woven fabric of polyamide or polyterephthalic acid diol esters

For all casings, a positive list of raw materials, production aids, and coatings is established. Length of the list differs with material type. Plastic casings are also subject to European legislation but concerning the use of additives, specific substances may be allowed. Residual content restrictions for individual substances are commonly set.

If colorants are used to color the artificial casings, they must comply with the requirements of Recommendation IX.

4.2.10
Materials for Coating the Outside of Hollow Glassware (BfR Recommendation XLVIII)

Substances allowed for coating the outside of glasses and bottles and the like is listed. There are no restrictions given, since when the coating is applied properly, only very small amounts, if any at all, find their way onto the inside of the glassware.

4.2.11
Soft Polyurethane Foams as Cushion Packaging for Fruit (BfR Recommendation IL)

Packaging for fruits may be manufactured from polyurethane foams. The manufacture of these foams is authorized by their monomers while additives authorized in the relevant commodity regulation are allowed along with some additional additives. The catalysts or activators needed in the polymerization process are regulated as well. Maximum use quantities may be established.

4.2.12
Temperature-Resistant Polymer Coating Systems for Frying, Cooking, and Baking Utensils (BfR Recommendation LI)

Polymers intended to be coated on metal substrates and used for baking, cooking, and frying of foods are listed. Also, binding agents, adhesion promoters, processing aids, and emulsifiers are included in the list. The recommendation requires instructions of cleaning before the first use. Some substances are subject to migration testing. For that purpose, migration conditions are prescribed. Other substances are limited by the maximum residue in the finished article. The specific migration limits and the residual content restrictions are expressed in mg/dm^2.

4.2.13
Fillers for Commodities Made of Plastic (BfR Recommendation LII)

Fillers are listed for plastics that include in this case also elastomers and coatings. Fillers are assumed to be used at a concentration of 5% and more, but which are not used as colorants. Various types of additives are included. Purity requirements are set on the extractability of some metal ions.

4.2.14
Absorber Pads and Packagings with Absorbing Function, in which Absorbent Materials Based on Cross-Linked Polyacrylates are Used, for Foodstuffs (BfR Recommendation LIII)

This recommendation deals with the acrylic acid-based cross-linked polymers. The substances used for cross-linking must be announced to the BfR. Production aids in compliance with Recommendation XIV may be used.

The absorbing capacity of absorber pads with respect to packaging systems must be sufficient to absorb the total amount of liquids released by the packaged foodstuffs.

References

1 Regulation (EC) No. 1935/2004 of the European Parliament and of the Council of October 27, 2004 on materials and articles intended to come into contact with food and repealing Directives 80/590/EEC and 89/109/EEC, see also Ref. [8].

2 Lebensmittel-, Bedarfsgegenstände- und Futtermittelgesetzbuch (Lebensmittel- und Futtermittelgesetzbuch – LFGB) LFGB Ausfertigungsdatum: 01.09.2005 Vollzitat: Lebensmittel- und Futtermittelgesetzbuch in der Fassung der Bekanntmachung vom 26. April 2006 (BGBl. I S. 945), zuletzt geändert durch Artikel 12 des Gesetzes vom 26. February 2008 (BGBl. I S. 215) *Stand*: Neugefasst durch Bek. v. 26.4.2006 I 945; Zuletzt geändert durch Art. 12 G v. 26.2.2008 I 215, see also Ref. [5].

3 Brauer, B., Schuster, R., and Pump, W. (2006) Lebensmittelbedarfsgegenstände. Chapter 38, in *Taschenbuch für Lebensmittelchemiker*, 2nd edn (ed. W. Frede), Springer-Verlag, Berlin, pp. 846–901, http://www.springer.com/life+sci/food+science/book/978-3-540-28198-6.

4 Bedarfsgegenständeverordnung BedGgstV Ausfertigungsdatum: 10.04.1992 Full citation: Bedarfsgegenständeverordnung in der Fassung der Bekanntmachung vom 23. December 1997 (BGBl. 1998 I S. 5), zuletzt geändert durch Artikel 13 der Verordnung vom 8. August 2007 (BGBl. I S. 1816) *Stand*: Neugefasst durch Bek. vom 23.12.1997: 1998 I 5: zuletzt geändert durch Art. 13 V vom 8.8.2007 I 1816, see also Ref. [5].

5 BfR web site: http://www.bfr.bund.de/cd/template/index_en and more specifically dealing with commodities: http://www.bfr.bund.de/cd/527.

6 Web site of BfR: Database Plastic Recommendations: http://bfr.zadi.de/kse/index.htm.

7 Note for Guidance http://www.efsa.europa.eu/EFSA/efsa_locale-1178620753812_1178620772989.htm.

8 EU food contact legislation http://ec.europa.eu/food/food/chemicalsafety/foodcontact/eu_legisl_en.htm.

9 Collection of analytical methods (web site available only in German) http://www.methodensammlung-bvl.de/.

10 UBA Drinking Water Guidelines http://www.umweltbundesamt.de/wasser-e/themen/drinking-water/epoxidharzleitlinie.htm.

11 German Drinking Water Ordinance http://www.eugris.info/displayresource.asp?ResourceID=3966&Cat=document.

5
The French Regulation on Food Contact Materials

Jean Gauducheau and Alexandre Feigenbaum

5.1
Introduction

The French regulation on food contact materials (FCM) dates back to 1905. Some texts in force even today refer to that law of 1905. Since then, the French lawmakers have expressed their constant concern for consumer protection, with food safety being a top priority. Progressively, laws and regulations came into force to take into account new concerns, facing new practical situations.

The French regulation is based on the principle of positive lists. These were constantly considered as the best way to protect the consumer's health and to ensure that none of the materials in contact with food contains any substance likely to endanger human health. Very early, official bodies were in charge of verifying the toxicological properties of substances before agreement. Many regulations, still in force, contain lists of approved plasticizers, of solvents, of colorants, and so on that had been accepted for the preparation of materials.

When the European bodies began to make laws in this area, the French regulation obviously became consistent with them. The directives were progressively introduced in the French regulation, not as simple translations, but by introducing in the form of French texts and guidelines, in compliance with the principle of "directives." Like other Member States, France is still free to make laws in areas that are not yet regulated at European level. In contrast to directives, European regulations are automatically in force in all member states, without the need of an adaptation into national laws.

The French regulation on FCM was built up progressively. As a consequence, it consists in a number of laws, sometimes independent of each other, sometimes connected, and sometimes overlapping. This gives a quite complex picture where it is difficult to find one's way. We will here give an overview and reclassify the texts, especially for areas that are not taken over by the European Commission.

Global Legislation for Food Packaging Materials. Edited by Rinus Rijk and Rob Veraart

ISBN: 978-3-527-31912-1

The French regulation on FCM deals with materials in contact with food and with feed. It covers

- materials destined to come in contact with foodstuffs: the responsibility of these materials lies with those who produce the materials
- materials already in contact with food: the responsibility that they are suitable for the foodstuffs in contact lies with those who put the packaged food on the market.

5.1.1 Basic Principles

The regulation includes recommendations on frauds on weight, volume, nature, and composition of foodstuffs. However, these are out of the scope of this chapter. The basic principle of FCM regulation, like for most national and EU regulations, is that the materials in contact with food must not

- endanger the health of human and animal consumers;
- modify in an unacceptable way the composition of the food;
- modify in an unacceptable way the organoleptic properties of the food.

To fulfil these requirements, France has elaborated various positive lists for different materials. The overview given in this chapter is not exhaustive.

In the European context, when a European specific regulation does not exist, or if it coexists with a national legislation, a material intended to come in contact with foodstuffs must be used in compliance with the law of the country where the food is packaged for the first time with this material, irrespective of the country the material has been produced in.

The Direction Générale de la Concurrence, de la Consommation et de la Répression des Fraudes (DGCCRF, Consumer, Competition, and Frauds Office, under the Ministry of Finances) is the agency to enforce the legislation. Since 2000, advice on food safety issues is given by the French Food Safety Agency (AFSSA).

5.1.2 Categories of Reference Binding Texts

To facilitate the retrieval of the reference texts needed, we use in this chapter the original French names.

5.1.2.1 Binding Texts: Lois (Laws), Décrets (Decrees), and Arrêtés (Orders)

These address a whole range of stakeholders. Laws, decrees, and orders are published in the Official Journal (JORF).

Lois (laws): They are prepared by ministers and adopted by the National Parliament and by the Senate. The President of the French Republic promulgates them.

Décrets: They are presented by the Prime Minister, on the basis of a report by one or several ministers. They are based on (or are consistent with) French laws, with European directives or regulations, or with other decrees, which they update.

Arrêtés (orders): They are decisions of one or several Ministers, in application of a law or of a decree.

5.1.2.2 Additional Information on the Texts

Circulaires, Lettres Circulaires

In a given area, these texts are adopted by a Minister or by a responsible authority for administration. These are intended for the agents of this administration in charge of the application of the law. If these are adopted by a Minister, the *Circulaires* are published in the *JORF*. If they are adopted by a responsible authority of an administration, they are not published. But they are usually available on the Internet (a convenient web site is that of the Laboratoire National d'Essais, LNE http://www.contactalimentaire.com/). Those concerning FCMs are compiled in the Brochure 1227, dedicated to FCMs, and edited by the *JORF*. The last update of this Brochure 1227 was published in 2002 (see below).

They are usually positive lists or addenda to positive lists.

Instructions

These texts are adopted by the responsible authority of DGCCRF and are sometimes cosigned by a minister. They are published in the *Bulletin Officiel de la Direction Générale de la Consommation de la Concurrence et de la Répression des Fraudes* (BOCCRF). See the DGCCRF web site: www.finances.gouv.fr/DGCCRF/.

Notes d'Information (Information Notes)

These are prepared by DGCCRF and contain information on legislative texts, work done by DGCCRF, and so on. Although these are in principle meant for the staff of DGCCRF, given their usefulness to the outside world, they are widely published, for example, on the Internet. See the DGCCRF web site www.finances.gouv.fr/DGCCRF/.

Advices, Scientific Opinions

The scientific opinion of the French Food Safety Agency is often sought by Ministers, by other administrations, or by other stakeholders (industry via DGCCRF, consumer associations, etc.). Before AFSSA, the *Conseil Supérieur d'Hygiène Publique de France* (CSHPF) was the agency responsible for this task. The opinions are usually endorsed into laws. Even if they are not endorsed, they are considered as reference documents and major recommendations. They are published in the Brochure 1227.

Brochure 1227 presents all the texts mentioned in Table 5.1 and is dedicated to materials intended to come in contact with food. It can be ordered from Les éditions des Journaux Officiels, 26 rue Desaix, 75 227 Paris cedex 15, France, www.journal-officiel.gouv.fr/.

Table 5.1 Reference texts in France: all of them are not laws, therefore we use the more general word "texts.

Type of text	Level of responsibility	Status
Laws	President of the Republic	Law
Decrees	Prime Minister	Law
Arrêté	Ministers	Law
Circulaire, letters circulaires	A minister or the head of the administration	Notes for implementation (for administration in charge of it)
Instruction	Head of DGCCRF (Consumer, Competition and Fraud Office)	
Information notes	A responsible authority of a given sector of DGCCRF	Summarizes the state of art
Opinions, advice	French Food Safety Agency	Scientific advice (of CSHPF and since 2001 of AFSSA)
Consumption code	Legal principles	

But these texts are usually binding.

The last edition was published in 2002. When the brochure is ordered, later texts are usually sent as additional leaflets. It is of general use in commercial recommendations to state in certifications "Compliant with Brochure 1227."

Texts of *circulaires*, instructions, notes d'information, advices can be found on the LNE web site: www.contactalimentaire.com/index.php.id=560.

5.1.2.3 The Consumption Code (Code de la Consommation)

This is a complete set of legal principles related to consumer protection. All decrees and arrêtés must be consistent with this code. Materials in contact with foods are covered under the following articles:

L214-1: characteristics of goods such as

- conditions of preparation, preservation, and storage of foodstuffs
- conditions of determination of the hygienic characteristics of foodstuffs
- conditions of labeling
- traceability.

L214-2: sanctions applicable

L214-3: taking into account EU regulations

So the decree 2007-766 of May 10, 2007 includes the EU Regulation 1935/2004 in Article L214-1 of the Consumption Code.

5.2 Integration of European Directives and Regulations on Food Contact Materials into the French Regulation

By law, regulations of the European Union are integrated as such into the French legislation. Directives are generally transposed, either as new laws or as modifications

Table 5.2 EU directives on FCMs that have not been transposed into the French regulation.

EU regulation that has not been transposed	Context
Directive 2001/62 on plastic materials and articles intended to come into contact with food	Repealed and replaced by Directive 2002/72, which has been transposed into a French regulation through the Arrêté of January 2, 2003
Directive 2002/17 on plastic materials and articles intended to come into contact with food	
Directive 2002/16 on some epoxidic compounds used for materials in contact with food	Repealed by Regulation 2005/1985

of the existing laws. A word for word translation is usually not required, but the principles and the outline of the directives are respected.

Some directives reported as such in the last edition of Brochure 1227 have not been adapted into the French regulation. These directives were later on repealed. These are presented in Table 5.2.

5.2.1
Main EU Regulations on Materials Intended for Contact with Food

This topic is dealt with in detail in Chapter 1 of this book. Let us recall major regulations:

Regulation 1935/2004 of 27 October 2004 on materials and articles intended to come in contact with foodstuffs.

Regulation 1895/2005 of 18 November 2005 concerning the limits of use of certain epoxidic derivatives for materials and articles intended to come in contact with foodstuffs.

Regulations 372/2007 of 22 December 2007 and *597/2008* of 24 June 2008 related to good manufacturing practices of materials intended to come in contact with foodstuffs.

Regulation 375/2007 of 2 April 2007 fixing provisional migration limits for plasticizers used in gasket seals intended to come in contact with foodstuffs.

Regulation 282/2008 on recycled plastic materials and articles intended to come in contact with foods and amending Regulation (EC) No. 2023/2006.

5.2.2
Transposition of the Directives on Materials Intended to Come in Contact with Foodstuffs

5.2.2.1 Regenerated Cellulose Films

Arrêté of 4 November 1993 related to materials and articles made of regenerated cellulose film in contact or intended to come in contact with food products, foodstuffs, and drinks. It transposes the directive 93/10.

Arrêté of 21 October 2004, modifying the arrêté of 4 November 1993.

This arrêté refers to directives 93/10, 93/11, and 2004/14.

One should notice the variations in the titles of the French arrêtés relative to the European directives

- *"in contact or intended to come in contact"* against *"in contact"* with foods in the EU regulation. The French regulation targets not only the responsibility of the materials' manufacturer but also that of the materials' user;
- *"with food products, foodstuffs, and beverages"* against *"with foodstuffs"* in the EU regulation. The French transposition clarifies the ambiguity of the term *"foodstuffs."*

The arrêté of 21 October 2004 has to be read together with that of 4 November 1993 since it presents only modified sentences and paragraphs. Directive 2007/42 of 29 June 2007, which repeals directives 93/10, 93/11, and 2004/14 has still not been transposed into the French regulation till the time this book went into press.

5.2.2.2 Plastic Materials

Arrêté of 30 January 1984, related to official methods of analysis, methods concerning the content of vinyl chloride. Two arrêtés signed on the same day incorporate the directives 78/142, 80/766, and 81/432.

Arrêté of 2 January 2003, related to plastic materials and articles intended to come in contact with food products, foodstuffs, and beverages. It transposes the Directive 2002/72 concerning materials and articles intended to come in contact with foodstuffs. It also includes in Chapter VI rules on food simulants and on migration measurements as set in directives 82/711 and 85/572.

Arrêté of 9 August 2005 modifying the arrêté of 2 January 2003 to take into account ER regulations 1935–2004.

Arrêté of 29 March 2005 modifying the arrêté of 2 January 2003 to transpose the Directive 2004/1, banning the use of azodicarbonamide as blowing agent.

Arrêté of 19 October 2006 modifying the arrêté of 2 January 2003 to take into account Directive 2005/79.

Arrêté of 19 November 2008 modifying tha arrêté of 2 January 2003 to take into account Directive 2008/39.

5.2.2.3 Ceramics

Arrêté of 7 November 1985, related to limitation on the quantities of lead and cadmium extractable from objects in ceramics intended to come in contact with food products, foodstuffs, and beverages. This transposed Directive 84/500 into a national regulation.

Arrêté of 23 May 2006, modifying the arrêté of 7 November 1985, including the provisions of Directive 2005/31.

5.2.2.4 Rubber

Arrêté of 9 November 1994, modified by Arrêté of 19 December 2006, related to materials and objects in rubber in contact with food products, foodstuffs, and beverages. This arrêté is much more general than Directive 93/11, which is restricted to teats and soothers. However, it takes over the terms of that directive concerning both the rules for determining *N*-nitrosamines and of *N*-nitrosable substances and the criteria applicable for their determination.

5.3 Specific French Legislation on Plastic Materials Intended to Come in Contact with Food

As we have seen in the previous section, most specific European directives on FCM dealt up to now with plastics. These directives have been incorporated in the French legislation. However, some issues concerning articles and packaging materials made from plastics have still not been regulated by the European Union. Some of these issues have been regulated in France, such as the selection of coloring agents. This specific legislation is mandatory in the French territory.

5.3.1 Reference Texts on Coloring Agents for Plastic Materials

Coloring agents, like any other additive used to manufacture food contact materials, are considered with the principle of positive lists: only authorized substances may be used, whether for coloring the core of a material or for a surface printing.

See Section 5.6; it deals in detail with colorants that can be used in particular for plastics in contact with food.

5.3.2 Reference Texts on Additives for Plastic Materials

They are referred to as arrêtés, instructions, or advice (see Table 5.1) following petitions by industry. Their technological function is often mentioned, at least for antioxidants. When such an additive becomes regulated by a EU regulation, this obviously supersedes the French regulation.

5.3.2.1 Information Note of DGCCRF 2003-27 of 24 March 2003

This note sums up in a list all additives authorized in France. Until that day, they were scattered in different texts compiled in Brochure 1227. For each additive, possible restrictions on use are specified.

5.3.3 Reference Texts on Recycled Materials

5.3.3.1 Avis of CSHPF Dated 1 June 1993

This text admits the process of production of PET from DMT obtained by methanolysis of recovered and collected PET.

5.3.3.2 Avis of CSHPF Dated 7 September 1993

This opinion of the High Public Health Council (CSHPF) deals with all recycled materials, including plastics. CSHPF "expresses its reservations about the use of recycled plastics in contact with foodstuffs when they do not present the same guarantees as the virgin materials they could replace." This applies to any recycled material and in particular to plastics.

5.3.3.3 Avis of AFSSA of 27 November 2006

This text gives general recommendations for PET recycling. It indicates that the processes have to be authorized and lays down guidelines for applicants. These guidelines also propose a useful approach for risk assessment of recycled PET.

In December 2007, AFSSA published draft guidelines for public consultation, with a broader scope than those of the EFSA, as they deal with all recycled plastics. The approach given for PET in AFSSA guidelines (published in English) describes a very precise and detailed risk assessment for PET, which is consistent with EFSA's approach.

See http://www.afssa.fr/. An English version of the guidelines is available on the web site of AFSSA.

5.4 Supplementary French Legislation on Materials other than Plastics Intended to Come in Contact With Food

5.4.1 Introduction

French regulation on areas not yet regulated by the European Union is still in force in France. These regulations concern

- actual materials: paper and board, tissues, silicone elastomers, rubber, various metals, wood;
- products to protect or to decorate the materials: coatings, varnishes, inks;
- substances and products that may come in contact with food deliberately (such as ion-exchange resins) or unintentionally (such as residues of dish washing products).

In addition, some reference texts present fundamental principles

- on the application of regulations in France;
- on the process of applications for approval of new substances to be used in specific areas falling exclusively under the scope of the French regulation.

5.4.2
Application of Regulations: the Decree 2007-766 of 10 May 2007

This decree is representative of the principles mentioned above. This reference text is named "application of the Consumption Code as regards the materials and articles intended to come in contact with foodstuffs." It refers in its preamble

- in priority to the EU Regulation 1935/2004 of 27 October 2004;
- to the Code of Consumption;
- to the law of 1 August 1905 on the suppression of frauds and to its application decree (no. 73-138 of 12 February 1973);
- to the decrees no. 92-631 of 8 July 1992, no. 99-242 of 26 March 1999, and no. 2001-1097 of 16 November 2001 on materials and articles destined to come in contact with food products, foodstuffs, and beverages for human and animals (food and feed products);
- to an advice of AFSSA.

This text emphasizes one of the basic principles: the EU regulation is included in the existing context of the French regulation, Code of Consumption and decrees with ad hoc modifications. The Regulation 1935/2004 is also included but only for its scope: "the arrêtés taken in application . . . of the decree of 12 February 1973 . . . and of 8 July 1992 remain valid as far as they do not contradict the provisions of the Regulation of 27 October 2004." Thus, "the objects which, by their aspect seem destined to come into contact with food, without being into the scope of the Regulation of 27 October 2004, must display, in a visible and indelible manner the mention or the symbol (fixed by law) . . . that they cannot be set in contact with foodstuffs."

The section also mentions sanctions incurred by offenders.

The European regulation is applied in France but through the French regulation. The French regulation cannot contradict the EU regulation, but it can be more precise or complement it.

5.4.3
French Regulation on Materials in Paper and Board Intended to Come in Contact with Food

Arrêté of 18 June 1912 related to coloring, storage, and packaging of foodstuffs and beverages. This has been updated by several reference texts, from 1952 to 1993. Article 7 deals specifically with paper.

> It is forbidden to place in direct contact with papers hand written or printed in black or in color foodstuffs intended for consumption other than roots, tubers, bulbs, fruit with a dry shell, pulses or vegetable with leaves. Furthermore it is forbidden to place any other paper than folding papers either white or straw-colored or colored by means of an authorized substance [the list is in annex of the arrêté] in contact with bread and with aqueous or fatty foodstuffs or with foodstuffs which may stick to the said paper.

The attached positive list includes two parts:

- Dyes that may be used for coloration in the core and on the surface of the paper. "Surface" refers to printing inks.
- Pigments "for the sole coloration of the surface" (inks); there is a very short list: calcium carbonate, titanium dioxide, iron oxides, aluminum, silver, and gold.

This text has been amended by the Circulaire 176 that authorizes other coloring substances in food packaging and by various advices of CSHPF. The major relevant advice is that of 7 November 1995, which recommends that *"the printed face of a packaging material, whether or not coated with a varnish, should not come in contact with foodstuffs."* Monitoring of migration is imposed by other reference texts, in particular, the EU Regulation 1935/2004.

This will be discussed in more detail in the chapter on coloring substances.

Circulaire No. 170 of 2 April 1955, concerning paper for packaging of foodstuffs. This reference text authorizes

- the use of authentic greaseproof paper (treated with sulfuric acid *"to give a low porosity, a great impermeability to water and to fatty substances and a great resistance at humid state"*);
- paper coated with the substances authorized for regenerated cellulose films.

Note d'information no. 2004-64, completing the note no. 155 of 26 October 1999, related to materials in contact with foodstuffs.

The chapter "paper and board" deals with

- paper and board based on natural fibers and which may also contain synthetic fibers;
- coated paper or paper having been submitted to a surface treatment;
- tissues and paper napkins.

The following papers are not in the scope of this note: papers coated with waxes and paraffins, complexes realized with other materials (see Section 5.4.7.4 on complexes), tablecloths, and tablemats.

1) Recycled paper and board that do not present the same guarantees as the virgin paper and board cannot be used in contact with foodstuffs (according to the advice of CSHPF of 7 September 1993 on recycled materials).
2) A positive list of authorized optical brighteners is attached to the advice of CSHPF of 13 October 1998.
3) Paper containing epoxidic compounds must fulfil requirements of the arrêté of 2 April 2003 (transposition of Directive 2002/16).
4) The information note also refers to the *"guide of good manufacturing practice for paper and board used in contact with foodstuffs."* This guideline has no legal value, but is a reference document since it has been approved by the High Public Health Council (CSHPF).
5) The printed face of paper and board must not be in contact with foodstuffs. In addition, printing must be in accordance with the advice of CSHPF of 7 November 1995 (see Section 5.6.4).

The information note also sets some migration limits

- for heavy metals: lead, cadmium, mercury, and chromium (VI);
- for some specific chemicals: polychlorophenols (PCP), polychlorobiphenyls (PCB), formaldehyde, and glyoxal.

Note of information no. 2006-156 of DGCCRF concerning paper and board coated with waxes, paraffins, silicones and polymeric emulsions.

5.4.4 French Regulation on Materials and Articles in Rubber Intended to Come in Contact with Food

Arrêté of 9 November 1994, related to materials and articles in contact with food products, foodstuffs, and beverages.

This has been already dealt with in Section 5.2.2. However, although it also deals with the release of nitrosamines and of *N*-nitrosatable substances from teats and soothers in elastomer or in rubber, this arrêté goes beyond the Directive 93/11.

- It deals with all the materials and articles in natural or synthetic rubber coming in contact with foodstuffs.
- It stipulates a positive list of authorized constituents: monomers and starting substances, accelerators, vulcanization agents, antioxygens, fillers, plasticizers, lubricants, emulsifiers, and so on, including their migration limits. The coloring substances authorized are the same as those authorized for plastic materials.

Arrêté of 19 December 2006 completing the previous one.

This reference text deals with acceptance of monomers, starting substances, and modifiers originating from other EU Member States as well as from Turkey or from a state belonging to the European market if they have been evaluated according to the guidelines of a European scientific body.

5.4.4.1 Note of Information 2004-64 of DGCCRF, Chapter "Rubber"

This note specifies how the regulation should be applied.

Under "rubber" fall elastomers, whether or not of natural origin, whether or not vulcanized. Rubber elastomers of silicone and gaskets for preserve cans are excluded from the scope of the regulation.

A table lists the acceptable limits for authorized substances used in the composition of rubbers, thus completing the arrêté of 9 November 1994.

5.4.5 French Regulation on Materials and Articles in Silicone Elastomers Intended to Come in Contact with Food

Arrêté of 25 November 1992 dealing with materials and objects made of silicone elstomers put or destined to be put in contact with food products, foodstuffs, and beverages.

The polymers used must be exclusively composed of organopolysiloxanes. The annex of the arrêté provides the positive list of authorized additives: fillers, scavengers, hardeners, cross-linkers; coloring matters must be selected among those authorized for materials and objects in contact with food.

5.4.6 French Regulation on Materials and Articles in Glass and Ceramics Intended to Come in Contact with Food

We have already mentioned the arrêtés of 7 November 1985 and 23 May 2006, dealing with transposition of Directives 84/500 and 2005/31, respectively.

The information note 2004-64 defines in the relevant chapters glass ceramics, vitreous ceramics, crystal, and enameled objects covered by this regulation. It refers to standards on these materials. It defines tolerable limits for lead, cadmium, and chromium (VI) and underlines conditions and methods for testing for controls.

5.4.7 French Regulation on Materials and Articles in Metals Intended to Come in Contact with Food

We will first review the general case of metallic materials and objects with emphasis on uses for packaging. In the second step, we will deal with household articles and industrial machinery coming in contact with foodstuffs.

5.4.7.1 General Case: Metallic Materials and Objects, Packaging

I-A Stainless Steel

Arrêté of 13 January 1973, related to stainless steel materials and objects in contact with foodstuffs: the text gives a list of elements that may be incorporated into the steel, with maximum concentration limits.

Information note 2004-64, chapter "stainless steel": it completes the arrêté with additional composition limits for other elements.

I-B Noncoated Steel (Black Iron)

Arrêté of 15 November 1945, fixing a list of materials that may be used without harm to public health for measuring instruments. Black iron is authorized only for contact with roots, tubers, bulbs, fruit with a dry shell, pulses, or vegetable with leaves and with foils.

Information note 2004-64, chapter on black iron.

The note refers to the guidelines on metals and alloys adopted by the Council of Europe. It confirms that the use of black iron should be restricted to fatty and dry foods. It also provides a list of elements that may be incorporated into the iron, with maximum concentration limits.

I-C Metal-Coated Steel (Tinned Iron)

Arrêté of 18 June1912, modified by the *arrêté of 5 July 1956* related to coloration, preservation, and packaging of foodstuffs and beverages. These texts are still part of the regulation; however, the limits have been modified.

Information note 2004-64, chapter on tinned iron.

Restrictions of use are specified for contact with highly acidic foodstuffs. The note also specifies composition limits for lead, cadmium, and arsenic in the coating. The steel itself must obey the rules set for black iron.

I-D Steel-Coated with Organic Coating

The steel must obey the same requirements as above.

We deal in a dedicated chapter with rules for coatings.

I-E Aluminum and Aluminum Alloys

Arrêté of 27 August 1987, related to materials and objects in aluminum or aluminum alloys in contact with food products, foodstuffs, and beverages.

- The metal must be composed of at least 99% aluminum. The maximum content of each possible impurity is specified.
- Coatings applied, if any, must comply with the regulation (see below).
- Anodization conditions are specified.

Information note 2004-64, chapter on aluminum.

The aluminum producer must give to the packaging manufacturer a certificate of compliance with the arrêté of 27 August 1987 along with analytical reports demonstrating compliance with the limits set in this arrêté.

I-F Tin and Tin Alloys

We deal here with tin used for tins and cans. We see below uses of tin in household articles.

Arrêté of 28 June 1912, already cited

> It is forbidden to place beverages or foods in direct contact with containers, utensils, tinned instruments coated or soldered with tin containing more than 0.5% lead or more than 3/10 000 arsenic or less than 97% tin, determined in metastannic state.
>
> It is authorized to proceed to solders on the outside of containers with tin and lead alloys, provided that there is no penetration of the lead containing alloy into the container, unless as excess material or incidents; penetration of the alloy must not be a result of the soldering process.

Information note 2004-64, chapter on tin.

Concerning the use of tin, for solders or as containers, the note specifies very low limits. These limits are much lower than those specified in the arrêté of 28 June 1912.

5.4.7.2 Household Metallic Articles and Parts of Industrial Materials

II-A Objects in Plain, Uncoated Steel

Information note 2004-64, chapter on objects in steel.

Objects made of plain steel must not be used in contact with acidic foodstuffs and beverages. A reference is given to the following standards:

- NF A 36-711 on stainless steel – except packaging – intended to come in contact with foodstuffs.
- NF A 36-714 related to flat articles made of steel.
- NF A 35-596 related to steel for cutlery.

A table specifies the maximum content of other elements in the composition of the objects. Tolerable limits are given for lead, cadmium, arsenic, and cobalt. The supplier of these objects must certify that the products comply with these specifications.

II-B Objects in Steel with Metallic Coating

Information note 2004-64, chapter on objects in steel with metallic coating.

The objects in steel coated with zinc or with zinc alloys must not be used in direct contact with food, with the exception of

- objects for manufacturing or storage of chocolate products or confectionery not containing liquid acidic substances;
- objects for distilleries and distilling operations;
- objects destined to store roots, tubers, bulbs, fruit with a dry shell, pulses or vegetable with leaves and with foils;
- coatings based on quasi crystals.

The maximum temperature of use must be specified on the labeling (e.g., 100 °C for zinc coatings).

The *arrêtés* cited for packaging do apply also for objects. In addition, the information note indicates tolerable limits for lead, cadmium, and arsenic impurities, as well as specific migration limits for nickel, chromium, and zinc.

II-C Objects in Steel with Organic Coating

Information note 2004-64, sheet on objects in steel with organic coating.

The rules are the same as for the similar packaging materials. Coatings are presented in a dedicated chapter.

II-D Objects in Cast Iron

Information note 2004-64, sheet on cast iron.

Acidic foodstuffs must not be stored in contact with cast iron utensils, whether before or after cooking.

Objects made of cast iron must not contain more than 0.05 % lead.

II-E Objects in Cast Iron with Metallic Coating

Information note 2004-64, chapter on objects in cast iron with metallic (nickel and chromium) coatings.

A reference is given to the guidelines of the Council of Europe on metals and alloys intended to come in contact with foodstuffs. The text defines maximum limits for lead, cadmium, and arsenic as well as specific migration limits for nickel and chromium. Rules for the use of food simulants are given.

II-F Objects in Cast Iron with Organic Coating

Information note 2004-64, sheet on objects in cast iron with organic coating.

The coatings are

- polymeric films (polyethylene terephthalate, silicones, etc.): they must comply with the corresponding regulations on these polymers;
- organic coatings: they must comply with the specific rules. This will be dealt with in a dedicated chapter.

The maximum temperature of use must be indicated by labeling each individual object.

Cast iron must comply with the specific requirements for cast iron.

If the cast iron is enameled, it must also comply with the specific requirements applying to ceramics.

The note also indicates methods for migration studies.

II-G Objects in Aluminum and Aluminum Alloys

Arrêté of 27 August 1987, related to material and objects in aluminum or aluminum alloys in contact with foodstuffs, food products, and beverages.

See Section 5.4.7.1.5

Information note 2004-64, sheet on aluminum.

The note differentiates

- objects in plain aluminum for single use;
- objects in plain aluminum for repeated use;
- objects in coated aluminum for single use;
- objects in coated aluminum for repeated use.

As a summary, in all cases

- the aluminum used must comply with the requirements of the arrêté of 27 August 1987, especially requirements for purity;
- the coating must comply with the regulation on coating;
- the aluminum supplier must provide to his customers a certificate of compliance with the arrêté of 27 August 1987, supported by an analytical report;
- the coating supplier must certify that the coating complies with the relevant regulation.

II-H Objects in Tin and Tin Alloys

Arrêté of 28 June 1912, related to coloration, preservation, and packaging of foodstuffs and beverages, which sets purity criteria for the used tin.

Arrêté of 15 November 1945, fixing a list of materials that may be used without harm to public health to produce measuring instruments.

Tin is authorized if it complies with the requirements of arrêté of 28 June 1912.

Information note 2004-64, sheet on tin.

The note deals with objects and household articles, measuring instruments, and food industries equipment of tin or tin alloys or coated partly or totally with tin.

- The objects must not be used in contact with strong acidic foods or strong basic foods or to heat foodstuffs at temperatures exceeding 150 °C.
- The storage of food in objects in tin, tin alloys, or tinned is not advisable.
- The tin content must be higher than 97%.
- Limits are set for lead, cadmium, arsenic, antimony, and copper.
- The specific migration limit for antimony is 0.01 mg/kg food.

II-I Objects in Zinc

Arrêté of 28 June 1912, related to coloration, preservation, and packaging of foodstuffs and beverages,

> It is forbidden to place any beverage or food in direct contact . . . with zinc.

An exception are all operations of manufacture or storage of chocolate products and confectionery not containing acid liquid substances and for distillery operations.

Arrêté of 15 November 1945, fixing a list of materials that may be used without harm to public health to produce measuring instruments.

Measuring instruments made of zinc are authorized for contact with roots, tubers, bulbs, fruit with a dry shell, and pulses or vegetable with leaves and with foils.

Information note 2004-64, sheet on zinc.

The note deals with objects for repeated uses, made exclusively of zinc and destined to come in contact with food and feed. Objects in steel coated with zinc are excluded (see Section 5.4.7.1.2).

For the uses authorized, the purity of zinc must be at least 99.85%. Limits for lead, cadmium, and arsenic are set.

The specific migration of zinc must not exceed 10 mg/kg food.

II-J Objects in Copper

Arrêté of 28 June 1912, related to coloration, preservation, and packaging of foodstuffs and beverages. It is forbidden to place any beverage or foodstuff in direct contact with copper, with the exception of

- operations of processing or storage of chocolate products and confectionery not containing acid liquid substances and
- distillery operations.

Arrêté of 15 November 1945, fixing a list of materials that may be used without harm to public health to produce measuring instruments.

Measuring instruments made of copper are authorized for contact with roots, tubers, bulbs, fruit with a dry shell, and pulses or vegetable with leaves and with foils.

Measuring instruments made of copper covered with fine tin are authorized for wine, alcohols, and alcoholic liquids.

Advice of AFSSA of 5 November 2001, related to an opinion request on the use of copper in direct contact with foodstuffs.

Since there is no complete overview on all the kitchen utensils made of copper and since the migration of copper depends strongly on the type of food and on the conditions of contact, the risk assessment has to be made on a case-by-case basis based on specific migration data and on the contribution to overall exposure.

The advice recalls that the tolerable daily intake of copper is 0.5 mg/kg body weight and that the dietary needs are between 1.5 and 2 mg/person/day.

II-K Objects in Whitened Metals

Information note 2004-64, sheet on whitened metals.

Whitened metals are metals coated with a light white layer of silver, nickel, tin, chromium, or any combination of these elements. They are used for tea and coffee sets, cake servers, tumblers, and so on. Silver-plated metals are excluded.

- Whitened metals must not be used in contact with acidic foods and beverages.
- Limits are set for lead, cadmium, and arsenic.
- Migration limits are set for nickel, chromium, zinc, lead, and cadmium. Conditions for the determination of migration are given, depending on the type of foodstuff.

5.4.7.3 French Regulation on Materials and Objects in Wood Intended to Come in Contact with Food

Arrêté of 15 November 1945, fixing a list of materials that may be used without harm to public health to produce measuring instruments.

Wood of the following trees may be used to make measuring instruments for solid foodstuffs: beech, elm, walnut, and poplars.

Lettre Circulaire of 28 October 1980

This text extends the previous authorized uses to containers for storage and preservation of solid foodstuffs.

Advice of CSHPF

Several opinions deal with the products for treatment of wood (September 8, 1992, October 5, 1993, July 11, 1995, and April 11, 2000): AZC, benzalconium, acarbenzadine, azaconazole, chlorothanonil, and so on (see addenda to Brochure 1227 for details).

Note of Information no. 2006–58 of DGCCRF concerning wood materials coated or not coated.

5.4.7.4 French Regulation on Materials and Objects as Complexes Intended to Come in Contact with Food

Information note 2004-64, sheet on zinc.

The note deals with complexes for which the layer in contact with food is made of plastic:

- complexes plastic/aluminum
- complexes plastic/paper/paper
- complexes plastic/paper (or board)/paper
- complexes plastic/paper (or board)
- complexes coating/plastic/paper (or board): the coating may be a hot melt or a thermosetting coating; the plastic layer can be a mealiest polyester.

The note draws the attention of suppliers and users to the need of compliance with the criteria of inertness:

- compliance with positive lists
- compliance with overall migration limits and with specific migration limits for monomers and additives
- compliance with the composition restrictions set for paper
- compliance with criteria for paper and board
- rules are set for the control of migration

5.5 French Regulation on Coatings Coming into Contact with Foodstuffs

These texts deal with

- mainly coatings to protect the inside of metal cans;
- technological coatings used on paper and board and packaging films coming in direct contact with food.

Coatings on a surface of packaging that do not come in direct contact with food are treated in the next chapter.

Several texts, spread all over the Brochure 1227 deal with these coatings:

Circulaire 159 of 23 June 1950, related to lacquers and coatings for containers.
Circulaires of 22 February 1966 and of 2 April 1969: they complete the previous one with lists of substances that may be used.

The regulation for plastics in contact with food applies to coating films that contain plastics. See the corresponding European directives and the specific French regulation, *Note d'Information DGCCRF 2003-37* (see above), which defines lists of additives for plastics pending a complete European regulation on this issue.

Regulation 1895/2005 applies to epoxidic derivatives if they are present in these coatings.

The *Arrêté of 4 November 1993*, modified by the *arrêté of 21 October 2004* on regenerated cellulose, transposing the corresponding directives, deals with lacquers used on these films.

The *Note d'Information 2004-64* refers to the Resolution AP (96) 5 of the Council of Europe, modified since by the Resolution AP (2004) 1.

The note specifies that

the coating supplier must verify that monomers and additives are included in the list of the arrêté of 2 January 2003, of the Information Note 2003-37, in the Brochure 1227 and in the Resolution of the Council of Europe;
the producers must verify that the finished materials comply with the overall and specific migration limits.

5.6 French Regulation on Coloring Matters Used on or Within Food Packaging in Contact with Food

5.6.1 Preliminary Remarks

Materials and articles can be colored either by printing with an ink or by incorporating in the mass of the material, which is possible for plastic materials and paper. The coloring matters are either dyes, which are soluble in the medium where they are incorporated, or pigments, which are insoluble in this medium and are used as dispersions.

5.6.2 Basic Principles

The French regulation relies on the principle of positive lists, which have to be updated from time to time. Only listed dyes and pigments can be used. The regulation is global:

> Materials and articles in contact with foods, food products and beverages must be elaborated exclusively with constituents the use of which is authorized following consultation of the CSHPF [nowadays AFSSA].

The positive lists apply to all uses of dyes and pigments on or within the material or article.

5.6.3
French Regulation on Colorants Used in the Mass of Paper or Plastic Materials

5.6.3.1 Texts and Regulations

Arrêté of 28 June 1912, related to coloration, preservation, and packaging of foodstuffs and beverages.

> It is forbidden to place any paper other than virgin paper either white or straw-colored or colored by means of one of the substances (list is in annex) in contact with bread or with any foodstuff humid or fatty likely to stick to the said paper, such as meat, poultry, fish, meat based preparations, butter, food fats, vegetables and fresh products, confectionery products and bakery.

The lists in annexes contain on the one hand the colorants for coloring the mass and on the other hand colorants for coloring the surface only.

Circulaire No. 176 of 2 December 1959, related to pigments and dyes for plastic materials and packaging.

The text concerns essentially the coloration of plastics in the mass and secondarily the printing inks.

- The only coloring materials authorized are those listed in an annex to the Circulaire. This list is divided in three parts: organic substances, colorants for cellulose films, and mineral substances.
- These substances must comply with purity criteria concerning their content in some metals and in aromatic amines.
- The benzene content of carbon black (determined by extraction) must not exceed 0.1%. This pigment must be free of benzo[*a*]pyrene.

5.6.3.2 Instruction of 2 August 1993

The Circulaire 176 concerns all colorants for printing and for mass coloration, for plastic materials, and for packaging. Following studies published by Hoechst showing that diazoic pigments derived from dichlorobenzidine decomposed above 200 °C, yielding toxic substances, it became obvious that these substances could no longer be used for mass coloration. However, their use for printing was still acceptable. This was endorsed by the Instruction of 2 August 1993.

It became obvious that the two uses should be separately regulated. Hence, the project of a new arrêté dealing specifically with the substances that can be used for the coloration of plastic materials in the mass. This new regulation has not been published when this chapter is printed.

5.6.4
French Regulation on Inks Used to Print Food Packaging

There is no specific arrêté dealing exclusively with printing inks. However, these are mentioned at different places in the regulation and this is the basis for their use in food packaging.

Arrêté of 28 June 1912, related to coloration, preservation, and packaging of foodstuffs and beverages (see Section 5.6.3).

This is the oldest text that ever deals with printing.

Circulaire No. 176 of 2 December 1959, related to pigments and dyes for plastic materials and packaging (see Section 5.6.3).

This text has been modified several times, but the *paragraph (e)* is still in the original version, as indicated in Brochure 1227:

> For inks used on the back side of transparent packaging films, when the print is protected by a protecting coating (solid), only the latter should satisfy with the requirements on materials intended to come in contact with food. It must effectively protect the food and not flake or peel off.

The concept of "protective coating" is similar to what is now known as functional barrier. It has been proved that a film of white ink or varnish does not protect the food from migration and hence is not an efficient functional barrier. The situation has therefore been reviewed by the following texts.

Arrêté of 4 November 1993, modified by the arrêté of 21 October 2004, on materials and articles made of a film of regenerated cellulose put in contact or intended to come in contact with food, foodstuffs, and beverages. This text lifts, for these materials, the ambiguity of the Circulaire 176:

> The printed layer of regenerated cellulose films must not be put in contact with foodstuffs, food products and beverages.

Avis of CSHPF of 7 November 1995 on the use of inks and coatings for the printing of packaging intended for foodstuffs.

This is an advice and not a law, and serves as a reference document for ink producers and users, for all applications in food packaging. Here are the main requirements for the general case of printing inks:

- the famous "ambiguity" of Circulaire 176 is cleared up: " the printed side of a packaging, whether or not over lacquered with a coating, should not come in contact with food";

- all the coloring matters authorized in the various regulations can be used for printing inks: those of the Circulaire 176 and those authorized for the mass coloration of plastics even after the advice;
- the authorization of new coloring substances requires an authorization procedure that should describe
  - compliance with purity criteria
  - migration tests
  - two mutagenicity studies (the SCF required three studies).

Dyes and pigments authorized following this evaluation procedure will be registered in a positive list. This procedure is slightly different from that applied to other constituents of the materials, as described in Chapter 7.

5.7 French Regulation on Requests for Authorization of Use of Constituents of Materials and Articles in Contact with Food

The positive lists that constitute the French regulation have to be updated from time to time.

Arrêté of 13 November 1986, related to applications for authorization of use of constituents of materials and articles placed in contact or intended to be placed in contact with food, food products, and beverages.

- An application must be sent to DGCCRF
- For each constituent, there must be a separate application dossier, describing
  - the intended use and the advantage of the new substance for the consumer
  - the innocuousness of the substances in the intended conditions of use
  - potential risks to environment
- The application dossier must include a set of data that are defined in detail:
  - data on physical and chemical behavior of the substance
  - toxicological properties of the substance

By derogation, a simplified dossier is requested for coloring substances.

Instruction of 27 November 1986, clarifying the application of the previous text (Arrêté of 13 November 1986) for toxicological data:

- acute toxicity data
- 90 day repeated dose toxicity study
- mutagenesis data

This instruction also applies to colorants. It specifies that the derogation established in the previous instruction of 13 November 1986 can apply only if the absence of migration in food products is demonstrated. There must be no migration, even if it is not visually detected.

5.7.1
Advice of CSHPF of 9 December 1997

This updates all the two previous texts.

- DGCCRF, which receives the applications, can authorize the substances only after consulting an independent expert committee: earlier the CSHPF, nowadays AFSSA.
- The application procedure is updated to take into account new knowledge and the guidelines for application of substances at the European Commission and at OECD.

The composition of the dossier is revised. It now includes

- general information on the substance: name, uses, conditions of use, maximum concentration in the material, technical argumentation on the interest of the substance, effects on environment, and so on;
- information on physical and chemical properties of the substance;
- inertness studies: overall and specific migration of the substance in the food simulants A, B, C, and D defined in the EU regulation;
- the migration data to determine a conventional exposure figure, called "Theoretical Exposure Level" (Niveau d'Exposition Théorique, NET). The NET is determined assuming a daily consumption of 1 kg foodstuffs where the substance has migrated
  - assuming that the diet is generally composed of 20 % fatty food and 80 % aqueous foods
  - each food is assumed to contain the substance at the level where it is measured in the corresponding simulant
  - NET (μg/person/day) $= 0.8 \quad ((MSA + MSB + MSC) \times 0.33) + 0.2 \quad MSD$, where MSA, MSB, MSC, and MSD are specific migration values determined in simulants A, B, C, and D, respectively, and expressed in μg/kg simulants;
- toxicological information, based on the conventional exposure defined as NET. The data requested are the same as in the EFSA guidance document, except that the migration values are replaced with NET values;
- a major difference with SCF (now EFSA) guidance document. "When the NET is below 0.5 μg/person/day, the applicants who can demonstrate the absence of carcinogenic potential by a recognized structure–activity relationship are exempted of studies demonstrating the absence of genotoxic potential";
- the case of the coloring matters is also treated in this text. Their migration has to be determined, either by analytical means or by visual methods. In the absence of a visible migration, the dossier must include at least a gene mutation test and a chromosomal aberration test.

The framework presented in the Advice of CSHPF of 9 December 1997 must be followed; it is next of the one of EFSA.

The request and the letters must be written in French, but the annexes can be written in English.

5.8 Conclusion

This chapter shows the evolution of the French regulation on materials in contact with food. Related issues such as cleaning of articles, ion-exchange resins, and ionization of packaged foods have not been included in order to focus on the issue.

Overall, it looks complex, as it is constituted of successive decisions on various components of the materials. The unity is found in the principle of positive lists.

The regulation is changing like all regulations of this type. A major reason for the changes is the incorporation of EC decisions.

This regulation is rather dense and complex. We hope that this chapter will help those who design new materials and articles to comply with the rules applicable in France. The structure of this chapter may have some repetitions, but it allows to follow in better way the driving thread of the French regulation.

Useful Links

Web site of the Laboratoire National d'Essais: http://www.contactalimentaire.com/.

DGCCRF web site: www.finances.gouv.fr/DGCCRF/.

Official journal web site: www.journal-officiel.gouv.fr/.

Glossary

AFSSA	French Food Safety Agency
Arrêté	decision of one or several Ministers, in application of a law or of a decree.
Avis du CSHPF ou de l'AFSSA	Advice from CSHPF or (after 2000) from AFSSA: scientific opinion of the French Food Safety Agency, often sought by Ministers, other administrations, or stakeholders (industry via DGCCRF, consumer associations, and so on). The opinions are usually endorsed into laws. Even if they are not endorsed, they are considered as reference documents and major recommendations.
BOCCRF	Official Bulletin of DGCCRF
BROCHURE No. 1227	Collection of French texts on the materials in contact with food. The last update was published in 2002. It is a basic document for certifications. Updates are available as separate prints, supplied with the Brochure, when ordered.

CSHPF	A higher French consultative body for public hygiene (was replaced by AFSSA in 2000 on several areas, including materials in contact with food)
DGCCRF	Consumer, Competition, and Fraud Office (Direction Générale de la Consommation, de la Concurrence et de la Répression des Fraudes)
JO	Official Journal, *Journal Officiel*
NET	"Theoretical Exposure Level" (Niveau d'Exposition Théorique)

6
Dutch Legislation on Food Contact Materials

Rob Veraart

6.1
Introduction

On December 28, 1935 a legislation was published known as the "Warenwet" (Food and Commodities Act). Translated to English this means "legislation on goods." This legislation has been amended many times and does cover products such as food, tattoos and piercings, cosmetics, food contact materials, and much more. The legislation covers only issues such as responsibilities for the legislation and enforcement, and fines.

For every area that is covered under the "Warenwet," a royal decision is published that describes the general definitions and requirements. A decision "Food Packaging and Utensils" has been published. This decision is comparable with the Framework Regulation (EC) No. 1935/2004. However, because the Framework Regulation is now a regulation, implementation in the national legislation is not allowed (this is in contrast to the Framework Directive 89/109/EEC that was in force before and that was implemented in the royal decision).

The positive lists and EC directives are implemented by the Dutch Packaging and Food Utensils Regulations (Food and Commodities Act) or in Dutch "Regeling Verpakkingen en Gebruiksartikelen," abbreviated as "RVG") under the powers of the Ministry of Health, Welfare, and Sports. Legislation is published in the official journal of the Netherlands, *Staatscourant*, which is published daily in the Netherlands. Information on the *Staatscourant* can be requested at:

Servicecentrum Uitgevers
PO Box: 20014
2500 EA The Hague
The Netherlands
Tel.: + 31 70 37 89 880
Fax: + 31 70 37 89 783
E-mail: sdu@sdu.nl
www.sdu.nl

Global Legislation for Food Packaging Materials. Edited by Rinus Rijk and Rob Veraart

ISBN: 978-3-527-31912-1

In total, over 50 decisions have been published since the framing of the legislation. Because it is quite difficult to obtain the status of a substance in a certain application (one has to review all publications), a (unofficial) codification is published by the SDU (in Dutch only, see http://www.sdu.nl/catalogus/VGBHW).

The Dutch Packaging and Food and Utensils Regulation, RVG, consists of the following items:

- A legislative part describing general issues and defining the 10 categories of materials that may be used for contact with food.
- Annex A that discuss in more detail the groups of materials (will be dealt with in more detail in the following sections):
  - Plastics
  - Paper and paperboard
  - Rubber products
  - Metals
  - Glass and glass ceramics
  - Ceramic materials and enamels
  - Textile products
  - Regenerated cellulose
  - Wood and cork
  - Coatings
- Annex B Chapter I in which the investigation of the finished products is discussed. This includes general items such as selection of simulants, time and temperature conditions, and calculations and provides for some analytical testings that are mentioned with some restrictions.
- Annex B Chapter II describes the investigation of raw materials and additives.

The food contact legislation in the Netherlands covers many areas including food contact materials that are nonharmonized at EU level. In most cases, this national regulation on food contact materials has positive lists of substances that may be used in the production of food contact materials. It is possible to add substances to these lists; a petition needs to be filed. In many cases, the petition is similar to the one that needs to be filed with the European Food Safety Authority (EFSA). The petition has to be filed on the EFSA format with

Ministry of Health, Welfare and Sport
Attn. Dr. H. Roelfzema
Tel: +31 703 406965
Fax: +31 703 405554
h.roelfzema@minvws.nl
Department for Nutrition, Health Protection and Prevention
Product Safety and Injury Prevention Division
Parnassusplein 5
PO Box 20350
2500 EJ The Hague
The Netherlands

The food contact legislation is enforced by the VWA (Food and Consumer Product Safety Authority).

The task of the VWA is to protect human and animal health. It monitors food and consumer products to safeguard public health and animal health and welfare. The Authority controls the whole production chain, from raw materials and processing aids to finished products and their consumption.

The VWA is an independent agency under the Ministry of Agriculture, Nature and Food Quality (LNV) and a delivery agency for the Ministry of Health, Welfare and Sport.

The three main tasks of the VWA are supervision, risk assessment, and risk communication. Other important activities are incident and crisis management and policy advice for the Minister of Agriculture, Nature and Food Quality. A significant part of its work involves liaising with other ministries. Maintaining international contacts is also an important part of its activities.

6.2 Plastics

6.2.1 Nonepoxy Plastics

In the Netherlands, the positive list for nonepoxy plastics is considered to be complete. The legislation can be found in Chapter I of the RVG, Sections 1 and 2. The positives lists as are present in Section 2 and the positive lists of Directive 2002/72/EC are implemented in Sections 2.1, 2.2, 2.3a, and 2.3b. Section 2.1 does include the monomers and other starting materials, as is present in Annex II, Section A of the 2002/72/EC. Section 2.2 does include the monomers and other starting substances that may continue to be used pending a decision on inclusion in Section A, as is present in Annex II, Section B of the 2002/72/EC.

The additives that are harmonized in the EU are implemented under Section 2.3a and 2.3b of Chapter I. In Section 2.3a of Chapter I, the positive list as given in Annex III, Section A, and in Section 2.3b of Chapter I, the positive list as given in Annex III, Section B, are included.

In addition to the harmonized EU legislation, an additional list of additives and aids to polymerization and/or their breakdown products is included in Section 2.4 of Chapter I. The additive list is considered to be complete and this means that additives used in nonepoxy plastics must be present either in the harmonized list (Sections 2.3a and 2.3b) or in the national list (Section 2.4). The aids to polymerization must be included in Section 2.4.

Restrictions for the finished products are identical to those noted in the Directive 2002/72/EC, and some additional components listed in Section 2.4 do have specific migration limits (SML), but restrictions established for plastics should be met using relevant factors for the surface to volume ratio in real use (see Section 13).

6.2.2 Epoxy Plastics

Epoxy plastics are excluded from the definition of plastics in the Directive 2002/72/EC. The Dutch legislation has a specific section (Section 3 of Chapter I) on epoxy plastics. Epoxy plastics are defined as "Products obtained by a polyaddition reaction of components with epoxygroups with components, that has two or more active hydrogen atoms per molecule or can form these during a reaction (hardeners), or with condensation products of polyols and isocyanates. The polymer part of the endgroup must contain at least 50% epoxy polymer."

Derivates of BADGE, BFDGE, and NOGE are excluded; they are regulated by Regulation (EC) No. 1895/2005.

No positive lists are present for the monomers. The positive lists for additives are present and are grouped in three types of additives:

- Lubricants and release agents
- Plasticizers
- Other additives

For the epoxy polymers, the following restrictions apply: (1) regarding the overall migration, the conventional 60 mg/kg limit must be used; (2) for the monomers, only a group residual content applies: epoxy groups 5 mg/kg as epoxy groups of Mw 43 (epoxydized soybean oil is excluded). No specific migration limits or residual contents are set for the additives.

6.3 Paper and Board

Two types of paper are distinguished in the Dutch legislation on food contact materials: paper for general purpose and paper intended for boiling, for packaging, and for contact with foodstuff above 80 °C.

6.3.1 Paper and Board for General Purpose

Paper and board intended for general use may be made from the following raw materials: vegetable fibers, recycled paper and board, fibers from plastics (in compliance with Chapter I of the RVG, plastics), whether or not provided with a coating (coating must be in compliance with Chapter X of the RVG, Section 3), fibers from regenerated cellulose (complying with Chapter VIII of the RVG), or fibers from textiles (complying with Chapter VII of the RVG).

In addition, processing aids may be added to the fibers. The processing aids are divided into different categories, each with a different technological purpose. However, except for some sections where it is explicitly mentioned, components may also be used for another technological purpose. The following technological functions are covered:

- Section 1.2.2a: basic processing aids
- Section 1.2.2b: precipitants, fixatives, retentive, and dehydrating agents
- Section 1.2.2c: slimicides, exclusively for use in process water
- Section 1.2.2d: bleaching agents
- Section 1.2.2e: dispersion, flotation, and antifoam agents
- Section 1.2.2f: fillers
- Section 1.2.2g: dyes and pigments
- Section 1.2.2h: sizes and fiber-binding agents
- Section 1.2.2i: paraffins and waxes
- Section 1.2.2j: moisture control agents
- Section 1.2.2k: preservatives exclusively for preserving paper coatings
- Section 1.2.2l: preservatives for the protection of packaged foods
- Section 1.2.2m: agents for improving wet strength
- Section 1.2.2n: macromolecular compounds
- Section 1.2.2o: plasticizers
- Section 1.2.2p: optical whiteners
- Section 1.2.2q: adhesives, solvents and inks
- Section 1.2.2r: remaining additives

In addition to the positive listing of the raw materials and processing aids, requirements on the finished product exist.

6.3.1.1 Overall Migration

If the paper has a coating, an overall migration limit of 60 mg/kg applies. If no coating is applied on the paper, a limit of 60 mg/kg applies to both the methylene chloride soluble part of the overall migration and to the nonsoluble part of the overall migration.

6.3.1.2 Specific Migration

A list of components with a specific migration limit is given, applying to the final article.

Migration is determined according to the conditions described in the EU Directive 82/711/EEC and 85/572/EEC as amended. This means that migration is determined using food simulants under standardized conditions of time and temperature. However, paper and board intended for contact with dry foodstuffs should be tested with the foodstuff itself as no simulants have been established for dry foods. In some cases, the use of modified polyphenylene oxide may be useful. To judge the final compliance with the regulation, the proper conversion factors should be applied as indicated in Section 6.13 of this chapter.

6.3.2
Paper for Filtering and Cooking Above 80 °C

Paper and board intended for filtering and cooking above 80 °C may be made from the following raw materials: fibers from plastics (complying with Chapter I, plastics, and they may not contain plasticizers) whether or not provided with a coating (coating must be in compliance with Chapter X of the RVG, Section 3), fibers from regenerated cellulose (complying with Chapter VIII of the RVG).

Additives must be on the positive list and only two categories are allowed:

- Section 2.2.2a: parchmentizers and neutralizers
- Section 2.2.2b: agents for improving wet strength

In addition to the positive listing of the raw materials and processing aids, requirements on the final product exist.

6.3.2.1 Overall Migration

An overall migration limit of 60 mg/kg applies.

6.3.2.2 Specific Migration

A list of components with a specific migration limit is given, applying to the finished article.

Migration experiments are the same as those for general-purpose materials subject to the applicable rules.

6.4
Rubber

Chapter III of the RVG deals with rubber articles. In this chapter, rubber is defined as "Products based on elastomers, to which one or more processing aids are added. The rubber products are obtained from mixtures of elastomers and additives as a result of crosslinking on a molecular scale, usually at elevated temperatures and with or without the application of pressure."

The method of producing rubber articles is provided in the policy statement on rubber of the Council of Europe [1] (Figure 6.1).

The manufacturing of rubber may start from natural source materials (natural rubber or latex) or from synthetic polymers. In the Dutch regulation, the natural sources are not mentioned and thus these types of products are not covered by the Dutch legislation.

6.4.1
Categories

The Dutch authorities do recognize that rubber into contact with foods and/or beverages in a wide variety of applications. In some cases, there may be contact with

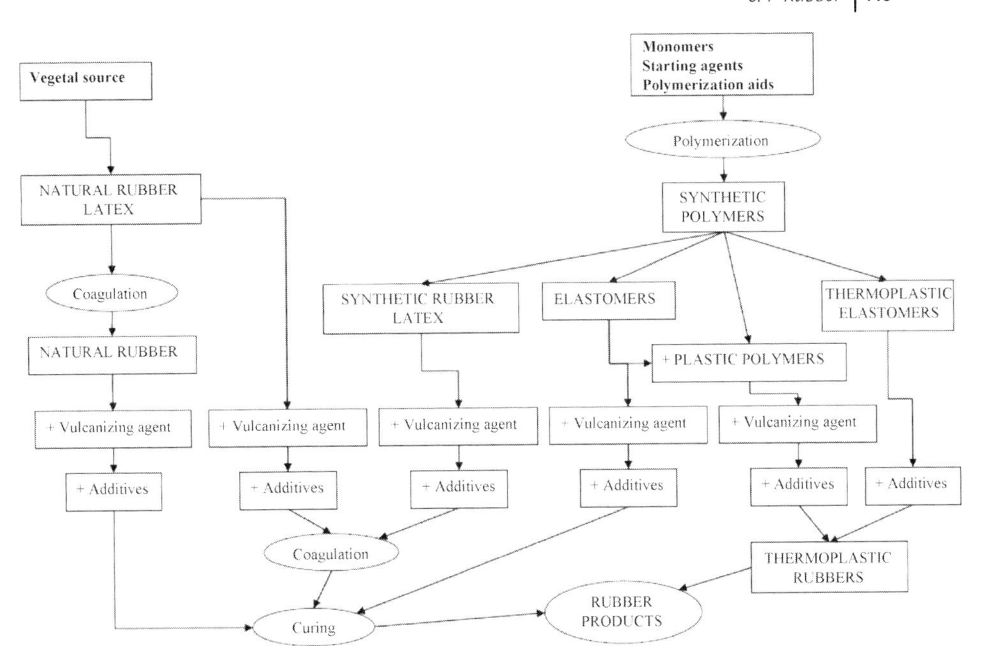

Figure 6.1 Scheme of manufacturing rubber products.

foods for vulnerable consumers, whereas other applications include contact with foods under conditions where migration is negligible. Therefore, three different categories of rubber materials, each with its own requirements and positive list, have been established.

Category I comprises rubber products requiring special attention because of their intended use, particularly for baby bottle nipples and articles intended to be taken into mouth by babies or young children or to come into contact with baby food.

To establish compliance with the specific migration limits of relevant substances, the assumption should be taken into account that a child uses five teats a day and, therefore, SMLs must be divided by 5 to check compliance with the migration obtained from one teat.

Category II and III concerns materials for which the migration may or may not be negligible. The criterion, in assessing if the migration is indeed negligible, is the result of the continued product of four factors R_1, R_2, R_3, and R_4 referring, respectively, to the relative contact surface, the contact temperature, the contact time, and the number of times that the utensil is used. Articles belonging to Category I are always exempted from this rule.

Category II comprises rubber products for which the continued product of the factors R_1, R_2, R_3, and R_4 is greater than 0.001, meaning that migration tests must be carried out.

Category III includes rubber products for which the continued product of the factors R_1, R_2, R_3, and R_4 is smaller than 0.001, meaning that no migration tests needs to be carried out.

The factors R_1, R_2, R_3, and R_4 are defined and determined as follows:

R_1 refers to the relative contact surface between rubber article and food or beverage, expressed in cm^2 of rubber surface per kg of food or beverage and is calculated with the following equation:

$$R_1 = \mathrm{RO}/100,$$

where RO is the area that is in contact with the food in cm^2.

For a relative surface smaller than or equal to 100 cm^2, R_1 has a value calculated according to the above-mentioned formula.

For a relative surface larger than 100 cm^2/kg, R_1 has always the value 1.00.

R_2 refers to the temperature during the contact between the rubber product and the food or the beverage. At a temperature lower than or equal to 130 °C, R_2 has a value calculated according to the formula:

$$R_2 = 0.05\,\mathrm{e}^{0.023T},$$

where "e" is the base of the natural or Napierian logarithms and T is the contact temperature, expressed in °C. For temperatures higher than 130 °C, R_2 always has the value 1.00.

R_3 refers to the time t, expressed in hours, during which a rubber product is in contact with the food or beverage. For a contact time shorter than or equal to 10 h, R_3 has a value calculated according to the formula:

$$R_3 = t/10.$$

For a contact time of more than 10 h, R_3 has the value 1.00.

R_4 refers to the number of times N that one and the same rubber article comes into recurrent contact with a quantity of food or beverage. For a number of times greater than 1000, R_4 is calculated according the formula:

$$10 \log R_4 = 6\text{-}2^{10} \log N.$$

For a number of times smaller than or equal to 1000, R_4 always has the value 1.00.

Example 6.1

A rubber ring is used as a sealant between a metal bottle and a plastic closure intended to hold cold drinks. The size of the bottle is 0.5 l and the area of the rubber part is 10 cm^2. This results in the following calculations:

$R_1 = \mathrm{RO}/100 = 10/100 = 0.1$
$R_2 = 0.05\,\mathrm{e}^{0.023T} = 0.05\,\mathrm{e}^{0.023\times 40} = 0.13$
$R_3 = t/10$ the time is difficult to estimate,
but a contact of longer than 10 h can be expected, and therefore, $R_3 = 1$
$R_4 = 1$, it is not expected that the bottle will last longer than 1000 times of use
$R = R_1 \times R_2 \times R_3 \times R_4 = 0.1 \times 0.13 \times 1 \times 1 = 0.013.$

Conclusion, the material is a category II rubber.

Example 6.2

A rubber closure is used in a coffee machine intended for cafeterias. The rubber ring has to be changed after 1 million servings. During the processing of the coffee, 1 cm^2 will be in contact with 150 ml of boiling water for 10 s.

$$R_1 = RO/100 = 1/100 = 0.01$$
$$R_2 = 0.05\, e^{0.023T} = 0.05\, e^{0.023\times 100} = 0.5$$
$$R_3 = t/10 = (10/3600)/10 = 2.8\, 10^{-4}$$
$$10 \log R_4 = 6-2^{10} \log N;\ 10 \log R_4 = 6-2^{10} \log 1000000;\ R_4 = 10^{-6}$$
$$R = R_1 \times R_2 \times R_3 \times R_4 = 0.01 \times 0.5 \times 2.8\, 10^{-4} \times 10^{-6} = 1.4 \times 10^{-12}$$

Conclusion, the material is a category III rubber.

6.4.2 Positive List

The components needed to make the rubber are covered by a positive list. All components listed may be used for category III rubber only. A restricted list of substances may be used also for category II rubber and a further restricted positive list of substances may be used to manufacture category I articles. Specific migration limits can be assigned for category I or category II rubber (no experimental testing is required for category III rubber). In most cases, the specific migrations assigned are different. Many times the specific migration for a category I rubber is 10 times lower than for category II rubber.

The positive lists can be divided into two groups: groups listed in Section 4.2.1 monomers and other starting materials. The monomers and other starting materials may be used to manufacture elastomers (no positive list is present for polymerization aids to make the elastomers). The other group includes the processing aids needed to convert the elastomers into a rubber. The processing aids are listed in Section 4.2.2 of Chapter III. The following groups of processing aids have been covered:

- 4.2.2a cross-linking agents
- 4.2.2b accelerators
- 4.2.2c retarders
- 4.2.2d activators
- 4.2.2e protective agents
- 4.2.2f plasticizers
- 4.2.2g fillers
- 4.2.2h emulsifiers and emulsion stabilizers
- 4.2.2i colorants and pigments
- 4.2.2j other auxiliary substances

6.4.3
Other Restrictions

In addition to the listing on the positive list and specific migrations as given in Section 6.4.2, other restrictions do apply to the finished product, including overall migration, specific migration on nitrosamines, nitrosatable amines, mercaptanes, dibenzylamine, benzothiazol, aromatic amines, and 6-aminohexanelactame.

Regarding the overall migration, it must be noted that the limits applied are somewhat different from plastics. For materials of category I, the limit is 20 mg/kg. For category II, a limit of 60 mg/kg does apply for water, 15% ethanol, and olive oil, and a limit of 100 mg/kg does apply for 3% acetic acid.

Finally, there are some labeling requirements for teats.

6.5
Metals

Chapter IV of the RVG describes the metallic packaging materials and utensils, provided or not with a coating other than of enamel. Two sections with positive lists are given: one for packaging materials and one for utensils.

After the manufacture, a protective coating may be applied to the finished product, but only in such a way that it can be removed in a simple manner before the finished product comes into contact with foods and/or beverages.

6.5.1
Metals Used for the Application of Packaging Materials

Three positive lists are included for three uses of metal in food packaging: as a metal (sheet), in solders, and as metallic coatings. In addition to the metal part, other additives are specified that may be used to produce the finished article. Seven groups of additives/items are specified:

- Greasing agents
- Rolling oils
- Lubricants for stamping and drawing
- Nonmetallic sealants for the seams
- Adhesive tape for covering the side seam
- Sealants for end double-seams
- Organic coatings

In addition to the coatings mentioned here, coatings mentioned in Chapter X of the RVG (see Section 6.11) may be used. The list of coatings mentioned in Chapter X is more extensive. Some ingredients that may be used to make a coating do apply to both lists, some of them do appear in one of the two lists. However, if a coating is used all the ingredients must be present in the same section.

6.5.2 Metals Used for the Application of Utensils

Three positive lists are included for four uses of metal in food utensils: as metal (sheet), in solders, as welding material, and as metallic coatings. In addition to the metal part, an organic coating may be applied.

6.5.3 Restrictions

Except for the sealants for end double-seams, the following restrictions do apply:

- The finished product must meet the overall migration and shall not exceed the value 60.
- A list of components with a specific migration limit is included.

6.6 Glass and Glass Ceramics

Chapter V of the RVG deals with the restriction of glass and glass ceramics. For glass and glass ceramics, requirements on the specific migration of 15 elements and the overall migration limits have been established in the Netherlands. The migration time and temperature conditions must be chosen as defined in the implemented Directive 82/711/EEC and amendments thereof. However, the testing should be done only with 3% acetic acid because this simulant can be regarded as the worst-case simulant. Elements for which limits have been established are: antimony, arsenic, barium, boron, cadmium, cerium, chromium, fluor, cobalt, lithium, lead, manganese, nickel, rubidium, and zirconium.

Glass products covered are industrial glass (for the transport and holding of food), glass used as packaging material, glassware, and crystals and articles made from glass that can be used at high temperature (in the oven or cooking). Although no positive listing is established, it is mentioned that mercury components may not be used in the production of glass and the use of lead(II) oxide may be used only for the production of crystal glass.

6.7 Ceramics and Enamels

Chapter VI of the RVG includes both ceramics and enamels because these materials are somewhat similar. Part 1 of Chapter VI describes the restrictions of ceramics and Part 2 describes the restrictions of enamels.

6.7.1 Ceramics

The ceramics chapter contains specifications on the migration of some elements only. In this chapter, the EU directives on ceramics are included (84/500/EEC [2] as amended by 2005/31/EC [3]). These directives describe the restrictions on lead and cadmium migration from articles made from ceramics and the way they should be determined. The migration testing of lead and cadmium must be performed using 4% acetic acid after contact for $22 \pm 2\,^{\circ}\mathrm{C}$ for $24 \pm 0.5\,\mathrm{h}$.

In addition to these EU restrictions, restrictions on the migration of 10 other elements have been included. The limits of the restrictions of these 10 elements are identical for enamels as discussed in the next section. These specific migration limits of the 10 elements must be determined using 3% acetic acid as simulant and using the regular temperature conditions as defined in the implemented Directive 82/711/EEC and amendments thereof. Furthermore, the finished article must meet the requirements on the overall migration by using 3% acetic acid only under regular conditions as defined in the implemented Directive 82/711/EEC and amendments. No other simulants has to be used for the compliance check of these 10 elements and overall migration because 3% acetic acid can be considered as the worst-case simulant.

6.7.2 Enamels

No EU legislation on enamel exists. In the Netherlands, however, legislation on these materials is established. The requirements are specific and overall migration to 3% acetic acid as simulant only under regular conditions are defined in the implemented Directive 82/711/EEC and amendments therefore. No positive list or requirement on ingredients is established. The following elements must meet the specific migration limits: arsenic, barium, boron, cadmium, chromium, cobalt, mercury, lithium, lead, rubidium, selenium, and strontium.

6.8 Textiles

In Chapter VII of the RVG, the legislation on the use of textiles in contact with food is described. As raw materials, a limited set may be used: regenerated cellulose (as described in Chapter VIII of the RVG), polypropylene and polyesters with terephthalic acids (which must be in accordance with Chapter I of the RVG), and vegetables fibers. In addition, positive lists for additives and processing aids are provided, covering the following groups:

- Preserving agents
- Processing and finishing agents
- Other additives

For the finished product, restrictions are the overall migration that has the conventional 60 mg/kg food limitation and a list of components for which a specific migration or a residual content restriction is present.

6.9 Regenerated Cellulose

Chapter VIII of the RVG describes the use of films made from regenerated cellulose. The film may be coated or uncoated. This chapter is equal to the implemented Directive 2007/42/EC [4].

6.10 Wood and Cork

Wood and cork may be used for contact with food and the legislation is provided in Chapter IX of the RVG. In the production of finished product, additives may be used as long as they are listed in one of the following groups:

- Preservatives
- Adhesives and binders
- Coatings
- Plasticizers
- Other additives

For the final product, restrictions are the overall migration that has the conventional 60 mg/kg food limitation and a list of components for which a specific migration or a residual content restriction is present.

6.11 Coatings

The legislation on coating is described in detail in Chapter X of the RVG. The following groups of coatings are defined in the Dutch legislation:

- Dispersions of macromolecular substances in water
- Dispersions of paraffins and waxes in water
- Dispersions of macromolecular substances in organic solvents
- Solutions in water
- Solutions in organic solvents
- Solvent-free material made from waxes and wax-like materials
- Other solvent-free materials
- Metallic coatings
- Polytetrafluoroethylene intended for use as coating for cooking, baking, and frying broilers and pans

Every group has a separate positive list for monomers and starting materials that may be used to make the coating, as well as a list of additives, aids to polymerization, and so on.

For the finished product, restrictions are the overall migration which has the conventional 60 mg/kg food limitation and a list of components for which a specific migration or a residual content restriction is present.

In addition to the above-mentioned list of coatings, additional coatings are listed in Chapter II of the RVG (wax and paraffins on paper) and Chapter IV of the RVG (coatings on metal).

6.12 Colored Materials

All food contact items are frequently colored. In the Dutch legislation, there are restrictions for both the colorant itself and the colored items. Although the requirements are mentioned in the section on plastic food contact materials (Chapter I Section 4 of the RVG), in many cases a reference is made to this section and therefore it applies to plastic, paper, rubber, textile, wood, cork, and coatings.

Regarding colorants, a maximum extractable amount is set for antimony, arsenic, barium, cadmium, chromium, mercury, lead, selenium, and primary aromatic amines. Purity requirements are set for carbon black.

Migration limits are present for colored finished products regarding aromatic amines, antimony, arsenic, barium, cadmium, chromium, cobalt, mercury, lead, manganese, nickel, selenium, and primary aromatic amines. In addition, there is a color fastness test to verify if colored items have the potential to release a color into the food.

6.13 Calculations

Calculations of the overall migration and specific migration of items that are harmonized at the EU and are implemented in the Dutch legislation must be done in the identical way to that described in EU directives. These items included nonepoxy plastics and ceramics.

Calculations regarding other items, such as epoxy plastics, paper, rubber, metals, glass, enamels, textiles, wood and cork, and coatings must be determined in the following way.

If the migration has been performed and the result is obtained in mg/kg simulant, the result must be converted to mg/kg food using the following equation:

$$M = 1000 \times m \times a_2/(a_1 \times q),$$

where M is the migration in mg/kg, m is the migration from the test sample in mg, a_1 is the area of the test sample that was in contact with the simulant during test, a_2 is the area of the finished product that will be in contact with the food during food contact, and q is the amount of food that will be in contact with the finished product (in grams).

Table 6.1 Values for conversion factor *f* as assigned in the Dutch legislation for non-plastics

Category	Description	Value of *f*
a	A material that covers the food completely or for a major part	6
b	A material that is in contact for a relative small area or for a very short time	0.5
c	Drums including storage tanks with a content of more than 25 l but less than 10 000 l	2
d	Storage tanks with a content of more than 10 000 l	0.2
e	Any material that is not covered by category a–d	3
f	Tubing and pipes normally used for transport of fluids	1

The conversion from a migration value determined in mg/dm^2 to mg/kg food has to be performed using the following equation:

$$\text{Migration (in mg/kg)} = f \times \text{migration(in mg/dm}^2\text{)}.$$

The value of *f* depends on the use of the food contact material. The more area of the food contact material is in contact with the food, the higher the value of *f*. The factors listed in Table 6.1 may be used to calculate compliance with the relevant restriction.

The above values are somewhat arbitrary but take better account of the real use than the factor 6 required by the EU directives.

References

1 Policy statement on rubber intended to come into contact with food of the Council of Europe, version 1 of 10-06-2004.

2 Council Directive 84/500/EECof 15 October 1984 on the approximation of the laws of the member states relating to ceramic articles intended to come into contact with foodstuffs.

3 Commission Directive 2005/31/EC of 29 April 2005 amending Council Directive 84/500/EEC as regards a declaration of compliance and performance criteria of the analytical method for ceramic articles intended to come into contact with foodstuffs.

4 Commission Directive 2007/42/EC of 29 June 2007 relating to materials and articles made of regenerated cellulose film intended to come into contact with foodstuffs.

7
National Legislation in Italy

Maria Rosaria Milana

7.1
Introduction

Describing in a few pages the Italian corpus of laws on materials and articles intended to be, or already, in contact with food is, of course, a challenging task.

In fact, the current Italian law on this issue was born in 1962 in the frame of Italian food law of 30 April 1962 no. 283 that laid down in its Article 11 the general principles for food contact materials and articles (FCM) that would not (a) endanger the human health and (b) alter the organoleptic characteristics. Since 1962, more than 50 legislative acts have been promulgated and the resulting frame is noticeably complicated. However, despite the difficulties to navigate through the different integrated decrees and amendments, the legislation shows a complete and in some cases detailed body of principles and rules applicable to FCM.

The idea behind the Italian legislation is that more knowledge on the components of food packaging implies less risk to consumers. Therefore, wherever possible, positive lists or characterization of allowed materials is established, with their migration limits and methods of analysis for enforcements. The concept is consistent with the Roman law approach, that is, the rules must be set in advance and must be known to and applicable to all stakeholders.

At present, FCMs are regulated in Italy under the framework of general principles applicable in all the cases and by specific rules in place for the most commonly used materials and articles. A synthetic scheme of the present laws is shown in Table 7.1.

This chapter is not a comment on all decrees that have been laid down but is an overview describing the structure and the main key points of the current laws on materials and articles in contact with foodstuffs. Only, the basic decrees and the amendments that introduced important changes have been mentioned, in the text, while the decrees that introduced only new substances or technical changes are listed elsewhere [1]. The following sections will illustrate the present situation, first the legislation on general principles and then the specific decrees with focus on each material.

Global Legislation for Food Packaging Materials. Edited by Rinus Rijk and Rob Veraart

ISBN: 978-3-527-31912-1

Table 7.1 Synthetic scheme of the present Italian legislation on food contact materials and articles

Material	Measure
General principles (all materials)	Legge no. 283 of 30/4/1962 Decreto Presidente della Repubblica no. 777 del 23/8/1982 Decreto Legislativo no. 108 del 25/1/1992 Regulation 1935/2004/EC
Plastics	Decreto Ministeriale 21/3/73 and amendments (including the EU directives)
Rubber	Decreto Ministeriale 21/3/73 and amendments (including the EU directives)
Paper and boards	Decreto Ministeriale 21/3/73 and amendments (including the EU directives)
Regenerated cellulose	Decreto Ministeriale 21/3/73 and amendments (including the EU directives)
Stainless steel	Decreto Ministeriale 21/3/73 and amendments (including the EU directives)
Glass	Decreto Ministeriale 21/3/73 and amendments (including the EU directives)
Ceramic	Decreto Ministeriale 4/4/85 (EU directive)
Tin cans	Decreto Ministeriale 18/2/84
Tin-free steel	Decreto Ministeriale 1/6/88
Aluminum	Decreto Ministeriale no. 76 del 18/4/2008
Recycled plastics	Regulation 282/2008/EC
GMP	Regulation 2023/2006/EC
Official control	Regulation 882/2004/EC

7.2 Decrees on General Principles

All materials and articles intended to come in contact with foodstuffs are regulated by general principles, in addition to specific rules, when existing, at the Italian and/or EU level. Italian decrees on general principles are the Decreto Presidente della Repubblica no. 777 of 23/8/82 (DPR 777) [2] and the Decreto Legislativo no. 108 of 25/1/92 (DL 108) [3]. These two decrees introduced in the Italian law the two successive versions of the EU Framework Regulations 76/893/CEE and 89/109/CEE. These two acts lead important concepts and harmonized general rules in the field of food packaging regulations:

1) essential requirements of safety
2) labeling
3) declaration of compliance for the specifically regulated materials

The first EU directive arrived when a relevant part of the Italian legislation was already in place and, therefore, the Italian transcription decrees contained not only concepts that were in common with other Member States of the EU but also national provisions applicable only in Italy that were the sanctions for the infringers, the declaration of compliance for all materials, and the principle of traceability.

The more recent EU Framework Regulation 1935/2004/CE, which repealed the previous and entered directly in force without transcription in national decrees, refined the above concepts and enriched the general rules with the essential requirements for the active and intelligent packaging, the introduction of the supporting documentation, and the obligation of traceability. It must be said that the new Framework Regulation, being in the frame of the EU Food Law (Reg. 178/2002/CE), also brought into this field at the national level concepts such as the official control and communitary and national reference laboratories that are fundamental to a harmonized enforcement action.

The present Italian situation is therefore the coexistence of the previous two decrees with the new Framework Regulation. In fact, it must be noted that the DL 108 amended the DPR 777 and did not repeal it and they both are still living. The parts treated in the new Framework Regulation automatically substituted what was already treated in the previous Framework Directives, but what was only at the Italian level and out of the field of the harmonization remains still valid.

This is the reason why in Italy the declaration of compliance is required for all materials.

In fact, the Framework Regulation 1935/2004/CE states that the declaration of compliance is obligatory for the materials specifically regulated, which at the EU level are plastic materials and articles (in the meaning of the Directive 2002/72/EC), ceramic, and regenerated cellulose materials and articles. For other materials, it is in the power of the Member States to lay down or to keep national rules. In Italy, the declaration of compliance is required in general for all materials and articles by the above-mentioned DL 108. More specifications for this obligation are in the DM 21/3/73 (plastics out of the EU directives, rubber, paper and board, stainless steel, and glass), in the DM 18/2/84 (tin plate), in the DM 1/6/88 (tin free steel), and in the Decree no. 76 of 18/4/2007 (aluminum).

The decrees on general principles are the reference legislation for all materials and articles that have no specific rules, as illustrated at the end of this chapter.

7.3 Decrees on Specific Materials

7.3.1 The Ministerial Decree 21 March 1973 and its Amendments

7.3.1.1 General Part

The Ministerial Decree 21 March 1973 (here referred to as *the decree*) [4] laid down by the Ministry for Health is the milestone of the whole Italian legislation on FCMs. In fact, it was promulgated in the frame of the Italian food law of 30 April 1962, no. 283, to lay down conditions, restrictions, and limitations in the field of FCMs.

The decree has been amended for 35 years, but it still keeps the original structure. It is composed of 41 articles and 4 technical annexes, with the following outline:

Art 1–8: General provisions, applicable to all materials listed in Articles 9–37 covered by this decree

Art 9/9 bis-14: Plastics
Art 15–19: Rubbers
Art 20–26: Regenerated cellulose
Art.27/27 bis-33: Paper and boards
Art 34–35: Glass
Art 36–37: Stainless steel
Art 38–39: Transitional measures and repealed decrees (obsolete)
Annex I: Scientific protocol to present the dossier for inclusion in the positive lists
Annex II: Positive lists for the materials listed in Articles 9–37
Annex III: Conventional classification of foods, food simulants, contact conditions for migration tests
Annex IV: Methods of analysis

Articles 1–8 of the decree contain important concepts applicable to FCMs regulated in this decree (plastics, rubber, paper and board, regenerated cellulose, glass, and stainless steel). As the other provisions of this general part, these apply only to the materials specifically regulated in this decree. Other materials (e.g., aluminum, tin-free steel, etc.) have their specific rules in other legislative acts. After the "field of application" and the definition of "food" and "food contact materials and articles" (Articles 1–3), the decree prescribes the obligation to comply with the positive lists and to follow rules to get new substances on these lists (Articles 3 and 4). In fact, to include new substances or materials in the positive lists, an application must be presented to the competent authority, that is, the Ministry for Health. The Annex I gives indication on the technical dossier that must be submitted for the evaluation by the competent health authorities. The dossier has the same approach as that adopted by the EFSA on the basis of the indications of the former DG Sanco Scientific Committee on Food.

Article 5 establishes the obligation to comply, when applicable, with overall and specific migration limits, but most important of all, it bears the concept of "barrier layer." In fact, the article reports that in the case of coupled or other complex materials, the compliant layer that is in direct contact with foods shall be capable to impede the migration of constituents from the layers that are behind. This performance must be demonstrated in the migration tests. Only this general concept, and no specific rules, is given to evaluate the barrier performance, this demonstration is the task of the industry. This concept not only refers to plastic materials and articles but is also valid for all materials when the layer in direct contact is one of those covered by this decree.

Also, Articles 6 and 7 are very important because they lay down the obligations of the declaration of compliance and the supporting documentation and, indirectly, of the traceability for each level of the production chain of FCMs. Labeling and instructions for proper use are indicated in Article 8 of this general part of the decree. It is worth to note that the ideas behind the content of Articles 6–8 are not substantially different from those of the most recent EU corresponding provisions (Articles 15–17 of the Regulation 1935/2004/CE).

The decree was amended by more than 40 decrees to introduce new substances, new methods, or in general new technical rules, or to transcribe EC measures. Therefore, all the EU directives on plastics have been introduced as amendments to the DM 21/3/73.

It must be stressed that the decree was born before the EU directives on FCMs and has a wider field of application. Therefore, only those parts of the decree that were covered by the EU harmonized legislation were amended by EU measures, the others remained at the national level. This will be better illustrated in the following sections, specific for each one of the materials covered under DM 21 March 1973.

7.3.1.2 **Specific Part**

Plastic

Plastic materials and articles in contact with foods (hereinafter *plastics*) are regulated by the general provisions of the above reported Articles 1–8 of the DM 21/3/73 and by the specific rules in this decree. To better explain the present situation, it is good to start with the beginnings. The original decree, in 1973, covered all kinds of materials and articles if the layer in direct contact with foods was plastic. Moreover, the decree covered not only plastics but also coatings, epoxy resins, and silicones. The decree laid down positive lists of resins (e.g., polyethylene, polyvinyl acetate, etc.) and additives with specific migration (SML), residual content (QM) restrictions, and conditions of use. Overall migration limit was set at 8 mg/dm^2 or 50 mg/kg. A five-type conventional classification of foods was established and simulants were also indicated: distilled water, 3% acetic acid, 15% ethanol, and olive oil. Methods of analysis for overall and specific migration in food simulants were given and rules for the contact conditions in migration tests were fixed.

Although the outline of the Italian legislation was not far from the EU approach to FCMs, the field of application was different and in Italy was wider in its scope than in EU where the harmonization deals only with plastic materials and articles composed of plastic layers. Therefore, because of the progressive introduction of the EC directives in the structure of the decree, the legislation has been amended as follows:

- The EU rules for the so-called "homogeneous plastics" [5] have been completely adopted for the positive lists of the monomers that are now fully harmonized.
- The Italian positive lists of resins for the other "heterogeneous" plastic materials and articles and other materials covered by the decree remain alive for these nonharmonized parts. However, the positive list of the EU monomers is also applicable.
- The EU list of additives has been adopted as positive list to include all materials and articles covered under the rules for plastic. The national positive list of additives has been integrated with the EU list and has followed a progressive harmonization process. The national list will remain valid until 31 December 2009 when the communitary harmonization will be complete.

- EU overall migration limits (10 mg/dm^2 or 60 mg/kg) have been completely incorporated and applied to both "homogeneous" and "heterogeneous" materials and articles covered under the rules for plastic. In addition, the EU method for overall migration, the simulants, and the contact conditions described in the Directive 82/711/EEC and amendments have been integrated fully into the Italian legislation.
- The EU conventional classification of the foodstuffs of the Directive 85/572/EEC substituted the national classification that, however, was not substantially different.
- The EU concept of functional barrier in the Directive 2007/19/EC with the specific rules established therein applies only to the field of "homogeneous" plastics, the others being under the general concept of barrier layer of the Article 5 of the decree, previously discussed.

Therefore, two parallel regimens exist for the harmonized and nonharmonized field of "only plastic" and heterogeneous plastic articles and other materials covered by the decree. They partially overlap because the harmonized parts on overall and specific migration, the conventional classification, and the simulants were extended at the Italian level to the whole field of FCMs covered by the decree.

Other points of the DM 21/3/73 must be highlighted to get a complete picture of the Italian approach to the legislation on plastic FCMs.

First of all, Article 10 deals with substances that are not listed but that can be present in the finished products. For example, "some monomers, low molecular weight substances, intermediates, catalysts, solvents, emulsifying agents." The idea behind Article 10 is that the migration of non specifically regulated components of the plastic material and articles, even though there are no positive lists, must not endanger human health. In practice, these substances must not migrate in an amount that can compromise the compliance with the general principles of safety that now are in Article 3 of the Regulation 1935/2004/CE but have been there in the Italian legislation since 1962 and before.

There is no positive list of colorants. As stated in Article 12, all colorants may be used provided they do not migrate into the food and they comply with the listed specifications concerning the amount of metals. Also, free primary aromatic amines must not be present in an amount above 0.05%. The method to determine the migration of colorants is described in Annex IV.

Some methods for specific migration are given in Annex IV. They deal with formaldehyde, chromium from coated aluminum or glass, trivalent chromium, lead, nickel, vinyl chloride, acrylonitrile, vinylidene chloride, and tin. Methods are also laid down, as limit tests, for the migration of some technological adjuvants (peroxides, aromatic amines, dithiocarbamates, thiourames, xanthogenates, mercaptobenzothiazol and zinc salt, benzothiazyl disulphide, phenols, and cresols) and for the purity of paraffins and microcrystalline waxes, Vaseline oil, and carbon black.

It must be underlined that even if no official methods, as the previous ones, are reported the specific migration or the composition requirements are obligatory when there is a corresponding restriction (SML or QM) defined in the positive lists.

A separate discussion must be carried out on recycled plastics. In fact, since 1973 Article 13 of the decree prohibits the use of used plastic objects or wasted plastic materials to prepare plastic objects intended to come in contact with foods. The unique derogations were laid down in 2005 and 2007 [6, 7], and at present it is allowed to prepare crates for fruits and vegetables using recycled polypropylene and high-density polyethylene provided the used materials and articles conformed to the applicable rules on food contact materials and were in contact only with foods. The list of fruits and vegetables that are allowed to be put in contact with such recycled crates and the geometrical characteristics of the crates are defined in the above-mentioned 2007 decree. At present, other plastic recycled materials and articles in direct contact with foods are not allowed by the Italian legislation. Obviously, the recent EU Regulation 282/2008/EC on recycled plastics in contact with foods will impact Article 13 of the Italian DM 21/3/73 that will be amended accordingly.

Rubber

The rules for rubbers since the beginning were parallel to those for plastics. At present, the existing differences are due to modifications introduced in plastics by the transcription of European directives, while the rubbers remain in the nonharmonized part, subject to national rules. Rubber material and articles in contact with foods (hereinafter *rubbers*) are regulated by the general provisions of the above-reported Articles 1–8 of the DM 21/3/73 and by the specific rules of the decree. The rules on rubbers are based on the three key points of the legislation: (1) positive lists, (2) migration limits, and (3) standardized migration test. In fact, there is a positive list divided into (a) elastomers and (b) additives, with SML, QM, restrictions, and conditions of use. The introduction of new substances in the lists has to follow the procedure described in the general part.

The conventional classification of foods is the same as laid down in European legislation for plastics, but the simulants are those originally listed in the DM 21/3/73: distilled water, 3% acetic acid, 15% ethanol, and olive oil. Test media substituting the simulant D (e.g., isoctane) are not applicable to rubbers. The worst-case contact conditions indicated by the law are 2 h at 80 °C or 10 days at 40 °C in the proper simulant. Methods of analysis for overall and specific migration in food simulants were given and the rules for the contact conditions in migration tests were fixed. The overall migration limit is 8 mg/dm^2 or 50 mg/kg.

Also, in the case of rubbers, there is an article on substances that are not listed but that can be present in the finished products and also in this case, in Article 16, "some monomers, low molecular weight substances, intermediates, catalysts, solvents, emulsifying agents" are mentioned as examples. As for plastics, the migration of nonspecifically regulated components of the rubber material and articles, even though there are no positive lists, must not endanger human health.

For rubbers, there is no positive list of colorants, and Article 18 says that all colorants can be used provided they do not migrate into the food and lays down the same specifications and the method of enforcement as defined for plastics.

The same methods as prescribed for plastics for specific migration of some substances (e.g., monomers, metals, etc.), some technological adjuvants (e.g., peroxides), and for purity of some additives (e.g., vaseline oil) are also applicable to rubbers.

The use of recycled rubbers in contact with foods is not allowed under Article 19, without derogations from this prohibition.

Regenerated Cellulose

This material was formerly regulated by the decree, but the introduction of EU directives drastically changed the situation. At present, the content of the decree is totally substituted by the harmonized EU legislation (Directive 2004/14/EC) and the relevant articles were repealed. The unique part that remains out of the harmonization process deals with synthetic casings. These objects may be softened only with glycerol that, when in contact with foods, shall not exceed the level of 13% (w/w) in the casing.

Paper and Board

Paper and board material and articles in contact with foods (hereinafter *P&B*) are regulated by the general provisions of the above-reported Articles 1–8 of the DM 21/3/73 and by the specific rules on P&B based on (1) positive lists, (2) compositional requirements, and (3) purity requirements. In fact, taking into account that the physicochemical properties of P&B do not allow contact with wet or liquid foods, compositional requirements were deemed more suitable than the migration tests in liquid simulants. Therefore, it must be highlighted that the overall migration concept is not applicable to P&B.

In the case of P&B, the positive lists are divided into the following parts:

Part A: Constituents (fibrous matter, fillers, auxiliary substances, and optical brighteners)
Part B: Processing aids

These lists are regularly updated with new substances that gain the approval following the procedure described in the General Part.

P&B must be manufactured according to good manufacturing practices and must comply with the following compositional requirements (Article 27):

1) In the case of contact with foods for which no migration test is prescribed in the EU conventional classification, at least 75% of fibrous matter, no more than 10% of fillers, no more than 15% of auxiliary substances.
2) In the case of contact with foods for which migration tests are prescribed in the EU conventional classification, at least 60% of fibrous matter, no more than 25% of fillers, no more than 15% of auxiliary substances.

Always bearing in mind that good manufacturing practices must be followed, the presence of traces of processing aids (e.g., reagents, dispersion agents, antimolds, etc.) is allowed.

The legal analytical methods to check compositional requirements are in Annex IV, Section 6 of the decree.

P&B must also comply with purity requirements, which are as follows:

1) The content of polychlorobiphenyls (PCBs) no more than 2 mg/kg P&B.
2) Lead extracted by a 3% acetic acid solution at 40 °C for 24 h: no more than 3 μg/dm^2.

These two requirements correspond in practice to the prohibition of recycled P&B for general use. In fact, recycled P&B are allowed only in contact with foods for which no migration test is prescribed in the EU conventional classification provided the finished products comply with the purity requirements described above (modification introduced by Annex to the Decree n 220 of 26/4/93) [8]. The only derogation to this provision is in Article 27 bis, for multilayer boards, with well-defined description in terms of weight per m^2 of the total sample and of the food contact layer. In fact, Article 27 bis describes multilayer boards with at least 200 g/m^2 weight and with at least three layers (cover, intermediate, and direct contact layer); the layer in direct contact with foods must have a weight at least of 35 g/m^2. In these products, the purity requirement of lead must be respected only for the layer in direct contact with food. However, these boards may be put in contact only with foodstuffs listed in Article 27 bis (e.g., cereals, dry pasta, sugar, salt, shelled fruit, etc.). It must be underlined that the description of the board for which this derogation applies does not fit with corrugated boards.

The legal analytical methods to control the purity requirements are in Annex IV, Section 6 of the decree.

Articles 29 and 30 deal with adhesives for paper and boards. Here is a distinction between the adhesives to couple the different layers and the adhesives for the edges of the finished products. In the first case, there is a positive list, while in the second case no positive list must be respected provided there is no bleeding from the edges to the layer intended to be in contact with foods.

Colorants that may be used for P&B are those listed in the Ministerial Decree 22/12/1967, Section C [9]. Other colorants may also be used provided they comply with the method for the "bleeding" of colorants that is described in Annex to the Ministerial Decree 22/12/1967. P&B colored or printed only on one layer, that is, not the one destined to be into contact with foods, are exempted from the bleeding test.

When printed, P&B may not be printed on the layer intended for contact with foods.

Optical brighteners that may be used are those listed in the positive list of the DM 21/3/73, amended for this point by the Decree no. 267 of 30/5/2001 [10]. Their total amount may not be higher than 0.3% w/w.

The decree contains also an article that requires to identify the layer intended to be in contact with foods, to allow a proper application of the relevant rules. When the indication is missing, both layers must comply with the rules.

The last article on P&B states an important concept: the P&B that do not comply with the specific rules for P&B are allowed to be used provided they comply with the specific rules on plastics. This implies that P&B, to be in compliance, should pass the overall and specific migration tests described for plastics.

Some methods for the specific migration of formaldehyde and trivalent chromium are given in Annex IV. Methods are also given, as limit tests, for the migration of some technological adjuvants (peroxides, aromatic amines, dithiocarbamates, thiourames, xanthogenates, mercaptobenzothiazol and zinc salt, benzothiazyl disulfide, preservatives, phenols, and cresols) and for the purity of paraffins and microcrystalline waxes, vaseline oil, and carbon black. As for plastics, it must be underlined that even if no official methods are reported, checking for compliance with the specific migration or the composition requirements is obligatory when there is a corresponding restriction (SML or QM) in the positive lists.

Stainless Steel

Stainless steel materials and articles in contact with foods (hereinafter *stainless steels*) are regulated by the general provisions of the above-reported Articles 1–8 of the DM 21/3/73 and by the specific rules of the decree. Also for these materials, the rules are based on the concepts of (1) positive lists, (2) migration limits, and (3) standardized migration test. The positive list of the admitted stainless steels, in Annex II, indicates the type of stainless steel individuated by international denominations according to the American Iron and Steel Institute (AISI), the Unification Italian Committee (UNI), or other international codes. The list is open to the entry of new alloys provided the above-described approval is obtained.

Overall migration limit is 8 mg/dm^2 or 50 mg/kg. Specific migration limits of 0.1 mg/l are established for both trivalent chromium and nickel, to be determined by the method described in Annex IV. Taking into account the fact that stainless steel objects are of course not for single use, the decree states that in the case of repeatedly used articles three subsequent "attacks" must be performed and the migration must be measured upon the third "attack."

The worst-case contact conditions, for objects intended to be used without limitations, are indicated by the law as follows:

1) Prolonged contact at room temperature: 3% acetic acid for 10 days at 40 °C.
2) Repeated use contact, short time at room temperature or above: 3% acetic acid for 30 min at 100 °C. Overall and specific migration on the third "attack."

Glass

Glass materials and articles in contact with foods (hereinafter *glass*) are regulated by the general provisions of the above reported Articles 1–8 of the DM 21/3/73 and by the specific rules of the decree. For this particular material, the approach was different from other materials, and although there is no true positive list with different chemical species or materials, a classification of the type of glass materials on the basis of their performance does exist.

Three categories are distinguished:

Category A: borosilicate and sodium-calcium glass, colorless or colored, destined to be submitted to general conditions, up to, and including sterilization conditions.
Category B: sodium-calcium glass destined to be submitted to conditions up to 80 °C.
Category C: lead crystal, for short and repeated contact.

In addition, for glass, migration tests are indicated. In fact, overall migration limit is 8 mg/dm^2 or 50 mg/kg, and specific migration limits of 0.3 mg/l are established for lead, to be determined in lead crystal objects (type C) by the method described in Annex IV.

The decree states that in the case of repeatedly used articles, three subsequent "attacks" must be performed and the measure of the migration must be performed upon the third "attack."

The worst-case contact conditions, for objects intended for use without limitations, are indicated in Article 35 as follows:

Category A: distilled water, 30 min at 120 °C, overall migration
Category B: distilled water, 2 h at 80 °C, overall migration
Category C:
(1) three successive attacks for 24 h at 40 °C with distilled water. Overall migration on the third attack
(2) three successive attacks for 24 h at 40 °C with 3% acetic acid. Migration of lead on the third attack

There is a specific point on the returnable glass container in Article 35. In fact, it is established that containers must be marked with the type of the category, A or B, they belong to and with the mark of the producer. This is to ensure a proper use when the returnable glass container is submitted to the second life cycle.

7.3.2 Ceramic

No Italian regulation is in force on ceramics, but the EU directives completely substitute the national provisions.

7.3.3 Tin Plate

Tin plate is regulated by the Ministerial Decree 18/2/84 [11]. To illustrate the approach of this decree, it is suitable to underline that this material is almost exclusively used for cans. The tin plate is constituted by a steel layer covered by a tin layer. To impede the contact between the canned food and the metal layers, an organic coating is applied on the tin layer, and this organic layer is in direct contact with foods; however, application of uncoated cans still exists. The technologies for cans are based on two- or three-piece assembly and welding by sealings or mouldings.

DM 18/2/84 contains in Annex I a list of prescriptions for the materials and the substances forming the finished tin plate objects. In fact, it defines compositional requirements for the steel layer, a purity requirement of at least 99.85% for the tin of the internal layer, a positive list, and a maximum amount for the allowed lubricants. The organic coatings that are allowed are those covered by the positive lists for plastics in the DM 21/3/73 that, as illustrated above, is applicable also to surface coatings on substrates other than plastic.

DM 18/2/84 defines maximum limits of total lead present in foods canned in cans with the edges welded with tin–lead alloy. For this purpose, a list of foods with the relevant total limits of lead is provided in Annex II. These limits were fixed taking into account the occurrence of lead in foods. Specific migration limits are settled also for iron (50 mg/kg food) and tin (150 mg/kg food). The relevant methods are given in Annex III.

With regard to the organic coating, a method for the control of the "organic overall migration" is laid down, in which the determination of organic migratable fraction is described. The applicable limits are those for plastics (10 mg/dm^2 or 60 mg/kg), and the contact conditions must be selected taking into account the actual conditions of filling, thermal treatments, and storage.

Decree n 405 of 13/7/95 amended the DM 18/2/84 in order to introduce new requirements for the foils of tin plate. In fact, to prevent the use of low-quality materials in food cans, technical requirements are imposed (defects, rust, discontinuities in tin layer, etc.).

Finally, the DM 18/2/84 recalls as applicable Articles 6 and 7 of DM 21/3/73 (see above) concerning the obligation of declaration of compliance, supporting documentation, responsibility of the producers, and technological suitability. The obligation on traceability is covered by Article 17 of the Framework Regulation 1935/2004/CE and the labeling issues by Article 15 of the same regulation.

7.3.4 Tin-Free Steel

Tin-free steel materials and articles (in this chapter, TFS) are regulated by the Ministerial Decree 1/6/88 [12]. The approach of this decree is analogous to that for tin plate, and also in this case, the material is almost exclusively used for cans, but it is suitable to underline that in this case the cans are always internally coated with organic layers.

DM 1/6/88 contains in Annex I the list of prescriptions for the materials and the substances forming the finished tin-free steel objects. In fact, it defines compositional requirements for the steel layer, the chromium-based layer, a positive list, and a maximum amount for the allowed lubricants, which are the same as for tin plate. Also in this case, the organic coatings that are allowed are those covered by the positive lists for plastics in DM 21/3/73, also applicable to surface coatings on substrates other than plastics.

DM 1/6/88 defines maximum limits of total chromium present in foods canned in TFS cans. Specific migration limits have also been established for iron (50 mg/kg food). The relevant methods are given in Annex III.

With regard to the organic coating, a method for the control of the "organic overall migration" is laid down, in which the determination of organic migratable fraction is described. The applicable limits are those for plastics (10 mg/dm^2 or 60 mg/kg), and the contact conditions must be selected taking into account the actual conditions of filling, thermal treatments, and storage.

Finally, DM 1/6/88 recalls as applicable Articles 6 and 7 of DM 21/3/73 (see above) concerning the obligation of declaration of compliance, supporting documentation, responsibility of the producers, and technological suitability. The obligation on traceability is covered by Article 17 of the Framework Regulation 1935/2004/CE and the labeling issues by Article 15 of the same regulation.

7.3.5
Aluminum

Decree no. 76 of 18/4/2007 [13] regulates the use of aluminum and aluminum alloys (in this chapter, *aluminum*) in materials and objects intended for food contact. This material is commonly used for both food packaging (foils, trays) and for food utensils (pans, parts of equipment, etc.). Decree no. 76 is applicable only to aluminum in direct contact with foods, not covered by layers of other materials. The approach to the protection of the consumer health adopted in this decree is based on (1) purity requirements of aluminum, (2) compositional requirements of aluminum alloys, and (3) limitations of use.

In this view, the aluminum must be more than 99% pure and maximum amounts of components of the alloys are indicated. Different tables list the compositional limits taking into account different technologies used to manufacture aluminum objects.

The conditions of use that are allowed for aluminum objects (Article 5) are

1) short contact (<24 h): no limitations of temperatures;
2) prolonged contact (>24 h): only refrigerated temperatures;
3) for prolonged contact, room temperature is allowed only for foods listed in Annex IV.

Foods described in Annex IV are not liquid foods and have a limited extractive power with respect to aluminum (e.g., cocoa and chocolate products, coffee, cereals, sugar, bakery products, etc.).

To allow the consumers to properly use aluminum objects, there are obligations for special labeling in Article 6. In fact, one or more of these indications must be reported:

1) not suitable for contact with strongly acidic or strongly salty foods;
2) destined to come in contact with foods at refrigerated conditions;
3) destined to come in contact with foods at not refrigerated temperatures, for times longer than 24 h;
4) destined to come in contact with foods listed in Annex IV at room temperature also for times longer than 24 h.

The identification of proper compositions and suitable indications and limitations of use prevent an excess migration of aluminum and make it possible not to prescribe migration tests. Therefore, compositional and purity requirements must be under control.

Finally, the obligation of declaration of compliance, supporting documentation, responsibility of the producers, and technological suitability are prescribed in Articles 8 and 9 of this decree. The obligation on traceability is covered by Article 17 of the Framework Regulation 1935/2004/CE and the labeling issues, in addition to those previously reported valid at the Italian level, by Article 15 of the same regulation.

7.3.6
Materials Without Specific Regulation

There are a number of materials for which no specific decree exists in Italy, for example, wood and cork, iron alloys, copper, and so on. In these and in other possible cases, the rules to be followed are the general rules given by the decrees on general principles, the DPR 777, the DL 108, and the Framework Regulation 1935/2004/CE. Therefore this implies that the business operator has to declare the compliance on the FCMs, without reference to positive lists and limits, by performing a self safety assessment to demonstrate compliance with the general principles laid down Article 3 of the Regulation 1935/2004/CE. Food simulants that were developed for specific materials such as plastic or stainless steel may not be appropriate for other materials and overall migration is not directly applicable to materials for which it is not specifically required by Italian laws. In general, the knowledge of the composition of the material is a first unavoidable step to highlight the possibility of migration of components or of their degradation products under conditions widely representing actual uses.

7.4
How to Get the List and the Text of the Italian Legislation

The full text of the Italian law is published in the *Gazzetta Ufficiale della Repubblica Italiana* (Official Gazette of the Italian Republic). It is only in the original Italian language and no official translations are available. Hard copies and online subscriptions can be requested to Istituto Poligrafico e Zecca dello Stato, P.zza G. Verdi, 10, 00100 Rome, Italy, Tel.: +39.0685082150, Fax: +39.0685082520, web site: http://www.ipzs.it.

Furthermore, a useful online source for the list of the Italian measures is made available by the Directorate General for Health and Consumers of the UE (DG SANCO), which keeps a regularly updated file with the citations of all decrees in force, their titles, and a brief summary of their content [1].

Note: The content of this chapter is the responsibility of the author and does not involve the ISS or any other italian public authority.

References

1 http://ec.europa.eu/food/food/chemicalsafety/foodcontact/docs/ReferencesEurNatLeg_20080625.pdf.

2 Decreto Presidente della Repubblica no. 777 of 23/8/82 on Gazzetta Ufficiale Repubblica Italiana No. 298 of 28/10/82,

Istituto Poligrafico e Zecca dello Stato, Rome, Italy.

3 Decreto Legislativo no. 108 of 25/1/92 on Supplemento Ordinario Gazzetta Ufficiale Repubblica Italiana No. 39 of 17/2/92, Istituto Poligrafico e Zecca dello Stato, Rome, Italy.

4 DECRETO Ministero della Salute 21 marzo 1973, Disciplina igienica degli imballaggi, recipienti, utensili destinati a venire a contatto con le sostanze alimentari o con sostanze d'uso personale. Supplemento Ordinario Gazzetta Ufficiale Repubblica Italiana No. 104 del 20 Aprile 1973, Istituto Poligrafico e Zecca dello Stato, Rome, Italy.

5 Guidelines of the Scientific Committee on Food for the presentation of an application for safety assessment of a substance to be used in food contact materials prior to its authorization (updated on 13 December 2001) SCF/CS/PLEN/GEN/100 Final 19 December 2001; http://efsa.europa.eu.

6 DECRETO Ministero della Salute 22 dicembre 2005, no. 299 Regolamento recante aggiornamento del decreto ministeriale 21 marzo 1973, concernente la disciplina igienica degli imballaggi, recipienti, utensili destinati a venire a contatto con le sostanze alimentari o con sostanze d'uso personale. Gazzetta Ufficiale Repubblica Italiana No. 37 del 14 Febbraio 2006, Istituto Poligrafico e Zecca dello Stato, Rome, Italy.

7 DECRETO Ministero della Salute 12 dicembre 2007, no. 270 Regolamento recante aggiornamento del decreto ministeriale 21 marzo 1973, concernente la disciplina igienica degli imballaggi, recipienti, utensili destinati a venire a contatto con le sostanze alimentari o con sostanze d'uso personale. Gazzetta Ufficiale Repubblica Italiana No. 33 del 8 Febbraio 2008, Istituto Poligrafico e Zecca dello Stato, Rome, Italy.

8 DECRETO Ministero della Salute 26. aprile 1993 no. 220 Regolamento recante aggiornamento del decreto ministeriale 21 marzo 1973, concernente la disciplina igienica degli imballaggi, recipienti, utensili destinati a venire a contatto con le sostanze alimentari o con sostanze d'uso personale. Supplemento Ordinario Gazzetta Ufficiale Repubblica Italiana No. 162 del 13 Luglio 1993, Istituto Poligrafico e Zecca dello Stato, Rome, Italy.

9 Decreto Ministeriale 22/12/1967 Disciplina dell'impiego e approvazione dell'elenco delle materie coloranti autorizzate nella lavorazione delle sostanze alimentari, delle carte e degli imballaggi delle sostanze alimentari, degli oggetti di uso personale e domestico, Gazzetta Ufficiale Repubblica Italiana No. 28 del 1 Febbraio 1968, Istituto Poligrafico e Zecca dello Stato, Rome, Italy.

10 DECRETO Ministero della Salute 30 maggio 2001 no. 267 Regolamento recante aggiornamento del decreto ministeriale 21 marzo 1973, concernente la disciplina igienica degli imballaggi, recipienti, utensili destinati a venire a contatto con le sostanze alimentari o con sostanze d'uso personale. Gazzetta Ufficiale Repubblica Italiana No. 155 del 6 Luglio 2001, Istituto Poligrafico e Zecca dello Stato, Rome, Italy.

11 DECRETO Ministero della Salute 18 febbraio 1984, Disciplina dei contenitori in banda stagnata saldati con lega stagno piombo e altri mezzi, Gazzetta Ufficiale Repubblica Italiana No. 76 del 16 Marzo 1984, Istituto Poligrafico e Zecca dello Stato, Rome, Italy.

12 DECRETO Ministero della Salute 1 giugno 1988 n 243, Disciplina dei contenitori in banda cromata verniciata destinati a venire in contatto con alimenti, Gazzetta Ufficiale Repubblica Italiana No. 153 del 1 Luglio 1988, Istituto Poligrafico e Zecca dello Stato, Rome, Italy.

13 DECRETO Ministero della Salute 18 aprile 2007 n 76, Regolamento recante disciplina igienica deei materiali e dgli oggetti di alluminio e leghe di alluminio destinati a venire a contatto con gli alimenti, Gazzetta Ufficiale Repubblica Italiana No. 141 del 20 Giugno 2007, Istituto Poligrafico e Zecca dello Stato, Rome, Italy.

8
Switzerland

Roger Meuwly and Vincent Dudler

The first federal food law promulgated in 1905 marked the beginning of the modern legislation on food safety in Switzerland at national level. However, food contact materials (FCMs) have been specifically regulated only since 1936 when the *Ordinance on Trade of Foodstuffs and Commodities* was published. This initial regulation reflects the safety problems encountered at that time, mainly heavy metal contamination from metallic and enamel utensils. The revision of this ordinance in 1964 took a more modern approach by regulating new materials such as plastic, varnishes, coatings, and waxes. The admissibility of plastic starting substances was then regulated by a positive list and migration maxima. In 1995, a major revision of the food legislation marked the beginning of the harmonization of the national legislation with the European legislation. At present, both regulations on FCMs are quite similar, but the Swiss regulation still presents some differences and particularities, mostly in domains that are not regulated by the European Commission.

8.1
Legislative System

Both food and food contact materials are governed by the same legislative act: the *Federal Law on Food and Commodities* of 1995 that establishes the scope, purpose, and principles of the regulation. The law also defines the enforcement system, the separation of duties between the federal government and the cantonal authorities as well as the responsibility of all those involved with FCMs. The two main objectives defined in the law are consumer protection against risks to health and the prohibition of deception or misleading practices. However, FCMs are excluded from this second objective. An important principle defined in the law is the "self supervision" (Article 23): anyone who manufactures, treats, supplies, imports, or exports FCMs must ensure that within the context of his/her activities, the goods are in conformity with legal requirements. He/she must analyze them or have them analyzed according to "good manufacturing practice." This principle is strengthened by the obligation on

Global Legislation for Food Packaging Materials. Edited by Rinus Rijk and Rob Veraart

ISBN: 978-3-527-31912-1

the FCMs manufacturers and traders to assist the authorities in their duty and to provide all relevant information (Article 25, Rights and Obligations of Manufacturers and Traders).

Two ordinances are ranked under the law: the *Ordinance on Food and Commodities*, which is a framework regulation and which also states the general requirements on FCMs (Article 34), and the *Ordinance on Materials and Articles*. The latter ordinance lays down all specific requirements for each type of material intended to come in contact with food.

Each legislative text can be amended following a revision process that depends on its position (importance) in the legal hierarchy. For a federal law, the Parliament should accept the amended text by vote. The ordinances are amended in a consultative procedure involving different stakeholders (government departments, enforcement authorities, industry, consumer organizations, etc.). The amended texts are published and all interested parties can comment on the proposals.

The Swiss Federal Office of Public Health (FOPH) has the responsibility of the regulation on food contact materials and acts as a legislator. The role of the FOPH is to elaborate the regulation based on latest scientific findings, to grant authorizations foreseen by the legislation, and to coordinate its enforcement. The FOPH also publishes directly some legal texts of less importance, the *Information Letters* and the *Directives* (in German: Weisung). *Information Letters* are recommendations given to all stakeholders. They specify the manner in which a legal text should be interpreted. The *Directives* are official instructions given to the enforcement authorities. In addition, the FOPH publishes the *Swiss Food Compendium*, a database that lists the recommended analytical methods and test conditions concerning food and FCM controls; more information on analytical methods can be found in Ref. 3.

The cantonal authorities carry out the enforcement of the legislation in Switzerland. At present, 21 official laboratories enforce the food and FCM legislation in the territory and at the border. The FOPH coordinates their activities when necessary.

8.1.1 Availability of Legal Texts and Official Documents

The Swiss legislation is directly accessible on the Internet. The amendments are first published in the *Official Compilation of Federal Legislation* [1]. Three to four months after coming into force, a consolidated version of the amended texts is made available in the *Classified Compilation of Federal Law* [2]. The documents can be downloaded in German, French, and Italian languages. The best way to find information on a regulated matter is to look through the systematic numbering (SR). The food and commodities are classified under the reference number SR *817*: The *Federal Law on Food and Commodities*, the *Ordinance on Food and Commodities*, and the *Ordinance on Materials and Articles* are listed under the systematic numbers SR 817.0, SR 817.02, and SR 817.023.21, respectively.

The FOPH *Information Letters* and *Directives*, as well as the *Swiss Food Compendium*, are official documents that are published directly by the FOPH and are downloadable from its web site [3].

8.1.2 Attestation of Conformity (Letter of No Objection (LNOs)

Upon request, as it is not a legal requirement, the Swiss authorities issue an "attestation of conformity" indicating that the material or the finished article may be lawfully used in Switzerland. The application procedure is available on the FOPH web site [4]. The attestation of conformity is generally requested for by the food industry and issued for the finished article in plastic, but it may also be available for products made from other groups of materials and for intermediates.

The attestation is given on the basis of the evaluation of the composition of the products according to the Swiss legislation. If no specific Swiss legislation exists, the compliance is evaluated on the basis of other legislations such as other international/national legislations in the EU, Council of Europe Resolutions, German BfR Recommendations, FDA (Food and Drug Administration of the United States), and so on. It is important to note that such attestations do not relieve the manufacturer, distributor, or the food sellers of their responsibilities for assuring safety in use. The practical requirements specified in the *Ordinance on Materials and Articles* have to be fulfilled and are controlled by the cantonal laboratories that are in Switzerland the enforcement organs.

The attestation remains valid for 5 years and is considered valid as long as the composition and intended use of the material remain as described in the original submission and as long as the legislation pertinent to the product is not amended. It is the responsibility of the manufacturer of the material or the article to follow the most recent development in the legislation. The principle of "self supervision" expressed in the *Law on Food and Commodities* clearly stresses the liability of the manufacturer, importer, seller, and so on the compliance of their products with the Swiss legislation.

8.1.3 Council of Europe

The documents elaborated by the Council of Europe (CoE) are intended for use as the basis for a transposition or preparation of a national law and could eventually serve as the basis for future EU legislation on these materials. Switzerland is an active member of the CoE partial agreement and has already transposed the following CoE Resolutions in the Swiss legislation: *Resolution AP (89) 1 on Colorants for Plastics* and *Resolution AP (2004) 5 on Silicones*. The *Resolution AP (2005) 2 on Inks* is in the process of implementation. In the absence of specific Swiss or EU regulations, it is fully accepted by the FOPH and by the food control authorities that the CoE Resolutions serve as reference to define the compliance of materials and objects in regard to the general requirements of Article 34 of the *Ordinance on Food and Commodities.*

8.2 Food Contact Materials and Articles

The general requirements of Article 34 of the *Ordinance on Food and Commodities,* which applies to all materials intended for food contact, states that the food contact materials and articles should transfer their constituents only in quantities that

- are not dangerous to human health;
- are technically unavoidable;
- do not change the composition of the food or their organoleptic characteristics.

The term "technically unavoidable" is used instead of the wording "shall be manufactured in compliance with good manufacturing practice" as in Article 3 of the EU Framework Regulation (EC) 1935/2004, although it expresses the same principle. It also covers the ALARA principle (As Low As Reasonably Achievable).

In case articles are composed of different layers, each of the layers of a combined article must comply with the corresponding legislation. For example, a plastic layer on a paper has to fulfill the requirement applicable to plastic materials and articles.

8.2.1 Plastic Materials and Articles

Plastic materials and articles intended to come in contact with food constitute one group of materials and articles, which are regulated by specific measures in the *Ordinance on Materials and Articles.* In spite of the fact that Switzerland is not a member state of the EU, the Swiss government intends to harmonize its FCM legislation with EU regulation. The Swiss requirements on plastic materials and articles generally refer to the EU Plastics Directive (*Commission Directive 2002/72/EC* and its amendments), although there are some exceptions.

The section on *Plastic materials and articles* of the ordinance applies to food contact materials and objects, and parts thereof that consist

- exclusively of plastics;
- of two or more layers of materials, each consisting exclusively of plastics, which are bound together by means of adhesives or by any other means;
- of plastic materials used as surface coatings and varnishing.

Contrary to the EU Plastics Directive, "materials and articles composed of two or more layers, one or more of which does not consist exclusively of plastics" are not excluded from the scope of the ordinance. The plastic layer of a composite material made of plastic and other materials (i.e., tetrabrik) has to fulfill by analogy the requirements applicable to plastic materials and articles.

The section on *plastic materials and articles* consists of a text with general requirements and an annex containing lists of permitted monomers and additives with specific limitations on the use of certain substances. The monomer list is a positive list; only those monomers and other starting substances listed may be used for the manufacture of plastic materials and articles subject to the restriction specified. The additive list is a nonexhaustive list of substances that may be used to achieve some specific and technical properties to the plastics or to provide a suitable medium in which polymerization occurs. Neither does it include substances that directly influence the formation of polymers (e.g., catalyst, chain-stoppers, etc.) nor does it include substances that are impurities of substances used, reaction intermediates, or decomposition products. No provisions are actually mentioned for dual-use additives (additives authorized as food additives and as plastic additives) like in Article 5a of the *Commission Directive 2004/19/EC*, except for additives used in active or intelligent materials. The date when the additive list shall become a positive list has not yet been set.

A third list deals with specific EU directives and some specific requirements on the purity of starting substances and contains, among others, articles on

- PVC and PVDC films: the use of phthalates as plasticizers in PVC and PVDC films for food contact materials is prohibited. The article establishes some restrictions for the use of plasticizers in PVC and PVDC films.
- BADGE and other epoxy derivatives: the sum of the specific migration of BADGE and some of its derivatives should not exceed the limit of 1 mg/kg in food or food simulants. The use of NOGE and BFDGE in the manufacture of cans is prohibited. Compared to the EU, Switzerland has not increased the specific migration limit (SML) to 9 mg/kg for the sum of BADGE and BADGE hydrolysis derivatives [5] as such an increase would have been in contradiction with the ALARA principle defined in the alinea b of the Article 34 of the *Ordinance on Food and Commodities.*
- Colorants: the rules on the use of colorants in plastic materials coming in contact with food is based on the CoE *Resolution AP(89)1*. The text specifies, among others, the purity of pigments and some migration limits.
- The prohibition of azodicarbonamide in plastic FCMs.

8.2.1.1 Limits on Migration

All plastic articles have to meet the overall migration limit (OML), which is set according to the type of surface in contact with food, of 10 mg/dm^2 of surface area of material or article or 60 mg/kg food or food simulants. They must also meet any applicable SML requirements that may be applicable to particular monomers, other starting substances, or additives. In addition to OML and SML values, QM restrictions (maximum permitted quantity of a substance in a finished material or article) may be required in some cases. In a case where SML and QM values are given for a substance, the QM value can be used for determining the compliance only when the SML value cannot be determined.

Verification of compliance with the migration limits shall be carried out in accordance with the rules laid down in Chapter 48 of the *Swiss Food Compendium*. The details of compliance testing, time and temperature conditions, and the choice of the food simulant, correspond in general to both, the *Council Directive 82/711/EEC* and *Council Directive 85/572/EEC*, as amended. The estimation of the specific migration level of a substance may also be established by using generally recognized diffusion models based on scientific evidence. The fact, as in EU legislation (Article 8 of the *Commission Directive 2002/72/EC*), that "the verification of the compliance with the specific migration limits shall not be compulsory, if it can be established that, by assuming complete migration of the residual substance in the material or article, it cannot exceed the specific limit of migration" is not explicitly mentioned in the Swiss legislation but is accepted. The noncompliance of a material or article has to be demonstrated by experimental testing. The finished article must comply with the migration limits (OML and SML) and with any other applicable specifications or conditions of use.

8.2.1.2 **Recycled Plastic Materials**

According to Article 10 of the *Ordinance on Materials and Articles*, recycled plastics can be used only for the manufacture of materials and articles intended for food contact with an authorization issued by the FOPH. In-house production scraps are excluded from the scope of this article. Petitioners should demonstrate that the recycled material meets the general safety requirement of the *Ordinance on Food and Commodities* and the specific requirements of the section on plastic materials and articles of the *Ordinance on Materials and Articles*. The submission shall include information on the feedstock logistic, recycling procedure, and the quality control of the recycled materials. Some recycling processes have already been authorized and are used to produce PET bottles from postconsumer bottle materials (recycling bottle to bottle).

Most PET beverage bottles consumed and collected in Switzerland are sorted and recycled domestically. According to the *Ordinance on Beverage Containers* (SR 814.621), at least 75% of all glass and PET bottles or aluminum cans sold are recycled. On behalf of the Swiss Federal Office for the Environment, PET-Recycling Switzerland (PRS) [6], a private association, is responsible for the collection of disposable PET beverage bottles. An advance recycling contribution has been charged since 1991 on PET single-use drinking bottles without deposit. The national recycling rate for PET bottles has just reached the target specified in the ordinance during the last few years.

8.2.2 **Regenerated Cellulose**

The use of regenerated cellulose (cellophane) articles in food contact materials is covered by Section 4 of the *Ordinance on Materials and Articles*. The requirements are based on the *Commission Directive 93/10/EEC* and its amendment (*Directive 2004/14/EC*). The section contains the scope, definitions, requirements, and annexes including positive lists of substances permitted for use in the manufacture of cellulose varnished, nonvarnished, and plastic-varnished regenerated cellulose films. If a plastic coating is to be applied to the film, only substances in the list of Annex 1 (Section 3: *Plastic Materials and Articles*) can be used.

8.2.3 Materials and Articles in Ceramic, Glass, Enamel, or Other Analogue Materials

This section specifies migration limits of cadmium and lead, both of which may be released from ceramic, glass, and enamel articles. Migration values are taken from the *Council Directive 84/500/EEC*, as amended by *Directive 2005/31/EC*.

8.2.4 Metals and Alloys

Metals and alloys are used as food contact materials, not only in the processing equipment, containers, and household utensils but also in foils for wrapping foodstuffs (aluminum foil). They are often covered by a surface coating, which reduces the migration of metal ions into foodstuffs. This section regulates, among others, the following:

- The use of articles containing lead, cadmium, zinc, or alloys of these metals is prohibited
- Copper articles
- Aluminum articles for juices, with a migration limit of 10 mg/l
- Requirements on metallic coatings

8.2.5 Materials and Articles in Paper and Board

This succinct section on paper and boards specifies, among other things, that materials and articles in paper and board are made of a quality that they will not stick to the food. The use of recycled paper as FCMs is limited to certain foods or is subjected to an authorization from the FOPH.

8.2.6 Active and Intelligent Materials and Articles

A new section covering active and intelligent materials and articles has been introduced recently in the *Ordinance on Materials and Articles* and is based on the provisions of Articles 2 and 4 of *Regulation (EC) 1935/2004*. Pending the adoption of additional rules, substances deliberately incorporated into active materials and articles shall be used only if they comply with the food legislation. Active materials and articles shall be adequately labeled with the indication of the active substance.

8.2.7 Paraffin, Waxes, and Colorants

Paraffin and waxes used in the manufacture of food contact materials must fulfill the requirements of the *Pharmacopoea Helvetica* and must be free from carcinogenic substances.

For the coloration of food contact materials, the following products can be used: colorants included in the *Ordinance on Food Additives* (SR 817.022.31), barium sulfate, barite derivatives, chromium(III) oxide, and copper.

8.2.8 Silicone Materials and Articles

In one of the last revisions of the *Ordinance on Materials and Articles*, a new section on silicone materials and articles has been introduced. This new regulation is based on the Council of Europe *ResAP (2004) 5 on silicones to be used for food contact applications*. Silicones constitute a group of polymeric chemical substances and preparations, all containing polysiloxanes, and include a range of products with a variety of properties and applications. The substances used for the manufacture of silicone materials and articles should be listed in the inventory list of Annex 5. The inventory list is divided into two parts: evaluated and nonevaluated substances for food contact materials. The nonevaluated can still be used pending their transposition in the list of evaluated substances and as long as the general requirement to ensure the health of consumers is fulfilled (RS 817.02, Article 34). The verification of the compliance with the migration limits should be conducted according to the test conditions for plastic materials. In addition, the silicone elastomers should not release more than 0.5% of volatile components during heating for 4 h at 200 °C.

8.2.9 Inks

This new section applies to printing inks and is based on the Council of Europe *Resolution ResAP (2005) 2 on packaging inks applied to the nonfood contact surface of food packaging materials and articles intended to come in contact with foodstuffs*. The revised ordinance came into effect on April 1, 2008, but with an enforcement time frame of 2 years. The actual list of starting substances given in Annex 6 is incomplete; it will be revised in 2009. Layers of packaging inks in direct contact with foodstuffs are excluded from the scope of this regulation. Inks used behind a functional barrier such as in glass bottles and metal cans are also excluded provided a set-off or transfer via a gas phase is ruled out.

8.3 Conclusions

The EU Commission has been very active over the years and many amendments to the legislation on food contact materials have been introduced. Some points of these modifications have not been actually transposed into the Swiss legislation, for example, the traceability of materials and articles, the declaration of compliance, the plastic functional barrier, the Fat Reduction Factor (FRF) for a more adequate estimation of exposure of the consumer, a new adequate simulant for some milk products, and so on. Future revisions of the Swiss legislation on food contact materials will take into consideration these points in order to increase the harmonization with the EU legislation. It is difficult to give a time frame for these modifications because the points of the modification have to be transposed into different laws or ordinances. It is estimated that all the points listed in the conclusion will be transposed before 2012 in the Swiss legislation.

References

1 Official compilation of federal legislations, http://www.admin.ch/ch/f/as/index.html.
2 Classified compilation of federal laws, http://www.admin.ch/ch/f/rs/rs.html.
3 Federal Office of Public Health, http://www.bag.admin.ch/index.html?lang=en.
4 Federal Office of Public Health, Procedure for the expertise of materials and articles in plastic, http://www.bag.admin.ch/themen/lebensmittel/04867/05348/index.html?lang=en.
5 Commission Regulation (EC) No. 1895/2005.
6 Verein PRS PET-Recycling Schweiz, www.prs.ch.

9
Legislation on Food Contact Materials in the Scandinavian Countries and Finland

Bente Fabech, Pirkko Kostamo, Per Fjeldal, and Kristina Salmén

9.1
Introduction

Scandinavian countries comprise Denmark, Norway, and Sweden. These countries have a long tradition of cooperation in the Nordic Council of Ministers with Finland and Iceland. With the exception of Norway and Iceland, they are member of the European Community and follow the rules on food packaging materials. The Nordic countries frequently combine their research sources in various projects that are then reported in the so-called "Nordic Report." In most cases, the Nordic countries have no or very limited national legislation on food contact materials (FCMs). Below summaries are given for Denmark, Finland, and Norway. Sweden has a system of authorization, which is explained in more detail.

9.2
Legislation in Denmark

The Danish legislation on food contact materials consists of an Order, a Circular, and the EU regulations on the area. The Order and the Circular implement the EU directives on food contact materials, and it is updated whenever new directives are adopted. Link to the regulation on the web (Danish only) is http://www.foedevarestyrelsen.dk/Foedevaresikkerhed/Materialer_genstande/Regler_for_materialer_og_g.htm.

Besides the implemented EU legislation, Denmark has national legislation on glass and ceramic products other than ceramics, for example, porcelain. This legislation is equivalent to the EU legislation on ceramics but covers migration of lead and cadmium also from the mouth rind.

Denmark has for many years requested trade and industry within the area of plastics, cellulose regenerates, and glass and ceramics to be registered for food inspection. These companies are inspected by inspectors. As a consequence of the

Global Legislation for Food Packaging Materials. Edited by Rinus Rijk and Rob Veraart

ISBN: 978-3-527-31912-1

EU regulation on GMP, this request has been extended to the food contact materials industry and trade covered by the regulation on GMP. The companies had to be registred at the Danish Veterinary and Food Administration by 31 December 2009. and will be covered by the public control.

Guidance to industry and trade is given on the web site of the National Veterinary and Food Administration (http://www.foedevarestyrelsen.dk/Foedevaresikkerhed/Materialer_genstande/Regler_for_materialer_og_g.htm).

The guidance covers checklists for in-house control and references to documents that can be used in relation to the assessment of specific materials such as paper and board (Nordic Report on Paper and Board Food Contact Materials, TemaNord 2008:515, ISBN 978-92-893-1657-6) and the Council of Europe guidelines on metals and alloys.

9.2.1
Public Control and In-House Documentation

Business operators, who have duties defined under the legislation on food contact materials, shall map possible hazards in connection to the safety of the materials and articles. Business operators are obliged to have control of points and processes that are critical. This is an essential point in the national control on food contact materials.

9.3
Legislation in Finland

The Finnish legislation on food contact materials is all in accordance with the corresponding legislation of the European Union. Regulations (EC) No. 1935/2004 and (EC) No. 2023/2006 are applied as such and all the directives have been implemented to the Finnish legislation. All the legislation is available on the web site of the Ministry of Agriculture and Forestry (only in Finnish and Swedish) http://www.mmm.fi/attachments/elo/kontaktimateriaalit/5FlcCtjLq/kontaktimateriaalit.pdf.

Besides the EU legislation, Finland has one national decree: the decree of the Ministry of Trade and Industry 268/1992 on heavy metals migrating from food contact articles. It is available only in Finnish and Swedish. This decree can be obtained via the above-mentioned web site. Our national decree on paper and board in food contact materials and articles was repealed in 2006.

The competent authorities who undertake the official control of food contact materials in Finland are the Finnish Food Safety Authority, Evira, and municipal food control units. The Finnish Customs Authority is responsible for the control of imports from outside the EU and partially also for the control of the EU internal trade. In 2008, there were some 190 municipal units in Finland. Evira is responsible for giving guidance on food contact materials to other competent authorities.

Evira gives guidance to competent authorities and business operators on its web site http://www.evira.fi/portal/fi/elintarvikkeet/valvonta_ja_yritt_j_t/kontaktimateriaalit/.

A guidance document for public control both in FCM production and in FCM usage in food packaging is under preparation. After the FCM control guidance is ready (in 2009), it will be available on Evira's web site. The guidance for business operators consists of the documents prepared in cooperation with the Nordic authorities and operators and published by the Nordic Council of Ministers (http://www.norden.org/pub/sk/index.asp?subject=LevnedsM):

- TemaNord 2008:517 ja 2008:709, food contact materials, in-house documentation and traceability, Nordic checklist to industry and trade
- TemaNord 2008:515 Report: paper and board in food contact materials

The Ministry of Agriculture and Forestry is responsible for the food contact material legislation in Finland.

9.4 Legislation in Norway

Norway is associated with EU legislation via the European Economic Area (EEA) Agreement. This means that legislation has to be accepted by the European Free Trade Association (EFTA) countries before the measures are implemented in national legislation in Liechtenstein, Iceland, and Norway. The EEA Committee formally decides if a legal document shall be implemented. The EFTA countries may request adaptations in the legal text; however, in practice, almost all EU measures on food contact materials are implemented in Norway.

It is important to have in mind that the EEA process sometimes may result in delays in the implementation of some measures in EFTA countries. The transitional period of routine amendments of directives or regulations is normally sufficient for having the same date of entry into force in Norway as in the EU.

Link to the regulation on the web (Norwegian only) is http://www.lovdata.no/cgi-ift/ldles?doc=/sf/sf/sf-19931221-1381.html.

9.4.1 The Packaging Convention of Norway and the EK Declaration

Emballasjekonvensjonen, EK (The Norwegian Packaging Convention), is a membership organization with cooperation between food retail chains, food industry, and suppliers of food contact materials. Importers and producers of food contact materials in Norway may be asked by the food industry to provide an EK declaration.

The EK declaration, itself, is not a legal requirement in Norway. However, the EK declaration is constructed according to the listed Compliance Declaration requirements of EU Directive 2007/19/EC, and it is based on supporting documentation. Documentation is controlled by Nofima Mat (the Norwegian Food Research Institute).

Most food industries require the EK declaration because then they are confident that food contact legislation in Norway is fulfilled. Declarations from the EK are also a guarantee that compliance declarations and supporting documentation are provided upstream from the suppliers of raw materials. For more information, contact Packaging Convention of Norway, EK: http://www.emballasjekonvensjonen.no/web/ek.nsf/ekFsetEng!Openframeset.

9.4.2
Paper and Board Food Contact Materials

The Norwegian Food Safety Authority expects that food contact materials made of paper and board are safe and produced in line with requirements in the report Paper and Board Food Contact Materials, TemaNord 2008:515: http://www.norden.org/pub/sk/showpub.asp?pubnr=2008:515&lang=3.

This report applies to food contact materials made of paper and board and the basis is the Council of Europe Resolution AP (2002) 1. Requirements are harmonized with EU Regulation (EC) No. 1935/2004, with respect to GMP, traceability, compliance declarations, documentation, and so on. Substances used in the production of food contact paper and board should have been evaluated by EFSA, BfR, or FDA. Specifications and conditions of use must be respected.

9.4.3
Control of Critical Points

Business operators that have duties defined under the Norwegian regulation on food contact materials shall map possible hazards in connection to the safety of the materials and articles. Business operators are obliged to have control of points and processes that are critical. Please find a translation of the complete legislative text below.

Section 4b. Control of critical points

Business operators that have duties defined by this regulation shall map possible hazards in connection to the safety of the materials and articles. Business operators are obliged to have control of points and processes that are critical. This control includes the following:

1) Evaluation of possible hazards in connection to the work processes in the enterprise
2) Point out stages in the work process where hazards may occur
3) Establish which of these points are critical for the safety of materials and articles (critical points)
4) Establish and implement effective routines for steering and control of critical points
5) When there is any change in the work process, the business operator shall evaluate possible hazards, critical points, and steering routines

Written documentation shall be available. Both those who are responsible for the business and those who participate in it have a duty to ensure that this requirement is fulfilled.

Business operators shall fulfill this requirement by October 27, 2006 at the latest.

Section 4c. Notification

All business operators who are about to start production, import, or sale of materials and articles must send a written notification to Norwegian Food Safety Authority (Mattilsynet).

The notification shall contain the following information:

1) Name or company name, address in Norway, and organization number
2) Name of the manager or other person responsible for the company
3) For importers: country of production and name of the producer
4) Activities of the business operator
5) Type material/article and range of use
6) Other information necessary to describe business operators' nature and scope

Mattilsynet shall immediately be informed if information as notified under this Section 1 is changed.

Existing business operators shall notify requested information latest by January 1, 2007.

9.4.4
Metals in Ceramics, Glass, Metalwares, and Nonceramic Materials Without Enamel

Norway has stricter restrictions on migration of lead, cadmium, and barium from ceramics, glass, metalwares, and nonceramic materials without enamel than all other countries in the EU/EEA area. When importing into Norway, please be aware of these stringent requirements, which may be even more especially in case of ceramic products. Please find a translation of the complete legislative text below.

Chapter VI. Metals in ceramics

Section 25. Maximum limits

Ceramic articles shall not release metals in amounts above the levels given in Annex IV. The migration of metals shall be tested according to methods laid down in Section 26.

Section 26. Methods of analysis

Analytical control of lead and cadmium that are released from ceramic articles shall be carried out according to EU directives 84/500/EEC, Annex I and II.

If an article consists of a vessel fitted with a ceramic lid, the same maximum limit (mg/dm^2 of mg/l) for metals applies as for the container alone. The container and the inside of the lid shall be tested separately and under the same conditions.

The method of analysis must have a detection limit of at least 50% of the maximum limit for released metals.

Annex IV. Maximum limits of metals released from ceramic articles.

Articles	Lead	Cadmium	Barium
that can be filled (hollowwares)	0.1 mg Pb/l	0.01 mg Cd/l	1 mg Ba/l
Articles that cannot be filled (flatwares)	0.02 mg Pb/dm^2	0.002 mg Cd/dm^2	0.2 mg Ba/dm^2
Drinking rim	0.02 mg Pb/dm^2	0.002 mg Cd/dm^2	0.2 mg Ba/dm^2

Articles that cannot be filled include products with inner depth, measured between the bottom and the horizontal projection of upper edge, equal to or less than 25 mm (flatwares).

Drinking rim is the part of the article that is intended for contact with lips or mouth. The drinking rim is 2 cm wide, measured from the upper edge of the vessel on both the inside and the outside. Articles made for drinking shall fulfill maximum limits both for release of metals from the vessel and for the drinking rim.

Chapter VIa. Metals in other materials than ceramics: glass, metalwares, and nonceramic materials without enamel

Section 26a

Glass, metalwares, and nonceramic materials without enamel shall not release lead and/or cadmium in amounts above levels given in Annex III. The migration of lead and cadmium shall be tested according to methods laid down in Annex III.

The method of analysis must have a detection limit of at least 50 % of the maximum limit for released metals.

Annex III. Maximum limits of metals released from articles made of glass, metals, or nonceramic materials without enamel/coating.

Articles	Lead	Cadmium
Articles that can be filled (hollowwares)	0.1 mg Pb/l	0.01 mg Cd/l
Articles that cannot be filled (flatwares)	0.02 mg Pb/dm^2	0.002 mg Cd/dm^2
Drinking rim	0.02 mg Pb/dm^2	0.002 mg Cd/dm^2

Articles that cannot be filled include products with an inner depth, measured between the bottom and the horizontal projection of upper edge, equal to or less than 25 mm (flatwares).

Drinking rim is the part of the article that is intended to come in contact with lips or mouth. The drinking rim is 2 cm, measured from the upper edge of the vessel on both the inside and the outside. Articles made for drinking shall fulfill maximum limits both for release of metals from the vessel and for the drinking rim.

9.5 Legislation in Sweden

In Sweden, the legislation on FCMs is all in accordance with the laws of the European Union. Apart from the Framework Regulation (EC) No.1935/2004, all directives in the food packaging area have been implemented.

Sweden does have one national law on FCMs. In the Ordinance LIVSFS 2005:20, there is one additional paragraph (§17) stating some provisions on what kind of materials should be used for cutlery and/or equipment when handling food (with provisions for zinc, lead, and cadmium).

The competent authorities who undertake the official control of FCM in Sweden are the National Food Administration and the municipalities (Sweden has some 290 municipalities). The National Food Administration is responsible of giving guidance on FCMs to other competent authorities.

9.5.1 Voluntary Agreement

In Sweden, Normpack, a neural and independent body, is active in the control of food contact materials. A trade and industry group, Normpack is a voluntary agreement in force since 1978 within the industry on a material code for food packaging with the Swedish National Food Administration.

Today, the Normpack has around 175 members representing all production stages in the value chain: producers of raw materials for materials and products to be used in contact with food, food industry, and wholesale/retail. The aim of the agreement is to promote the interest of the industry, trade, and consumers in Sweden in product safety of food packaging materials by

- establishing a Swedish system of specifications regarding the characteristics of packaging materials;
- creating a contact forum for the public, authorities, and trade;
- maintaining and increasing the knowledge and skill in this field of the parties concerned.

The Normpack-Norm is based on present Swedish and EC regulations and directives, as well as legislative instruments such as Warenwet (the Netherlands), FDA (the United States), and BfR (Germany).

In some cases, these are temporary solutions. A Temporary Norm has been agreed upon with the Swedish National Food Administration and is in force till actual EC regulations covering the specific situation are published.

More information can be obtained from www.stfi-packforsk.se/np.

Swedish Material Norm for Materials and Articles in Contact with Foodstuffs

The Normpack-Norm, Status January 2008

§1 Materials and articles in contact with foodstuffs shall meet the demands laid down in the Swedish Food Act (SFS2006:804) and the European Directive No. 178/2002/EC, the Food Decree (SFS 2006:813), the Regulations (EC) No. 2232/1996 and 1935/2004, the ordinance for materials and articles intended to come in contact with foodstuffs (LIVSFS 2003:2, updated by LIVSFS 2006:20), the ordinances on nutritional supplements (LIVSFS 2003:9), on food additives (LIVSFS 2003:9, updated by LIVSFS 2003:20), on foreign substances in food (SLV FS 1993:36 with amendments).

The following decrees and recommendations that are in force, specific for materials and products for food contact, are included in the Normpack-Norm:

> EU-*Ordinances* 2232/96, 178/2002 (the Food Act), 1935/2004 (the Frame Regulation), 1895/2005 (epoxy derivatives), 2023/2006 (GMP), 372/2007 (plasticizers in lid gaskets) and the EU-*Directives* 78/142 (VCM), 80/766, 81/432, 86/572 (food simulants). 82/711 (test conditions, amended by 97/48), 2004/24 (cellophane), 2007/42, 2002/72 (plastics), 2004/19, 2005/79, and 2007/19 (plastics and lid gaskets).

§**2** For materials not covered in detail in Swedish food legislation, one of the following regulations shall be invoked:

- The Dutch Packaging and Food-Utensils Regulation (Warenwet), Holland.
- Kunststoffe im Lebensmittelverkehr, Empfehlungen des Bundesinstitutes für Risikobewertung (BfR), and Bedarfsgegenständeverordnung published in Bundesgesundheitsblatt, Germany.
- Code of Federal Regulations, 21, Food and Drugs, §§ 174, 175, 176, 177, 178, 180, 181, 182, 184, 186, and 189 (FDA), USA.

§**3** To prevent incorrect usage of materials and articles in contact with foodstuffs, suppliers and buyers at all manufacturing and handling levels shall consult on the suitability of the product for the intended purpose.

§**4** Temporary Norm for cellulose-based materials (except cellophane) in contact with foodstuffs

4.1 Fibrous raw materials may be used for materials in contact with fruit and vegetables that are usually washed and/or peeled provided that the material meets the general purity demands according to BfR, Empf. XXXVI.

4.2 Fibrous raw materials may be used for materials in contact with dry food products provided that the material meets the general purity demands according to BfR, Empf. XXXVI. Only such recycled fibers may be used as are defined in BfR, Empf. XXXVI A.I.4 and with special consideration to footnotes 2 and 2a. "Untersuchungen von Papieren, Kartons and Pappen für den Lebensmittelkontakt, Wiedergewinnung von Papierfasern," Section 2, "Begriffe."

4.3 Fat or wet foodstuffs should not be in direct contact with materials containing recycled fibers. If the material contains recycled fibers, there should be a functional barrier between the foodstuff and the material (recommendation from the Swedish Food Administration).

4.4 As the BfR Empf. XXXVI does not refer to materials with a functional barrier, the migration of foreign substances is not allowed to exceed 10 mg/dm^2 according to the Warenwet legislation, Chapter II.

4.5 BfR Empf. XXXVI/I, XXXVI/2 and XXXVI/3 shall be followed without remarks according to the BfR recommendations.

4.6 The material shall always maintain a high microbiological purity adapted to the foodstuff in which it will come in contact.

§**5** Temporary Norm with amendments for additives with limits in plastic materials: (EU) Directives 2002/72, 04/19, 05/79, 07/19, EU ordinances 2023/2006 and 372/2007. If there is a limit regulating materials and articles, the following alternative methods are suggested to establish whether the product meets the demands of 2002/72/EC with subsequent amendments:

1) Measurements using standardized methods.
2) Measurements using fully validated or recommended methods.
3) Mathematical calculations.
4) If methods 1, 2, and 3 are not available, a method of measurement working satisfactorily with reference to the fixed limit value can be used until a fully validated method of measurement has been established.

§**6** Temporary Norm for printing inks: description of legal status of printing inks in contact with food.

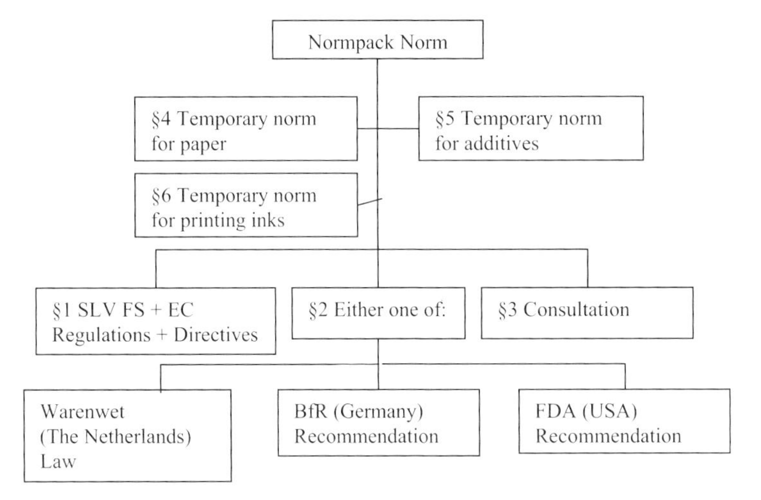

10
Code of Practice for Coatings in Direct Contact with Food

Peter Oldring

10.1
Introduction

In the EU, harmonized legislation for direct food contact materials (FCMs), other than the Framework Regulation (EC) No. 1935/2004 [1], exists only for a few classes of food contact materials, such as plastics, ceramics, regenerated cellulose (RCF), rubbers, and elastomers. Today, this can be considered, by some, as an unsatisfactory situation, not only for industry but also for authorities and control laboratories. Furthermore, even when there is a harmonized European legislation in some cases as is the case for plastics, it is still incomplete (Directive 2002/72/EC [2] and amendments thereof). A clear understanding and workable approach of how the safety of food contact materials can be demonstrated is needed for all stakeholders involved.

There are some substance-specific regulations or directives that apply to non-EU regulated food contact materials containing EU regulated substances. Two such examples are the VCM (vinyl chloride monomer) Directive 78/142/EEC [3] and the Epoxy Regulation (EC) No. 1895/2005 [4]. These regulations apply to all food contact substances wherever these substances could possibly be present. For example, VCM could be present in either plastics or coatings in direct contact with food, and epoxy mainly in coatings but sometimes also in utensils such as tabletop.

The Good Manufacturing Regulation (EC) No. 2023/2006 [5] also applies to all food contact materials, including all types of food packaging. It encompasses the surface in contact with foodstuffs and any transfer from the nonfood contact surface comes within the scope of this regulation. This would include transfer of substances from the external (nonfood contact surface) to the internal (direct food contact surface), whether it is by

- set-off – direct transfer as a result of contact between two surfaces (internal and external), normally on a coated roll of material or a stack of coated substrate.
- vapor-phase transfer – normally only relevant for systems where heat is used to cure or dry one or more components, such as an ink or coating.
- through migration – only relevant where the substrate is not an absolute barrier, such as a metal can.

Global Legislation for Food Packaging Materials. Edited by Rinus Rijk and Rob Veraart

ISBN: 978-3-527-31912-1

- transfer of external to internal from manufacturing equipment, used in some of the processes, which has become "dirty."

Coatings in direct contact with food are not within the scope of harmonized European legislation. Therefore, national legislation applies. This has caused the coating industry and its customers to face a number of issues. In many cases, national legislations differ between countries. In some cases, they contradict one another. This raises the question as to how one can demonstrate European-wide compliance and more importantly compliance with Article 3 of the Framework Regulation (EC) No. 1935/2004.

The can coatings industry, in particular, has suffered from many food scares since 1996 due to species migrating from coatings into foodstuffs. It initially started when BADGE (bisphenol A diglycidyl ether) was detected in cans of sardines and anchovies. Switzerland, not part of the EU, interpreted the limits differently from the EU. This cost industry millions of Euros with toxicological testing and numerous reformulations being needed. Eventually, after 10 years, the use of BADGE at levels significantly (9 mg/kg for the sum of BADGE, BADGE$\cdot H_2O$, and BADGE$\cdot 2H_2O$ and 1 mg/kg for the sum of BADGE$\cdot$HCl, BADGE$\cdot$2HCl, and BADGE$\cdot HCl \cdot H_2O$) above those provisionally authorized (1mg/kg for the total sum excluding BADGE$\cdot 2H_2O$) was permitted in Epoxy Regulation (EC) No. 1895/2005. Switzerland, however, only raised its SML to 1 mg/kg for BADGE and all of its derivatives. Other issues such as NOGE followed and the can coating industry and their customers struggled to find a way out of the confusion that reigned. Cans of foodstuff were destroyed by some EU member states but were allowed to remain on sale in others. At CANCO (2001) [6], the can coating industry (CEPE), their suppliers (various trade associations), and their customers, the canmakers (EMPAC), and their customers, the food industry (CIAA), all stated that they wanted harmonized and workable EU legislation for can coatings.

The coating industry through CEPE had been working for a number of years with the Council of Europe to update Resolution AP (96) 5. This resulted in AP (2004) 1 [7] – the resolution on coatings in direct contact with foodstuffs. However, the industry felt that in the form the resolution was published it was unworkable due to unrealistic deadlines for transfer of substances not on SCF Lists 0–4 to SCF Lists 0–4, particularly bearing in mind the likely workload on EFSA as a consequence of the additives list for plastics becoming a positive list and the recent demands of REACH Regulation (EC) No. 1907/2006 [8] requiring resources for toxicity testing to be focused on this regulation. In the latest revision of AP (2004) 1, these deadlines have been removed. Of more concern were the undefined guidelines on how to comply with the Framework Regulation (EC) No. 1935/2004. AP (2004) 1 only laid down principles rather than provide details for compliance, thus even with the deadlines for the transfer of substances from Lists B and D to A and C, respectively, removed industry strongly believes that more guidance was required. The Code of Practice (C-o-P) attempts to address this shortfall.

Thus, industry took the initiative and started working on a Code of Practice for coatings in direct food contact. All in the supply chain were involved from raw material suppliers to the food industry. The activity was coordinated by CEPE, initially through the metal coating sector, but then expanded to include heavy-duty, flexible coatings, and so on. The process was not without its problems, and understandably the perspectives at either end of the supply chain differed. Notwithstanding the difficulties encountered, agreement was reached, albeit a compromise, and the first version of a Code of Practice was published in September 2006. Updated versions are periodically published on the CEPE web site [9]. The supporting trade associations are listed on this web site and in the Code of Practice. It should be noted that under EU law the Code of Practice is of a voluntary nature and following it or not is the decision of individual companies. In many instances, business-to-business agreements may exceed some of the general principles in the Code of Practice.

The Code of Practice combines elements from the Framework Regulation (EC) No. 1935/2004, the Plastics Directive 2002/72/EC and its amendments, and the Council of Europe Framework Resolution AP (2004) 1 on coatings intended to come in contact with foodstuffs.

It introduces some new concepts such as the use of exposure and structural alerts to estimate risk from migrating species and the "no-migration" principle. Following the Code of Practice should enable all in the supply chain to have a standardized means of demonstrating compliance with the Framework Regulation (EC) No. 1935/2004, and more important, it will enable the food industry to fulfill their legal obligations as far as coated packaging is concerned.

10.2 Contents of the Code of Practice

The following is an overview of the contents of the Code of Practice. It was structured to reflect the approaches used in other food contact article legislation, namely, an introduction, followed by Articles and finally Annexes, some of which explain some of the specific features surrounding coatings compared to plastics, for example. Unlike traditional EU legislative documents, there are no "whereas," but there is a glossary. Furthermore, the "main text" contains explanatory notes in order to communicate the intent to the reader. These are in italics in text boxes.

	Content of Code of Practice
	Introduction
Article 1	Subject matter and scope
Article 2	Good manufacturing practice (GMP)
Article 3	List of substances authorized
Article 4	Incomplete list of monomers and other starting substances
Article 5	Incomplete list of additives
Article 6	Substances having multiple functions
Article 7	Specific restrictions of substances (SML)
Article 8	Overall migration limit (OML)
Article 9	Rules for multilayer coatings

(continued)

	Content of Code of Practice
Article 10	Labeling requirements, declaration of conformity, and supporting documents
Article 11	Demonstration of compliance with OML, SML, and QMA and evaluation of results
Annexes	
Annex I	*Overview of how coated food contact articles are manufactured*
Annex II	*Monomers and other starting substances* *List A: monomers and other starting substance assessed by SCF/EFSA* *List B: monomers and other starting substances not assessed by SCF/EFSA*
Annex III	*Additives* *List C: additives assessed by SCF/EFSA* *List D: additives not assessed by SCF/EFSA*
Annex IV	*Generic description of resins used in food contact applications and their components*
Annex V	*Food and food simulants to be used for testing coatings in contact with foods*
Annex VI	*Risk assessment for migrants from coated articles in contact with foodstuffs*
Annex VII	*Basic rules for demonstrating compliance with the OML*
Annex VIII	*Basic rules for demonstrating compliance with SMLs*
Annex IX	*Glossary*
Annex X	*References to GMPs*
Annex XI	*List of substances considered as dual additives*

10.3 Main Points of the Code of Practice

10.3.1 Scope of the Code of Practice: Article 1

1) This Code of Practice shall only apply to the food contact surfaces of the following:
   a. Coated light metal packaging up to a volume of 10 l.
   b. Coated metal pails and drums with volumes ranging from 10 to 250 l.
   c. Coated articles with volumes 250–10 000 l.
   d. Heavy-duty coated articles with a volume >10 000 l.
   e. Coated flexible aluminum packaging.
   f. Printing inks and coatings in direct food contact.
2) Those sectors to be incorporated at a later date when more details are available
   a. Coatings primarily used to seal food packaging.
   b. Coatings for flexible packaging.
   c. Coatings and inks for paper and board.

3) The Code of Practice does not apply to
   a. Repeated use nonstick coatings, which remain regulated by the specific chapters of BfR [10], RVG [11], and FDA [12] applicable to them.
   b. Extrusion coated materials or articles where the extrusion coating, being a plastic, should comply with the provisions of Directive 2002/72/EC as amended.
   c. Laminated packaging articles or components where the food contact layer, being a plastic, should comply with the provisions of 2002/72/EC, as amended.
   d. Printing inks and coatings applied to the nonfood contact surface of food packaging materials and articles intended to come into contact with foodstuffs.
   e. Adhesives.
   f. Coatings on paper and board that remain regulated by specific chapters of BfR, RVG, and FDA applicable to them.
   g. Coatings on regenerated cellulose that are covered under Commission Directive 93/10/EEC and its amendments.
   h. Can end sealants based upon rubbers and elastomers that remain covered by rules applicable under national legislation.
   i. Tin coatings.
   j. Wax coatings.
   k. Gaskets for metal closures that are covered by the Plastics Directive 2002/72/EC fourth amendment, namely, 2007/19/EC [13].
4) Article 1 gives brief descriptions of different types of direct food contact coatings and inks, while Annex 1 provides an overview of how coated food contact articles are manufactured. It is divided into sections:
   a. Coated light-metal packaging.
   b. Drums and pails.
   c. Coatings for articles with volumes 250–10 000 l; however, these are not applicable.
   d. Heavy-duty coatings.
   e. Coated flexible aluminum packaging.
   f. Coated and printed plastic packaging.
   g. Coated and printed paper and board.
   h. Coated and printed flexible aluminum packaging.

Annex IX is a glossary as much of coating terminology is confusing to those not skilled in the art and often the same word has different meanings in different parts of the world or between two companies.

10.3.2
Good Manufacturing Practice: Article 2

All coatings must be manufactured and applied according to good manufacturing practice. This is in line with the Framework Regulation (EC) No. 1935/2004 and the GMP Regulation (EC) No. 2023/2006. CEPE has published its GMP and many other trade associations along the supply chain have GMPs and web links for these are on the CEPE web site. More details of web links are given in Annex X.

10.3.3
Substance Lists: Articles 3, 4, and 5

Like all other forms of current food contact regulation, the C-o-P works on the basis of lists of approved materials. Substances are divided into four categories:

(A) Monomers and other starting substances fully assessed (SCF Lists 0–4)
(B) Monomers and other starting substances partially assessed
(C) Additives fully assessed (SCF Lists 0–4)
(D) Additives partially assessed

The lists are in fact the lists in Technical Document No. 1 and its amendments in AP (2004) 1. The lists were compiled by industry, with the assistance of the Council of Europe, and are based upon raw materials actually used and with a national food contact approval somewhere. Working along the supply chain, coating manufacturers supplied CEPE with a list of raw materials they used and CEPE forwarded this list to the relevant trade associations. The suppliers were contacted by their association and broke the raw materials into substances. This approach ensured confidentiality was maintained along the supply chain. The lists are updated periodically by CEPE or Cefic FCA and the revisions are submitted to the Council of Europe for inclusion as an amendment to Technical Document No. 1 of AP (2004) 1. The Council of Europe critiques the updates and periodically may revise the lists, for example, by removing generic substances. Since many substances are used only for coatings and were never fully assessed by the SCF due to the priority of approving substances for plastics, many dossiers were never submitted to enable a full assessment by SCF (now EFSA). In other cases, data requested by the SCF have not been submitted. Lists B and D contain substances with at least one national approval, including compliance with USFDA. Thus, there has been some form of approval process for all substances listed even if it has not been to the latest SCF/EFSA protocols. The number of substances is in excess of 1100 with a significant number (about 500) not on SCF Lists 0–4. This was prior to the last submission, and it is believed that recent dossiers for additives for plastics, which can also be used in coatings along with generic substances being amalgamated, will reduce this number to about 400. The Code of Practice proposes a system of prioritizing those substances that are only partially assessed as it is impractical for EFSA to review 500 + dossiers in a short period of time, particularly as these substances are not "new" for food contact having already been authorized somewhere. The tonnages, REACH deadlines, and exposure are used in the proposed prioritization process. In addition to the substances listed, the coating industry reserves the right to use any substance on SCF Lists 0–4 provided relevant restrictions, if any, are met.

Annexes II and III give more details about the lists of monomers and other starting substances and additives. A criticism labeled at the coating industry was that the lists of monomers and starting substances did not reflect what was actually used. Thus in association with the Spanish authorities and delegation to the Council of Europe a generic description of resins, derived from substances on lists A and B (monomers and other starting substances), used in food contact applications and their components was developed and is described in Annex IV – Generic description of coatings typically used in food contact applications and their components. This Annex also gives representative functional groups that could reasonably be expected to be present to enable toxicologists to take a view on their suitability for direct food contact applications. The uses of these resins along with additives in representative types of coatings are also given. Annex IV is subdivided as follows:

Annex IV.I – Description of coatings.
Annex IV.II – Overview of different types of resins typically used in food contact coatings.
Annex IV.III – Generic descriptions of composition of resins typically used in food contact coatings and potential functional groups which could be present.
Annex IV.IV – Examples of typical food contact coatings.

A very significant development in food contact approaches is the use of the "no-migration" principle, which is described in Article 3. The "no-migration" principle facilitates the use of substances that have not been assessed by EFSA and that are not CMRs classes 1 and 2 to be used provided the following conditions are met:

- Migration is nondetectable at a detection limit of 10 μg/6 dm^2.
- The Declaration of Compliance states that the no-migration principle has been used for compliance.
- On request the identity of the substance along with an analytical method will be divulged by the coating manufacturers. This will enable the converters to confirm that they comply with the "no-migration" principle.

The use of CMRs class 3 – suspected carcinogens – is still under debate, but the food industry has by the above mechanisms the right to accept or reject the use of a CMR Class 3 substance, therefore the coating suppliers see no valid reason why CMRs Class 3 should be rejected for inclusion in the "no-migration" principle.

10.3.4
Dual Use Additives – Article 6

The use of substances which are also food additives, known as dual use additives, will be declared in the Declaration of Compliance. On request and under suitable confidentiality agreements their identities will be declared, although the most efficient mechanism still has to be finalized. The foodstuffs industry (CIAA) has provided a list of what they consider to be dual use additives and this is given in Annex XI, as well as the CEPE web site. It has been agreed, between CEPE and CIAA, that the use of any dual use additive in an additive role will always be declared. However some of the dual use additives are in fact used to manufacture resins, for example benzoic acid. In these cases declaration of the presence of dual use additives will only be made if the migration exceeds 10 μg/6 dm^2 the rationale is that when used as an additive there is always the potential for all of the substance to migrate, but when used as a "chain stopper" most of the dual use additive is "tied in" to the resin.

10.3.5
Restrictions of Substances and Testing – Articles 7, 8 and 11

The original concept of the EU Authorities was to authorize substances for food contact applications irrespective of which food contact materials they were used in. This is unlike the FDA system where specific uses are frequently listed. The EU system has the advantage that once approved the substance can be used in any food contact material provided any restrictions are met. The disadvantage is that the exposure approach used by the FDA is negated, resulting in an assumption that everyone eats 1 kg food per day packaged in the same material and the migration of any substances are at their limits. It is argued by many that this is over conservative, but it is an inheritance of today's EU system of controlling food contact materials. Therefore in the Code of Practice, the SMLs given in 2002/72/EC and its amendments are respected, even though these only apply to plastics. This is covered in Article 7. Any restrictions given in the "Synoptic Document" are also respected. The OML of 10 mg/dm^2 (60 ppm) is also respected in Article 8. Simulants specified for plastic materials are referenced, however there are exceptions for heavy-duty and light metal packaging. CEN 14 235 [14] addresses polymeric coatings on metal substrates and the issues associated with corrosion of the substrate by using 3% acetic acid. Article 11 describes how compliance with SML, OML and QMA along with the evaluation of the results can be done. Annexes V, VII and VIII describe the use of simulants and methods for OML and SMLs. The treatment of migration both SML and OML is significantly different for heavy-duty coatings, particularly as they are on multi-use food contact materials and articles.

10.3.6
Multilayer Coatings – Article 9

Rules are given in Article 9 to cover situations when two or more layers of coating are applied to the food contact surface of the substrate. In essence all components of all layers should be compliant with rules governing direct food contact and thus those in this Code of Practice.

10.3.7
Declaration of Conformity – Article 10

The Declaration of Compliance follows that proposed in the draft fourth Amendment to 2002/72/EC, that is, 2007/19/EC, plus additional information about the use or nonuse of the no-migration principle and, on request, a statement whether monomers and other starting substances or additives not on SCF Lists 0–4 are present, as well as the absence or presence of dual use additives, along with their identities if requested, obviously under suitable confidentiality agreements. In addition adequate information for the user of a material or coating should be transferred down the supply chain. This is interpreted as substances with SMLs or other restrictions being declared under suitable confidentiality agreements. The above will enable downstream users to improve their in-house risk assessments required for the supporting documentation. Clarification is still awaited from the Commission on the exact requirements for supporting documentation, but as in 2007/19/EC it will be made available to authorities on request and without unreasonable delay. The Declaration of Compliance will list the regulations for which the coating or material applies, according to their place in the supply chain. Reference to surface area is in 2007/19/EC, but in practice it is envisaged that coating suppliers will quote migration per square decimeter or six squared decimeters. The uses should also be listed, but for coating suppliers migration into simulants will be all that can be offered downstream, as in many cases some of the end uses of the supplier's coatings in food packaging are unknown to the coating suppliers. The converters are better suited to quote actual surface areas and end uses than their suppliers.

The coating supply chain has met numerous times to agree the principles of a Declaration of Compliance. The following is a summary of what has been agreed, although it will not be implemented immediately. In essence it consists of a two part document. The first readily available to all in the supply chain, states that the coating and coated article are suitable for food contact in the EU. The second part is disclosed on a named person to person basis. This document will contain identities and other proprietary information, hence the need for confidentiality agreements and the transfer only to named persons. Unlike most nondisclosure agreements, it is necessary to transfer this information downstream, albeit among other components (see later). A summary of the contents of both documents agreed between EMPAC and CEPE is given below:

Document 1: Widely available to all

- Statement of compositional compliance declaring that all constituents are listed in the Code of Practice (or by reference, AP2004/1).
- Statement that when applied and cured under the recommended conditions of use, the coatings will comply with the requirements of EU Regulation (EC) No. 1935/2004.
- Statement of Compliance with appropriate specific legislation
  - 1895/2005 [15]
  - 78/142/EEC
  - National Member State legislation as appropriate
  - and so on.
- Statement that when applied and cured under the recommended conditions of use, the coatings will comply with all relevant restrictions (QMA, SML, OML) set out in the Code of Practice.
- Statement defining any limitations on compliance (food types, process times and temperatures, markets, etc.).

Document 2; Only given to named recipients under a confidentiality agreement. This is a highly confidential document supplied on request by the coating suppler to the canmaker.

- Identification of all constituents with a restriction on use.
- Notification or identification of any constituent that is not yet fully evaluated (i.e., non 0–4) with an indication of steps in progress to ensure either that it becomes fully evaluated or that it will be substituted with a fully evaluated material.
- Identification of all constituents which are also authorized direct food additives.
- Identification of any constituents not listed in the Code of Practice but used on the basis of the "no-migration" principle.

The envisaged mechanism for confidential information transfer between named persons is as follows:

The coating suppliers contact their raw material suppliers and obtain lists of the substances in the raw materials they purchase. The amounts of each substance and how it is used to make the raw material are not divulged. The coating supplier compiles an aggregated list of all substances present in each of the coatings supplied to their customers. The identity or supplier of the raw materials containing those substances are not divulged, neither are the amounts or process used to make the formulation. The converter transfers to the packer, on request, a list of substances present in all of the food contact coatings on that packaging. This process is shown schematically in Figure 10.1. It should be noted that not all players in the supply chain will want this level of detail and it is arguable whether some of the companies are able to digest, fully understand and act on all the information.

All from named people to named people

Raw material supplier to coating manufacturer

Raw material X contains substances a, b, c
Raw material Y contains substances a, d, e
no amounts, no PPAs, no processes, etc. transferred.

Coating manufacturer to converter

Coating Z contains two raw materials X & Y
List of substances present in coating a, b, c, d, e
no trade names, no amounts transferred

Converter to food industry

Packaging has two coatings Z & W
Coating Z contains a, b, c, d, e
Coating W contains b, c, g, h
List substances present in packaging a, b, c, d, e, g, h
no trade names, no amounts transferred

Figure 10.1 Transfer of information along the supply chain.

10.3.8
Risk Assessment: Annex VI

Annex VI describes treatments that can be used to assess any risk associated with migrants originating from the coating on food contact articles. In addition to known substances, this treatment is applicable to products formed during the manufacture of resin or during the curing process, known as NIAS (nonintentionally added substances), albeit with modifications and guidelines anticipated from the Commission, EFSA, and ILSI (International Life Science Institute). This has been developed because full characterization of all individual peaks is not practical. It uses the concepts of exposure and structural alerts. This enables the estimation of a limit of migration equal to an exposure of < 1.5 μg/person/day using, for example, probabilistic modeling to demonstrate compliance with Article 3 of the Framework Regulation (EC) No. 1935/2004. The level of toxicological concern (TTC) is an approach that the EU regulators, and their independent adviser, EFSA, still have to accept. The level for the TTC is set at 1.5 μg/person/day provided the migrant is not a genotoxin. There is a debate among the scientific community as to whether any or only some genotoxins can have a threshold. In other words, exposure to one molecule of a genotoxin could cause cancer. On the other hand, it is recognized that many people are exposed to genotoxins daily, for example, cigarette smoke, and yet they do not develop cancers.

If the level of 1.5 μg/person/day is exceeded, other considerations using internationally recognized techniques should be applied, such as SARs (structure–activity relationships) and Cramer classes for toxicological threshold. Cramer class 3 permits exposure to a level of 1800 μg/person/day. ILSI have a task force to develop this approach to migrants from food contact materials.

Many of the newer approaches to managing risks from migrants originating from food packaging, such as TTC, rely upon an estimate of the actual exposure to those migrants, both as an individual and as a population. Today, an approach, acceptable to the authorities, for estimating actual exposure does not exist. Understandably, a *per capita* approach, which is arguably the one used by the USFDA, is not acceptable to the Commission or member states. The industry through various projects (CSL stochastic model, Matrix project) has tried to develop acceptable methodology. After a number of years, DGRESEARCH is funding a project under the seventh Framework. The project is known as FACET and more details are given in Chapter 11.

10.4
How can the Code of Practice be Applied?

The first step is to ensure that all components of the coating are approved for food contact applications. Only substances that remain in the "dried film" need to be assessed for food contact compliance. Solvents and volatile neutralizing amines, for example, should evaporate during the film formation stage or curing in some cases. Residual solvent levels should be checked by the converter. In order to assess the suitability of the film-forming components of a coating, start with monomers and other starting substances and additives.

Initially, each component of the food contact material should be compared against the Directive 2002/72/EC, the EFSA Register of opinions and statements in the area of food contact materials [16], Synoptic Document, or CEPE Lists. If listed and/or on SCF 0–4, there should not be an issue with that component unless a restriction is breached. In most cases, this would be a migration limit (SML), extraction limit (QM or QMA), or an end-use restriction (e.g., not for fatty foods).

For the remaining components, it is necessary to refer to national legislation. This significantly varies for each member state. For example, paper and board and nonstick coatings would refer initially to German recommendations (BfR), while can coatings would initially most likely refer to Dutch legislation (Warenwet) as this is the most comprehensive regulation for that particular application. A possible and not unusual situation is that while many may be listed, they all may not be listed. Then, it is necessary to search for other national legislations. Issues can arise when one member state specifically lists a substance, but another does not, and by default it is assumed that it is not allowed, particularly if that member state has a "positive list." Here, the "Treaty of Rome" principle can be applied. Unless a member state can justify prohibition on toxicological grounds, the substance is permitted in all member states. Again, the CEPE lists can facilitate this process.

If there are any remaining monomers, starting substances, or additives that are not listed, recheck whether they are on the CEPE Lists. If they are not, then determine whether they meet the criteria for the no-migration principle. If they do not, the coating does not conform to the Code of Practice.

It is always necessary to determine if any of the substances present are dual-use additives and whether they meet the criteria necessary for declaration (Articles 6 and 11). The CEPE web site can be used for this purpose.

Almost always, catalysts are used to manufacture the resins, prepolymers, and polymers that are present in coating formulations. In some cases, such as thermoset coatings, catalysts are used to facilitate the cross-linking reactions. The catalysts used for either role should be listed in the Synoptic Document or Council of Europe Resolution AP (92) 2 [17] on the control of aids to polymerization (technological coadjuvants) for plastic materials and articles intended to come in contact with foodstuffs.

Finally, it is necessary to test the coating to ensure that it meets any restriction listed for the substances. This testing should be under conditions that mimic industrial application and cure or drying as close as possible. Normally, laboratory conditions are used to "launch" new products, but the results need confirmation from production material.

A declaration of compliance is then needed according to the criteria in Article 11. It should be borne in mind that the above process involves close communication with suppliers and customers, and difficulties surrounding transfer of sensitive proprietary information should not be underestimated. One of the outstanding, and probably major, challenges for the application of this Code of Practice is making this exchange as issue free as possible.

10.5
Conclusions

The whole supply chain for coated materials in contact with food along with the food industry is confident that the Code of Practice is a significant improvement over today's situation. Based on principles derived from the Council of Europe Resolution AP (2004) 1 and the approach used in the Plastics Directive 2002/722/EC and its amendments, it offers a workable and pragmatic approach to demonstrate compliance with Article 3 of the Framework Regulation (EC) No. 1935/2004. Industry will be working with member states for assessing the workability of the Code of Practice and giving them the chance to critique its use in practice. The supply chain believes that the Code of Practice offers regulators and member states a rapid and efficient mechanism to develop a coating regulation, particularly as a consequence of the process whereby member states can critique it before it is debated among them more widely.

List of Abbreviations

Cefic FCA European Chemical Industry Council Food Contact Additives; http://fca.cefic.org/
CEPE European Council of Producers and Importers of Paints, Printing Inks and Artists' Colors; www.cepe.org
CIAA Confederation of the Food and Drink Industries in the EU; http://gda.ciaa.eu/asp/welcome.asp
EMPAC European Metal Packaging
ILSI International Life Science Institute

Acknowledgments

The author wishes to acknowledge the significant contribution of Dr J. Guerrier (CEPE) in the preparation of this Code of Practice, CEPE for facilitating the process, and all trade associations and their representatives involved in agreeing on the Code of Practice.

References

1 Regulation (EC) No. 1935/2004 of the European Parliament and of the Council of 27.10.2004 on materials and articles intended to come in contact with food and repealing Directives 80/590/EEC and 89/109/EEC. *OJ* L338, 2004, 4.
2 Commission Directive 2002/72/EC of 6 August 2002 on plastic materials and articles intended to come in contact with foodstuffs. Corrigendum *OJ* L39, 2003, 2.
3 Council Directive 78/142/EEC of 30 January 1978 on the approximation of laws of member states on materials and articles that contain vinyl chloride monomer and are intended to come in contact with foodstuffs. *OJ* L 44, 1978, 15.

4 Commission Regulation (EC) No. 1895/2005 of 18 November 2005 on the restriction of use of certain epoxy derivatives in materials and articles intended to come in contact with food. *OJ* L 302, 2005, 28.
5 Commission Regulation (EC) 2023/2006 of 22 December 2006 on good manufacturing practice for materials and articles intended to come in contact with food. *OJ* L 384, 2006, 75.
6 CANCO: Ensuring the safety of consumers: can coatings for direct food contact. Project QLAM-2001-00066. Final report, 2002. P. Boogaard, A. Feigenbaum (coordinator), K. Grob, J. Guerrier, P. Oldring, R. Rijk, D. Scholler, C. Simoneau, D. Smith, Representative of DG research: A. Boenke. http://crl-fcm.jrc.it/files/canco_report.pdf (accessed December 2009).
7 Framework Resolution AP (2004) 1 on coatings intended to come in contact with foodstuffs (superseding Resolution AP (96) 5) http://www.coe.int/T/E/Social_Cohesion/soc-sp/Public_Health/Food_contact/.
8 Regulation (EC) No. 1907/2006 of the European Parliament and of the Council of 18 December 2006 concerning the Registration, Evaluation, Authorisation and Restriction of Chemicals (REACH), establishing a European Chemicals Agency, amending Directive 1999/45/EC and repealing Council Regulation (EEC) No. 793/93 and Commission Regulation (EC) No. 1488/94 as well as Council Directive 76/769/EEC and Commission Directives 91/155/EEC, 93/67/EEC, 93/105/EC, and 2000/21/EC. *OJ* L 396, 2006, 1.
9 http://www.cepe.org; then under publications, coatings, food contact)[0].
10 Bundesinstitut für Risikobewertung Recommendation LI. Temperature Resistant Polymer Coating Systems for Frying, Cooking and Baking Utensils. http://bfr.zadi.de/kse/faces/resources/pdf/510-english.pdf.
11 Regeling verpakkingen en gebruiksartikelen (Warenwet) Chapter X Coatings. Koninklijke Vermande/SDU Uitgevers, P.O. Box 20025, 2500 EA Den Haag, The Netherlands.
12 Food and Drug Administration (FDA) Code of Federal Regulations, §21Part 177.1550, Perfluorcarbon resins. http://www.access.gpo.gov/nara/cfr/cfr-table-search.html.
13 Commission Directive 2007/19/EC of 2 April 2007 amending Directive 2002/72/EC on plastic materials and articles intended to come in contact with food and Council Directive 85/572/EEC laying down the list of simulants to be used for testing migration of constituents of plastic materials and articles intended to come in contact with foodstuffs. *OJ* L 97, 2007, 50.
14 CEN/TS 14235:2002 Materials and articles in contact with foodstuffs. Polymeric coatings on metal substrates. Guide to the selection of conditions and test methods for overall migration, ISBN 0 580 41153 2, Published 2003.
15 Commission Regulation (EC) No. 1895/2005 of 18 November 2005 on the restriction of use of certain epoxy derivatives in materials and articles intended to come in contact with food. *OJ* L 302, 2005, 28.
16 http://www.efsa.europa.eu/EFSA/ScientificPanels/AFC/efsa_locale-1178620753812_Opinions425.htm.
17 Resolutions AP (92) 2 Available from http://www.coe.int/t/e/social_cohesion/soc-sp/public_health/food_contact/COE%27s%20policy%20statements%20food%20contact.asp#TopOfPage.

11
Estimating Risks Posed by Migrants from Food Contact Materials

Peter Oldring

11.1
Introduction

One of the issues facing regulators and industry is how one can demonstrate the safety of packaging of foodstuffs. In the EU, this is reflected by the need for compliance with the Framework Regulation (EC) No. 1935/2004, particularly Article 3. In the EU, the current approach is arguably driven by the hazard (toxicity) of any migrating species and not the risk. A fundamental principle of toxicology is that a biological effect increases as the dose increases, in other words, the use of the oft-stated phrase "the dose makes the poison" attributed to Paracelsus.

The rapid advances in analytical techniques and equipment over the past two decades have resulted in detection limits for known migrating substances being substantially reduced and many substances being detected, which were hitherto unexpected, primarily because earlier analytical techniques never detected them. As result, this has resulted in food scares and questions being raised about the safety of food contact materials with respect to substances that migrate into the foodstuff. However, even toxic substances cannot endanger human health if they are not consumed or are only consumed at very low levels, the "*de minimus*" principle. Risk of a contaminant in food, irrespective of its source, is a combination of the toxicity of the substance (hazard) and how much of that substance is consumed (exposure).

$$\text{Risk} = \text{hazard} \times \text{exposure}$$

Thus, to ensure the safety of food packaging, is it necessary to consider not only the toxicity of any migrating species (hazard evaluation) but also how much is present in the foodstuffs (occurrence estimation) consumed and how much of it is consumed (consumption estimation). Today, it can be argued that in the EU a hazard-based approach is the legal basis to risk management. Many including industry would like the EU to have legislation based on an exposure approach. Recent developments have seen DGSANCO state at a member state expert group meeting that they would consider exposure-based legislation when the

Global Legislation for Food Packaging Materials. Edited by Rinus Rijk and Rob Veraart

ISBN: 978-3-527-31912-1

necessary tools are available. This chapter gives some background to different exposure approaches, culminating in a description of a 4-year seventh Framework DGRESEARCH (about €6 Mio) project, FACET, to derive such an exposure tool, which commenced on September 1, 2008.

11.2
Hazard Assessment

The approach for assessing the toxicity when the toxicity or structure of the substance is known is different from that when it is not. When the structure is known and sufficient quantities are available to facilitate toxicity testing, it is possible to apply a "traditional toxicological assessment." This would follow the approaches used in the EU by the European Food Safety Authority (EFSA), formerly SCF, and in the United States by the FDA. They are well documented but differ in one fundamental respect. The EU toxicity testing requirements are dictated by the level of migration, whereas in the United States the exposure dictates the testing required. Typically, animals (rats) are used. Various doses of the chemical are administered, and normally at the lower dose levels no adverse affects are observed. At higher dose levels, adverse effects may be found. In these cases, the dose or doses that cause adverse effects (effect level, EL) can be determined. Typically, the dose for which no observed effects are found is determined and is known as the No Observed Effect Level (NOEL). An alternative is to express the NOEL as NOAEL (No Adverse Effect Level), but the difference is that NOAEL differentiates between an observed effect that is adverse and an effect that is not necessarily adverse. As outlined by Barlow, toxicity data can be used to either

- predict safe levels of exposure for humans or
- predict potentially harmful levels of exposure and the likely nature of the harmful effects.

The acceptable daily intake (ADI) or tolerable daily intake (TDI) can be derived from the NOEL as follows:

$$\text{ADI or TDI} = \text{NOEL}/100.$$

The factor 100 represents a 10-fold safety factor for different levels of sensitivity between humans and another10-fold factor to allow interspecies differences [1]. The latter means that a factor is built in to allow any difference between rats and humans.

From the ADI for a potential migrant, it is possible to derive a level of migration above which human health could be endangered. In order to do this, it is necessary to have a level of exposure to the migrant and the body weight of the consumer. In the EU, the convention is to assume that ever person weighs 60 kg and eats 1 kg of food packaged in the same material and the migration is at the maximum level.

When no toxicity data exist, then other approaches can be used to assess the risk. If the structure is known, it is possible to use structural activity alerts. Munro *et al.* [2] developed a threshold approach according to their chemical structures. They used the structural classes initially developed by Cramer *et al.* [3] to develop a "decision" tree.

11.2.1
Cramer Class I

Substances with simple chemical structures and for which efficient modes of metabolism exist, suggesting a low order of oral toxicity.

11.2.2
Cramer Class II

Substances that possess structures that are less innocuous than Class I substances but do not contain structural features suggestive of toxicity like those in Class III.

11.2.3
Cramer Class III

Substances with chemical structures that permit no strong initial presumption of safety or may even suggest significant toxicity or have reactive functional groups.

It should be noted that this approach was developed for food flavors and its suitability to migrants from food packaging needs to be assessed and confirmed by toxicologists.

From the NOELs of a substantial group of such substances, the fifth percentile of the lowest NOELs for each group was selected, multiplied by 60 (for the 60 kg body weight assumption), and divided by 100 safety factor as described above to give an exposure level below which it was considered that no additional toxicological data were required for any substance meeting the criteria for that Cramer Class. See Barlow for more details. The levels corresponding to each Cramer class are

Cramer class	Human exposure threshold (g/person/day) (TTC)
I	1800
II	540
III	90

The term threshold of toxicological concern (TTC) was used for this threshold. This approach has been further extended by Barlow [4] and Kroes *et al.* [5]. The decision tree is shown in Figure 11.1. It can be seen that an exposure below 0.15 μg/person/day of any substance except the "cohorts of concern" (compounds, such as aflatoxin azoxy compounds, nitroso compounds, TCDD, and steroids) does not cause any concern. If there are no structural alerts for concerns over potential genotoxicity, then this threshold is increased to 1.5 μg/person/day. There are various steps in the process with different thresholds. Overviews of the approach for thresholds are given by Barlow [4], Cheeseman *et al.* [6], ILSI [7], Kroes *et al.* [5, 8], Munro *et al.* [2], and the US FDA [9]. In order to effectively use this approach, it is necessary to be able to assess the potential exposure.

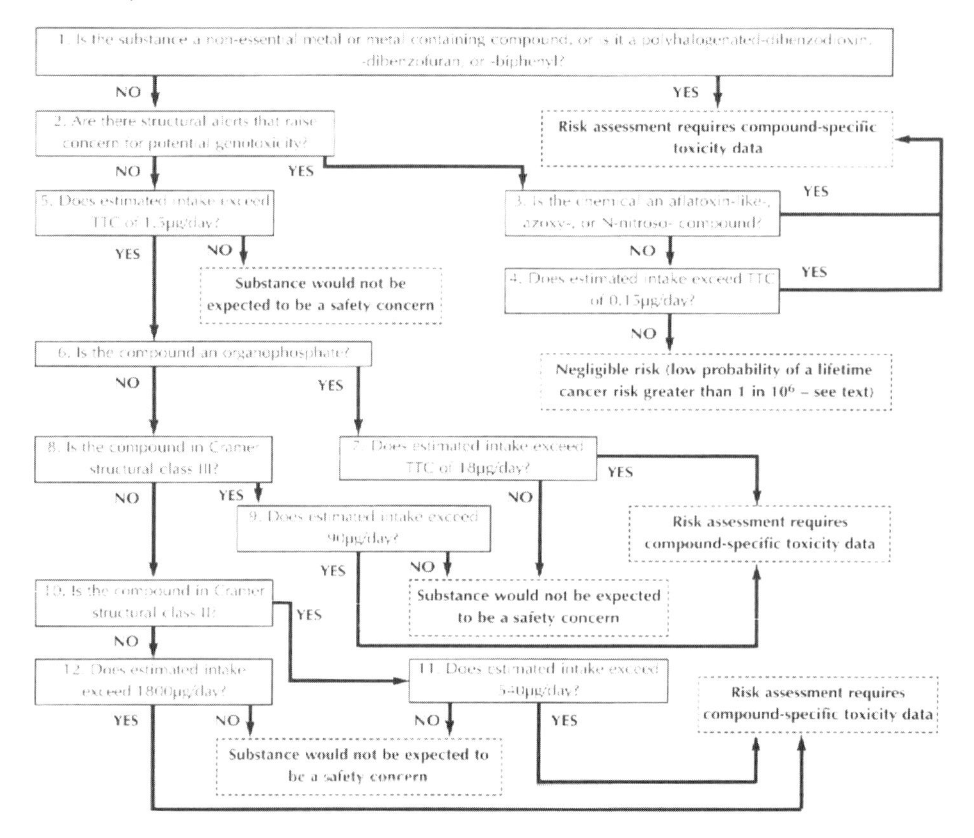

Figure 11.1 Decision tree proposed by ILSI Europe to decide whether substances can be assessed by the TIC approach. (From Ref. [5].)

11.3
Exposure Assessment

11.3.1
Introduction

Today, it is not possible to accurately calculate the exposure to migrants from food packaging due to lack of data. The prime purpose of food consumption surveys is for nutrition. In fact, it is unlikely that we will ever be in a position to know the exposure exactly except perhaps in a very few special cases. Hence, it is necessary to estimate exposure, and various approaches can be used each with it own benefits and drawbacks. Approaches can range from very crude to very refined. It could be argued that exposure to migrants from food packaging materials is more difficult to realistically estimate than that from food additives or other contaminants, such as pesticides, because in many surveys of the consumption of foodstuffs, the food is described but not its packaging.

Determining, estimating, or guessing the packaging of each and every foodstuff consumed adds another dimension to an already difficult problem. Many of the exposure approaches (e.g., Rees and Tennant [10, 11], Parmar *et al.* [12], and Kroes *et al.* [13]) were developed for estimating exposure to food additives, flavors, or contaminants, but not contaminants arising from packaging of the foodstuffs.

Different approaches may be used, ranging from the EU assumption of 6 dm^2/kg, to that of the US FDA, which in essence uses a per capita approach based upon industry statistics, to the latest probabilistic (Monte Carlo) modeling techniques. In this approach, the gaps in the data are addressed by using a random number generation approach. Here, the most likely values for different parameters affecting exposure to any migrant from packaging and others are chosen at random and used to construct an exposure scenario. Probabilistic modeling is ideally suited where there are data gaps, allowing confidence limits to be put on any exposure estimate.

Initially, it is necessary to consider what exposure is and what data are required. While it is necessary to consider all sources for a substance to be found in foodstuffs for a full exposure assessment, this chapter only considers the contamination of food from its packaging.

11.3.2
What is Exposure?

Exposure to a substance found in foodstuffs, irrespective of source, is derived as follows:

For each individual,

- exposure for food item = concentration of the substance in a food item × weight of item consumed
- exposure for meal = sum of exposure of all items consumed during that meal
- exposure over lifetime = sum of exposure for all meals

For the population,

- exposure for population = distribution of exposure for every individual obtained by repeating the above three steps for the entire population

The exposure intake for any population is a sum of the products of the concentration (c) in mg/kg or μg/kg of the migrant within the food item eaten and the weight (w) in kilograms of that item and is expressed as [14],

$$\mathrm{DD}_{jk} = \frac{1}{W_j}\sum_{l=1}^{n(k)} w_{jkl}c_{jkl}.$$

Here, DD_{jk} is the daily dose for any individual j on day k consuming up to $n(k)$ items on that day. W_j is the weight of the individual j and c_{jkl} is the concentration of the migrant in the food item l, while w_{jkl} denotes the weight of item l on day k eaten by individual j.

11.3.3
What Data are Needed to Estimate Exposure?

In order to estimate exposure to a given migrant(s), the following data are required:

1) Weight of foodstuffs consumed
2) Concentration data of migrant(s) in the foodstuffs consumed

For packaged foodstuffs, it is necessary to know or to estimate the packaging of the foodstuffs consumed in order to derive concentration data. Obtaining these data is normally not straightforward and many assumptions are required. While all EU member states conduct food consumption surveys, few contain detailed information on the packaging of the foodstuffs consumed. The EU Member States have nutritional surveys of varying quality. Their main purpose is to enable authorities and other stakeholders to determine the "nutritional health" of their population. These surveys today have to be used as the basis for assessing the consumption of foodstuffs.

Packaging may be defined for some of the food items recorded in some survey data, but this is not the norm. The packaging of nondefined items has to be assigned on the basis of their estimated market shares, any available marketing data, and/or expert judgment. Thus, it is necessary to make assumptions based on available information and expert judgment, but in so doing the impact of the assumptions on the estimate needs to be evaluated.

Concentration data for a given migrant can be obtained from simulant studies or by determining its concentration in the foodstuffs of concern or by mathematical modeling.

11.3.4
Who Should be Protected in an Exposure Assessment?

The protection of the high consumer from contamination by migrants from the packaging of the food they are consuming is important not only for governments and their advisory bodies, such as EFSA, but also for industry. Obtaining a realistic estimate of exposure to a particular migrant(s) for the high consumer is always an issue owing to the lack of data. However, per capita estimates normally include nonconsumers, which reduces the exposure to the consumer population. They cannot be used to express exposure for nonaverage consumers, particularly the high consumer.

Although exposure to a given amount of a given contaminant may not present an issue for one individual, it may be a serious issue for another. Thus, the questions, which must be considered in any exposure assessment, include the sensitivity of the individual, in the sense that an infant has a much lower body weight than the conventional assumption used in the EU that a person weighs 60 kg. Also a child, while having a lower body weight, consumes a relatively higher proportion of foodstuffs/kg of body weight, which results in a higher exposure in terms of mg or μg/kg body weight [15].

In addition, exposure of other vulnerable groups such as low-income socioeconomic groups must be considered.

The percentile to use for the "protection" of any population from any risk is a political and not a scientific decision. The regulators and politicians have to be realistic as 100% of the population can never be protected all the time. The definition of the high consumer varies, but for food contact materials, the 90th, 95th, or 97.5th percentile is typically used for the high consumer [13] rather than the maximum value.

11.3.5 Factors to Consider in an Exposure Assessment

A factor which could result in nonaverage exposure is loyalty, either to a brand or to a type of packaging, that could impact the estimate of exposure. Packaging loyalty is when a consumer will always drink a can (or glass bottle or PET bottle) of beverage irrespective of the brand, as distinct to a brand loyal consumer who will always drink the same brand of beverage irrespective of its packaging. Packaging loyalty or brand loyalty will skew any exposure estimate. Packaging loyal, high consumers are subjected to a higher exposure (if the migrant originates from that packaging) than a nonpackaging loyal consumer if there is more than one form of packaging for that foodstuff.

The exposure for a loyal consumer could significantly vary if there were, for example, a different food additive present in one compared to another. Today, dietary intake studies generally contain inadequate data to accurately determine the exposure of brand or packaging loyal consumers.

One of the simplest approaches for packaging loyalty [14] is to assume that if a consumer initially consumes a foodstuff in a particular type of packaging, the consumer will always consume that item in the same packaging (100% loyalty), whereas the nonpackaging loyal consumer will consume that foodstuff item with its packaging randomly selected in proportion to the market share of the packaging for that type of foodstuff.

11.3.6 Estimating Exposure to Migrants from Food Contact Articles

An overview of obtaining estimates of exposure to migrants from food contact materials is available [27]. Today, there is no universally recognized single approach for estimating exposure, particularly to migrants from food packaging. ILSI [16] conducted a workshop on determining exposure to migrants from food contact materials and recommended that a tiered approach be used.

Exposure assessments can be determined in a number of ways that can be summarized as

- *Simplistic* (or simplified/straightforward) normally using a worst-case assumption, as is the case in the EU.

- *Deterministic* where a fixed value of consumption (of a given foodstuff or family of foodstuffs) normally at the high end is combined with a high or most likely the highest level of migration found.
- *Probabilistic* where statistical modeling is used to predict those values related to the unknown inputs required to obtain a more refined estimate of exposure.

11.3.6.1 Simplistic

Today, for risk assessment in the EU it is assumed that each, and every, person eats 1 kg of food, packaged in the same material with a surface area-to-volume ratio of 6 dm^2/kg, every day of their life. Allocating migration limits for a substance, it is assumed that the substance always migrates at the highest level, corresponding to its TDI, for all packaging and all foodstuffs, thus

$$\text{migration at 1 mg/kg} = \text{exposure at 1 mg/person/day.}$$

$$\text{Thus : exposure} = \text{hazard}$$

This has the advantage that exposure to a substance for any migrant is independent of its packaging, with a few exceptions, unlike the US FDA, where the use of a substance can be restricted to a particular application, including type of package or foodstuff. The major disadvantage is that in most cases, different types of packaging contain different substances and their migration behavior frequently depends upon the foodstuff and any processing. Therefore, the current EU approach arguably overestimates exposure to migrants and applies stricter limits, which do not necessarily improve consumer safety, but could restrict consumer choice. There are some counterarguments that this approach underestimates the exposure for infants [15], but in reality there are a number of projects that may clarify this situation.

While most people recognize the shortcomings of the current EU approach for food contact materials, there is no simple solution, as the data required with the necessary detailed information do not exist.

11.3.6.2 Deterministic

The deterministic approach is better suited to food additives or contaminants and not migrants from FCMs. Here, both the types of food that could contain the additive and the amount consumed are known. The concentration in the foodstuff is either legally limited or can be measured. Using a high consumption of the foodstuffs containing high levels of additives with a relatively high concentration value enables an exposure assessment to be made. If it is below the TDI (TTC), then there is no cause for concern and no further efforts should be devoted to improving the estimate of exposure. The simplistic approach described above could be considered as a simplified deterministic one.

11.3.6.3 Probabilistic (Stochastic) Modeling

Probabilistic (stochastic) modeling can be used for estimating exposure to migrants from the packaging of foodstuffs. Not all exposure assessments need the refined approach of probabilistic modeling. However, as a tool it is gaining greater acceptance for assessing exposure where there are data gaps. Probabilistic modeling has been used by Lambe *et al.* [17] to assess the intakes of flavors. Petersen [18] compared theoretical and practical aspects of probabilistic modeling.

Probabilistic modeling overcomes the lack of data not only by estimating the most likely exposure to a given migrant(s), using input data with uncertainties, but also by deriving confidence limits for any assessment. Treating data with uncertainties is one of the strengths of probabilistic modeling, dealing with data-rich and data-poor inputs. Variability is a factor and should be separated from uncertainty giving rise to one- or two-dimensional probabilistic models, with the one-dimensional model combining uncertainty and variability and the two-dimensional model propagating them separately [14].

Probabilistic modeling uses statistics based upon the Monte-Carlo approach, repeatedly (typically >1000 iterations) calculating the exposure by obtaining estimates for the mean and the uncertainty for any given percentile using different input parameters, some of which are randomly generated. Where there is uncertainty, lower and upper limits can be set, around a most likely value, and the model randomly generates input parameters between the lower and the upper limits for each iteration, with the majority of the values being distributed about the most likely value, using whatever distribution centered around the most likely is considered appropriate. A more detailed description of such a model is given in Holmes *et al.* [14]. It is recognized that a number of groups are working in this area including, for example, CSL, Crème, and Rikilt. There are a few models, but the differences in use are in the data input and the results obtained.

Clearly, not all the data required are readily available and, in most cases, very limited data exist. Thus, it is necessary to make assumptions, but in so doing the impact of the assumptions on the estimate needs to be evaluated. Probabilistic modeling facilitates this requirement.

Inputs can be considered as being either fixed or variable being varied between upper and lower limits. Where some input data have considerably greater uncertainty or variability than others, it is questionable if the amount of effort and treatment required for the treatment of the less uncertain or less variable parameters is justified. The case of migrants originating from packaging is a good example. Today, for most cases, significantly less is known about the packaging of all the foodstuffs consumed than the types and amounts of foodstuff consumed; thus, uncertainty arising from the accuracy of the amount of foodstuff consumed could be considered insignificant compared to what it was packaged in.

It is possible to perform probabilistic calculations with input distributions based on small data sets or expert judgment, but distributions derived from small data sets or expert judgment are likely to be very uncertain. However, if these uncertainties can be adequately represented within the probabilistic assessment, or dealt with by making conservative assumptions for the affected inputs, probabilistic methods should still provide a useful refinement.

11.3.7 The US FDA Approach to Estimating Consumer Exposure to Migrants from Food Contact Materials

The US FDA approach to assessing exposure to migrants from FCMs is explained in CFSAN/Office of Food Additive Safety, April 2002 and is available on their web site (http://www.cfsan.fda.gov/). It describes the use of exposure estimates in Food Contact Notifications (FCNs) that would be normally based upon simulant rather than food migration data, as is the case for new materials. In the US FDA approach, a consumption factor (CF) is combined with a food distribution factor and concentration data to derive an estimate of exposure from all food types and all FCMs containing the substance of interest.

The CF describes the fraction of the daily diet expected to contact specific packaging materials and represents the ratio of the weight of all food contacting a specific packaging material to the weight of all food packaged. To account for the variable nature of food contacting each food contact article, the FDA has calculated food-type distribution factors (f_T) for each packaging material to reflect the fraction of all food contacting each material that is aqueous, acidic, alcoholic, and fatty. Tables for both factors are supplied by the US FDA. This is then combined with concentration data to obtain an exposure estimate, assuming a daily consumption of food and drink of 3 kg per person per day. This gives an estimated daily intake (EDI) for a substance per source of packaging. If there is more than one source, the EDIs are combined to give a cumulative estimated daily intake (CEDI).

The concentration of the substance in the food contacting the food contact article, $\langle M \rangle$, is derived by multiplying the appropriate f_T values by the migration values, M_i, for simulants representing the four food types. This, in effect, scales the migration value from each simulant according to the actual fraction of food of each type that will contact the food contact article.

$$\langle M \rangle = f_{\text{aqueous and acidic}}(M_{10\%\ \text{ethanol}}) + f_{\text{alcohol}}(M_{50\%\ \text{ethanol}}) + f_{\text{fatty}}(M_{\text{fatty}}),$$

where M_{fatty} refers to migration into a food oil or other appropriate fatty food simulant.

The concentration of the substance in the diet is then obtained by multiplying $\langle M \rangle$ by CF. The EDI is then determined by multiplying the dietary concentration by the total weight of food consumed by an individual per day, assuming that an individual consumes 3 kg of food (solid and liquid) per day.

$$\text{EDI} = 3\ \text{kg food/person/day} \times \langle M \rangle \times \text{CF}$$

A concentration in the daily diet of 1 ppm corresponds to an EDI of 3 mg substance/person/day. This approach is designed to deal with single use (e.g., food packaging) rather than repeated use (e.g., nonstick frying pan) FCMs.

11.3.8 Approaches to Determining Exposure to Migrants from Packaging of Foodstuffs that could be Used in the EU

There are a number of different approaches [10, 12, 13] that can be used in order to obtain estimates of consumption of different foodstuffs and any exposure to chemical contaminants in the foodstuffs. In essence, they fall into three categories [11] that are

- per capita
- model diets or worst-case scenarios
- duplicate diets

Per capita estimates of food chemical intake can be made for virtually every European country. There are two basic approaches for undertaking this estimate.

- Multiply the average food consumption of the whole population by anticipated or actual levels of the migrant.
- Divide the total available food chemical by the number of individuals in the population. Obviously, this approach is unsuitable for migrants, but it is ideally suited for food additives.

An advantage of the per capita approach is that it is cost-effective and relatively straightforward. If the estimated levels of exposure are significantly lower than those that could cause concern, then arguably further refinement of the estimate of exposure is unjustified. Wherever possible, per capita data should not be averaged, but a range retained in any subsequent calculations. The "better" the manufacturing data and demographic data the "better" and more reliable the per capita estimate.

The total diet method uses data on food purchases based on household budget surveys. Food groups that contribute the major sources of exposure for a given migrant can be identified. If the migrant is restricted to one type of packaging, it is necessary to select foods packaged in that packaging. This does not facilitate estimating exposure for individual consumers as the household budget surveys are frequently for families that may or may not be consumers of all items. The model diet, based on consumer statistics, is a proposed diet that models the diet of the average or possibly (with adequate data) the nonaverage consumer.

One of the most sophisticated approaches is the duplicate diet method where for every item consumed an identical amount of the item is put aside for analysis at a later date. This approach enables all sources of the substance of interest to be estimated, but it is expensive.

11.3.8.1 Overview of Dietary Surveys

To obtain estimates of exposure, it is necessary to obtain information about the different foodstuffs consumed, with details about their packaging. However, data with this extent of detail hardly exist; thus, the first step is to obtain data about the consumption of different foodstuffs. Today, there are three broad categories of sources of data for food consumption, covering the food supply chain, households, and individuals, namely,

- food balance sheets (FBS), which outline food availability and the market supply situation
- household budget surveys (HBS) that record the food brought (but not necessarily consumed by an individual) into the household
- food consumption or dietary surveys that try to capture the foodstuffs consumed by an individual during a specified period. Some workers [13] further divide this into duplicate diets and individual consumption surveys

Food balance sheets consider raw food commodities and do not reflect the packaging of the foodstuffs consumed. They are useful in indicating trends that can be used in any exposure assessment.

Some countries undertake shopping basket or household budget surveys, which involve recording the contents of a shopping basket for a known family size. However, this has many shortcomings. Furthermore, the packaging is not recorded, but there are a few surveys of packaging of foodstuffs in retail outlets [19], which could be used to give an indication of the most likely packaging or its market share. If no allowance is made for waste food, compared to that consumed, then this will increase estimates. Normally, foodstuffs not consumed at home are not recorded, resulting in a potential underestimate of exposure.

Different approaches for determining the foodstuff consumption of individuals can be broadly subdivided into record keeping or recalling the food items consumed. An advantage of surveys of individuals is that additional data about the individuals, such as age, actual body weight, gender, socioeconomic status, ethnicity, and so on, can be used in combination with their dietary habits enabling a more complete picture of exposure to be obtained, particularly when subgroups of the population could be of special interest. Any survey cannot feasibly survey every person in the population; thus, it must be representative of the whole population under consideration.

In record keeping methods, consumers record all the items and amounts consumed over a period of time, normally over 1–7 days. The weight of nonconsumed food should be deducted in order to allow for wastage. Foodstuffs consumed outside of the home should also be included.

In recall approaches, a trained interviewer asks individuals what they ate in the immediate past, typically the preceding 24 or 48 h. A major disadvantage is that it relies on the memory of the interviewees.

Food frequency questionnaires determine the frequency with which certain foodstuffs are consumed over a given period. Thus, it is necessary to predefine the foodstuffs of interest and these may be targeted to a nutrient(s) or food(s) of specific interest.

A short-term dietary survey being used to assess exposure from migrants from the packaging of the foodstuffs consumed cannot represent the longer term picture for at least two reasons. First, the packaging of a foodstuff in many instances will (almost certainly) change during the consumer's lifetime of the consumer; for example, the growth in PET for beverages has been enormous and will impact any exposure assessment. Second, the eating habits of the consumer will change during the consumer's lifetime.

Under and overreporting are potential sources of errors and there are different approaches to make allowances for these phenomena. However, in the case of estimating exposure to migrants from food packaging, it is considered that the sources of error from uncertainty about the packaging of many of the foodstuffs consumed far outweigh errors due to under or overreporting.

11.3.8.2 Concentration Data

The actual concentrations of any substance migrating from the packaging into each and every foodstuff are uncertain. In order to evaluate the exposure to any substance, it is necessary to determine the amount of each and every foodstuff consumed, which may have been in contact with packaging containing the substance. Surveillance surveys do not measure the concentration of a substance in every foodstuff, but typically target more those foodstuffs in which the substance(s) being surveyed is (are) considered to have their highest levels, and a data could be skewed to a higher concentration level and hence an unrealistically high exposure assessment could result.

An approach to give concentration data with reduced uncertainty limits is surveillance surveys. In these cases, there is no doubt about the concentration of a migrant if a foodstuff for those foodstuffs tested and, in some instances, a range of values likely to be found can be used rather than a single value as would be obtained by using a simulant. However, it is still necessary to allocate migration concentration values to those foodstuffs that were not part of the survey. This can be achieved by either assuming certain foodstuffs are similar in migration characteristics to those that have been tested in the survey or by using simulant migration data. However, surveillance surveys need to be run in conjunction with consumption surveys to maximize the accuracy of the estimates.

In many cases, concentration data in real foodstuffs are unavailable, particularly for new substances. Thus, simulant data for migration are typically used.

One of the main issues with concentration data is how the nondetectable (ND) values are treated. In many instances, the substance(s) of interest is nondetectable in either food simulants or real foodstuffs. If the migrant is not present in particular food packaging materials, then the foodstuffs consumed in that packaging will have a true zero concentration and can be excluded from any estimate of exposure. However, if the migrant species could be present in the packaging of that foodstuff, even if it cannot be detected in the foodstuff, then it is necessary to make allowances for its presence. It is clear that if the value is ND, it cannot be assumed that the value is zero and, on the other hand, it is reasonable that it cannot always be at the limit of detection (LOD). Therefore, it is necessary to try and use more realistic values, for which there are many approaches; one is to use a value between zero and LOD [13], with normal or Gaussian distributions between zero and LOD often being used, giving a mean value for the ND of half the LOD. The emerging use of probabilistic modeling of exposure to migrants enables NDs to be handled in a number of different ways, with values between zero and the LOD being statistically generated according to whatever distribution is required or considered appropriate. Improving the detection limit, for those substances where ND is the only concentration value, in many cases is one of the most efficient ways to demonstrate lower exposure.

Food surveillance surveys may contain a range of values for the concentration data for a given food or group of foods. Some may be ND while others may be above the LOD. For exposure assessments, it is possible to use various approaches to utilize this data, by assuming that the migrant is always present at the highest level recorded or the average level or the mean and so on. Another approach is to fit the actual data to a statistical distribution, for example, normal or lognormal. This enables a more representative value for migrant concentration to be used. An arguably improved treatment would be to use a probabilistic modeling to randomly select concentration values in the statistical distribution range, with possible weighting around the mean, in order to better represent the realistic concentration of the migrant in a food or range of foodstuffs.

Food surveillance surveys give concentration values in either μg/kg (ppb) or mg/kg (ppm). However concentration data derived using simulants normally give results in $\mu g/dm^2$ or mg/dm^2; thus, in order to relate these values to concentrations in foodstuffs, it is necessary to know the actual surface to weight (volume) of the packaged foods. In practice, this is seldom known and in the EU the factor typically used is 6 dm^2/kg. Data from Bouma *et al.* [19], Holmes *et al.* [14], and ILSI [20] indicate that in practice 6 dm^2/kg is too low by a factor of at least 2. However, this is compensated by the overassumption of 1 kg of the foodstuff always being packaged in the same material of 6 dm^2. To compound this dilemma is the apparent growth in single person consumption.

It is generally recognized that values measured in simulants are normally worst case as the simulant normally extracts more of the substance than the foodstuff. Yet another approach is to use a strong solvent, such as acetonitrile, and extract all the substance that could potentially migrate. If the estimate of exposure does not give cause for concern, then there is no need to conduct simulant studies.

An area that still needs resolution is the lag time for multilayer packaging. Species that could migrate and are not in the food contact layer may over a period of time diffuse through the layers and eventually enter the foodstuff for products with long shelf lives or through thermal treatments.

Another approach to obtain migration data particularly for some plastic materials is the use of modeling. Today, this approach is suitable only for certain materials. Diffusion within, and migration from, food contact materials is a predictable process that can be described by mathematical equations. Mass transfer from a plastic material, for instance, into food simulants obeys Fick's laws of diffusion in most cases. Physicomathematical diffusion models have been established, verified, and validated for migration from many plastics into food simulants and are accepted in the United States and in the EU.

Because of the complex, heterogeneous, and variable nature of foods, compared to simple food simulating liquids (simulants), no general tools for modeling migration into foods are yet available. An EU project with the acronym "FOODMIGROSURE" (www.foodmigrosure.com) with the objective of developing a migration model for estimating mass transfer from food packaging plastics into foodstuffs by extending the existing model for food simulants to more complex foodstuffs as contact matrices has recently finished. These models probably represent the only practical way that the complete combination of relevant parameters, including variable food composition, in-pack processing, and storage times and temperatures can be encompassed when compiling concentration data sets large enough to accurately describe the foodstuffs as eaten by European consumers. For more details on the use of modeling to predict migration from food contact materials consult Refs [21–23].

11.3.8.3 Packaging of Foodstuffs Containing the Migrant(s) of Interest

Closely related to concentration data is the type of packaging from which the migrating species can originate. For the purpose of estimating exposure to migrants from food packaging, the focus has to be on the primary packaging. This would be, for example, the bottle for a beverage, whereas secondary packaging would be the plastic wrapper holding 12 bottles in the pack. If, however, knowledge would indicate that substances present in the secondary packaging may penetrate primary packaging and enter the foodstuffs, then the secondary packaging has to be considered.

In order to obtain an estimate of exposure, it is necessary to combine data derived from surveys of the food consumption with data derived from surveys of food packaging. Even then there could be issues as the food packaging may be identified as plastic without identifying the plastic. In some instances, the packaging may be multilayer, and expert knowledge or analysis would be the only certain method of determining the food contact layer.

One of the most straightforward approaches with the lack of packaging data is to use the total production of packaging materials for different foodstuffs, with corrections for imports and exports and divide it by the population. This is in essence the per capita approach and this was undertaken for canned foods and beverages [24]. This has the disadvantage that it will underestimate the exposure due to the nonconsumer.

In one of the most recent food consumption surveys [25, 26], the actual items of food packaging were collected and identified, thereby becoming the first food nutritional survey to determine the packaging of the food consumed, albeit for a limited number of children (about 600). In a project sponsored by the FSA (Project A03051) with Newcastle University, packaging of the foodstuffs consumed by children of different age groups has been identified wherever possible. European industry is also undertaking projects to improve the knowledge of the packaging of the foodstuffs consumed.

Bouma *et al.* [19] undertook a survey of types of packaging for foodstuffs in the Netherlands. Polyolefins (polypropylene, 27% and polyethylene, 34%) accounted for the majority of the packaging, with polyvinyl chloride, polystyrene, polyethylene terephthalate, and paper and board being the next most frequent forms of packaging. However, even knowing that the packaging is derived from a particular polymer may still be inadequate for a more refined exposure assessment. While it is adequate for assessing the exposure to the monomers of the polymer, it will not necessarily help with the additives. The surface area ratios ranged from 6 to 95 dm^2/kg; however, the higher values were for dried herbs. Consult Oldring [27] for more details on obtaining packaging information.

11.4 FACET: Overview of Food Contact Material Work Package and Interactions with Other Work Packages

FACET – Flavors, Additives (Food) Contact Materials Exposure Task – consists of 20 partners addressing an FP7 DG research call. It will provide the Commission, authorities, and industry with a free publicly available PC-based risk management tool to estimate exposure to substances in the foodstuffs originating from food contact materials, as well as food flavorings and food additives. The 4-year project started on September 1, 2008. The partners (with those relevant to food contact materials in bold) are listed in Table 11.1.

Industry has formed a consortium FIG – FACET Industry Group – to participate in the project as a full partner (No. 4, Table 11.1). Today, FIG consists of 12 trade associations and their members. Membership covers raw material suppliers, converters, and the food industry. More details are given in Table 11.2. In addition to industry supplying expertise and resources to the project, it is partially funding some of the activities up to a maximum of €500 000 in cash. The structure of the FIG is available upon request. The fact that industry is investing heavily, both in terms of finances and in terms of resources, in this project demonstrates the importance which they place on a realistic assessment of exposure to migrants from FCMs. With the present EU regulatory system and analytical techniques improving rapidly, resulting in detection limits dropping almost daily, traditional toxicological approaches are struggling to cope with the results. Therefore, for the future, industry strongly believes that a new exposure-based approach is necessary to ensure a viable and innovative food packaging industry.

There are 10 Work Packages, which are listed in Table 11.3. WP 4 is primarily concerned with food contact materials, although there will be interactions between many of the work packages. There are eight partners in Work Package 4, namely, CSL (#3), FIG – CEPE (#4), FABES (#9), Fraunhofer (#10), STFI (#12), USC (#16), NIRDIM (#18), and JRC (#21).

WP 4 is split into three parts.

- WP4.1 (led by CSL): the necessary data for the exposure and migration modelers to use will be collected.
- WP4.2 (led by Fraunhofer): the migration modeling for migrant concentration data will be applied to multilayers and paper and board. Coatings may or may not be incorporated.
- WP 4.3 (led by CSL): will apply QSAR approaches to migrants from modeling.

Table 11.1 List of partners.

Partner number	Name	Short name	Country
1 (Coordinator)	University College Dublin	UCD	Ireland
2	University of Ulster	UU	UK
3	**Central Science Laboratory**	CSL	UK
4	**European Council of the Paint, Printing Ink and Artists Colours Industry (Legal name for FACET Industry Group – FIG)**	CEPE	Belgium
5	Food Chemical Risk Assessment Ltd	FCRA	UK
6	Agence Fran aise de Sécurité, Sanitaire des Produits de Santé	AFSSA	France
7	National Institute for Food and Nutrition Research	INRAN	Italy
8	Technical University of Munich	TUM	Germany
9	**FABES Ltd**	FABES	Germany
10	**Fraunhofer Institut für Verfahren-stechnik und Verpackung**	Fraunhofer	Germany
11	National Public Health Institute	KTL	Finland
12	**STFI-Packforsk**	STFI	Sweden
13	Central Food Research Institute	CFRI	Hungary
14	Faculdade de Ciências da Nutri ão e Alimenta ão da Universidade do Porto	FCNA	Portugal
15	CREMe Software Ltd	CREME	Ireland
16	**University of Santiago de Compostela**	USC	Spain
17	National Food and Nutrition Institute	IZZ	Poland
18	**National Institute for Research and Development of Isotopic and Molecular Technologies**	INCDTIM	Romania
19	Confédération des Industries agro-Alimentaire	CIAA	Belgium
21	**Joint Research Center (Ispra)**	JRC	Italy

WP 4.1 and WP 4.2 are each split into five parts.

- WP 4.1.1: Compiling an inventory list of substances used to make food packaging materials.
- WP 4.1.2: Occurrence and concentration data for substances in packaging materials.
- WP 4.1.3: Listing different foodstuffs with the packaging materials that are used for them.
- WP 4.1.4: Linking packaging materials and their substances with different foodstuffs and concentration data.
- WP 4.1.5: Extending the databases to represent all EU member states.

Table 11.2 Trade associations participating in FACET.

To date 13 associations have signed up FIG	
APEAL	Steel
CEFIC-FCA	Additives for packaging
CEPE/EuPIA	Coatings and inks
CEPI	Paper and board
CIAA	Food industry
EAA	Aluminum
EMPAC	Canmakers
EUPC	Plastic converters
EWF	Wax federation
FEICA	Adhesives
FPE	Multilayer plastic converters
Plastics Europe	Plastic suppliers

- WP 4.2.1: New thermodynamic classification of foods/food groups based on solubility properties (log $P_{O/W}$ versus $K_{P/F}$ studies).
- WP 4.2.2: Study of effective diffusion properties of foods/food groups considering as a basis for a "A_F concept" for foods.
- WP 4.2.3: Parameters for multilayer/multimaterial migration modeling (reference partition coefficients).
- WP 4.2.4: Migration modeling for multilayer/multimaterial packaging in contact with foods – "$(D/K)_n D_F$" model.
- WP 4.2.5: Probabilistic modeling of concentration of FCM constituents in packed foods and link to exposure modeling in WP8.

Under the leadership of CSL and a steering group of FIG, industry will compile an inventory list of materials used in food contact packaging, including nondirect food contact uses such as adhesives between multilayers and inks. This should result in a "super" Synoptic Document. Typical concentrations of substances will be supplied to

Table 11.3 List of work packages.

Work package no.	Work package title	Lead partner	Person-months
1	Project management	1	40
2	Flavorings	7	95
3	Additives	6	76
4	Packaging	3	283
5	Food intake	2	73
6	Chemical occurrence	1	113
7	Regional modeling	5	116
8	Data bases and modeling	15	127
9	Concentration data	19	18
10	Dissemination	1	19

migration modelers. Where data are missing, expert judgment will be used to fill the gaps. Between the different partners in the supply chain, the likelihood of occurrence of packaging and substances for different foodstuffs will be estimated.

While FIG's main focus is WP4.1, FIG also has expertise in QSAR, stochastic modeling, and exposure assessments. Thus, industry will actively participate in the following work packages:

- WP 4.2 migration modeling
- WP 4.3 QSAR
- WP 5 food intake (coding issues)
- WP 6 chemical occurrence
- WP 7 regional modeling
- WP 8 databases and exposure modeling
- WP 10 dissemination particularly to industry

In WP 5, coding of foodstuffs consumed, it is likely that there will be three different codes for each item consumed in order that the relevance to the task in hand can be most efficiently used. The food consumption databases of eight countries will form the basis of a partially harmonized European food consumption database. WP 7 will then fill the data gaps by modeling and sampling.

The results of existing approaches that have been pioneered by industry, such as the Matrix project, CSL exposure model, and STFI project, will be incorporated into FACET, particularly WP 4.1.

While the initial objectives of the FCM part of this project will be to assess the exposure to known substances originating from the FCMs, either directly or indirectly (e.g., set-off), with the introduction of a TTC, the approach could be extended to deal with NIAS (nonintentionally added substances). In this case, estimates of actual exposure are essential and this risk management tool will provide that.

The industry is convinced that an exposure-based approach for assessing and managing any risks from food contact materials is the most effective solution for both regulators and industry, thus industry fully supports the FACET initiative.

11.5 Conclusions

It is necessary to consider how the shortfall of consumption data with packaging information can be addressed. The European Food Safety Authority has initiated a database of food items consumed (EFSA colloquium, 2005), but in the short-to-medium-term packaging of individual food items will not be identified.

As can be seen, there is no single method for determining exposure to migrants from food packaging. The availability of the necessary input data and the accuracy of the estimate dictate the methods that can be used. It is strongly recommended that a tiered approach is used for assessing exposure, starting with the simplest, and only if the result gives a cause for concern should more refined approaches be used. For others to understand the approach used for any estimate, all assumptions and sources of data should be clearly spelled out.

It is believed that the potential offered by probabilistic modeling to overcome the inadequacies in the data will start to be realized in the coming years, particularly with the FACET project. The issue could be explaining the use of statistics.

The need for exposure assessments will increase as concerns about migrants being found increase. It is also of great value for applying the threshold of toxicological concern and demonstrating compliance with the Framework Regulation 2004/1935/EC.

In the longer term, it should be possible to use such processes for a number of applications including

1) assessment of exposure to an unexpected substance being found
2) estimating the most likely exposure to any migrant "x"
3) establishing risk management options for DGSANCO or national authorities
4) demonstration of compliance with the Framework Regulation 2004/1935/EC
5) in support of dossier submissions to EFSA for new substances

References

1 Renwick, A.G., Dorne, J.L., and Walton, K. (2000) An analysis of the need for an additional uncertainty factor for infants and children. *Regulatory Toxicology and Pharmacology*, **31**, 286–296.

2 Munro, I.C., Ford, R.A., Kennepohl, E., and Sprenger, J.G. (1996) Correlation of a structural class with no-observed effect levels: a proposal for establishing a threshold of concern. *Food and Chemical Toxicology*, **34**, 829–867.

3 Cramer, C.M., Ford, R.A., and Hall, R.L. (1967) Estimation of a toxic hazard: a decision tree approach. *Food and Cosmetics Toxicology*, **5**, 293–308.

4 Barlow, S. (2005) Threshold of toxicological concern (TTC): a tool for assessing substances of unknown toxicity present in the diet, ILSI Monograph.

5 Kroes, R., Renwick, A.G., Cheeseman, M., Kleiner, J., Mangelsdorf, I., Piersma, A., Schilter, B., Schlatter, J., van Schothorst, F., Vos, J.G., and Wurtzen, G. (2004) Structure based thresholds of toxicological concern (TTC): guidance for application to substances present at low levels in the diet. *Food and Chemical Toxicology*, **42**, 65–83.

6 Cheeseman, M.A., Machuga, E.J., and Bailey, E.B. (1999) A tiered approach to threshold of regulation. *Food and Chemical Toxicology*, **37**, 387–412.

7 ILSI Monograph (2000) Threshold of Toxicological concern for chemical substances present in the diet, August 2000.

8 Kroes, R., Galli, C., Munro, I., Schilter, B., Tran, L.-A., Walker, R., and Wurtzen, G. (2000) Threshold of toxicological concern for chemical substances present in the diet: a practical tool for assessing the need for toxicity testing. *Food and Chemical Toxicology*, **38**, 255–312.

9 US Food and Drug Administration (1995) Food additives: threshold of regulation for substances used in food contact articles; final rule. *Federal Register*, **60**, 36582–36596.

10 Rees, N. and Tennant, D. (1993) Estimating consumer intakes of food chemical contaminants, in *Safety of Chemicals in Food* (ed. D. Watson), Ellis Horwood Series, pp. 157–181.

11 Rees, N. and Tennant, D. (1994) Estimation of food chemical intake, in *Nutritional Toxicology*, Raven Press, New York, pp. 199–221.

12 Parmar, B., Miller, P.F., and Burt, R. (1997) Stepwise approaches for estimating the intakes of chemicals in foods. *Regulatory Toxicology and Pharmacology*, **26**, 44–51.

13 Kroes, R., Muller, D., Lambe, J., Lowik, M.R.H., van Klaveren, J., Kleiner, J., Massey, R., Mayer, S., Urieta, I., Verger, P., and Visconti, A. (2002) Assessment of intake from diet. *Food and Chemical Toxicology*, **40**, 327–385.

14 Holmes, M.J., Hart, A., Northing, P., Oldring, P.K.T., Castle, L., Stott, D., Smith, G., and Wardman, O. (2005) Dietary exposure to chemical migrants from food contact materials: a probabilistic approach. *Food Additives and Contaminants*, **22** (10), 907–920.

15 Castle, L. (2004) Exposure to constituents of food contact materials. TNO Symposium (Zeist).

16 ILSI Monograph (2002) Exposure from food contact materials, November 2002.

17 Lambe, J., Cadbey, P., and Gibney, M. (2002) Comparison of stochastic modelling of the intakes of intentionally added flavouring substances with theoretical added maximum daily intakes (TAMDI) and maximised survey-derived daily intakes (MSDI). *Food Additives and Contaminants*, **19**, 2–14.

18 Petersen, B.J. (2000) Probabilistic modelling: theory and practice. *Food Additives and Contaminants*, **17**, 591–599.

19 Bouma, K., Stavenga, K., and Draaijer, A. (August 2003) Domestic use of food packaging materials in the Netherlands, Report NDFCM 010/01 available at http://www.vwa.nl/download/rapporten/Voedselveiligheid/20041019_food_%20packaging_materials.pdf.

20 ILSI Monograph (1996) Food consumption and packaging use factors, July 1996.

21 Brandsch, J. and Mercea, P., Ruter, M., Tosa, V., Piringer, O. (2002) Migration modelling as a tool for quality assurance of food packaging. *Food Additives and Contaminants*, **19**, 29–41.

22 Reynier, A., Dole, P., and Feigenbaum, A. (2002) Integrated approach of migration prediction using numerical modelling associated to determination of key parameters. *Food Additives and Contaminants*, **19**, 42–55.

23 O'Brien, A. and Cooper, I. (2002) Practical experience in the use of additive models to predict migration of additives from food contact polymers. *Food Additives and Contaminants*, **19**, 56–62.

24 Dionisi, G. and Oldring, P.K.T. (2002) Estimates of per capita exposure to substances migrating from canned foods and beverages. *Food Additives and Contaminants*, **19** (9), 891–903.

25 Duffey, E., Hearty, A.P., Gilsenam, M.B., and Gibney, M.J. (2006b) Estimation of exposure to food packaging materials 2: Patterns of intake of packaged foods in Irish children aged 5–12 years. *Food Additives and Contaminants*, **23**, 715–725.

26 Duffey, E. *et al.* (2006a) Estimation of exposure to food packaging materials 1: Development of a food packaging database. *Food Additives and Contaminants*, **23**, 623–633.

27 Oldring, P.K.T. (2007) Exposure: the missing element for assessing the safety of migrants from food packaging materials, in *Chemical Migration and Food Contact Materials* (eds K.A. Barnes, R. Sinclair, and D. Watson), Woodhead Publishing, pp. 122–157.

12 Compliance Testing, Declaration of Compliance, and Supporting Documentation in the EU

Rob Veraart

12.1 Introduction

Both by the EU and the Member States, legislation has been developed for food contact materials (FCMs) to ensure that no components can migrate to the food during contact with FCMs in an amount endangering human health. In this chapter, information on tests is provided that can be used to demonstrate that a certain food contact material is suitable for a specified application. These tests can be initiated by the producer of a food contact material, by a user of a food contact material, or by an enforcement authority. In some cases, the person who wants to demonstrate compliance may hire specialized companies or laboratories to perform the tests specified.

The tests needed to be considered can be one or a combination of the following tests: the overall migration; describing how much is migrating in total, the specific migration; describing the amount of a specific (group of) component(s) migrating into the food during contact, the residual amount; describing how much of a component is present in the finished product (FP); determination of not-intentionally added substances (NIAS), which may migrate to the food; organoleptic testing; and some other tests.

Demonstration of compliance is obligatory for FCMs regulated by EU directives on plastics [1], regenerated cellulose [2], and ceramics [3]. For that purpose, a supplier should provide a declaration of compliance for these FCMs while he should have available documentation that proves the validity of the declaration. Substances excluded from EU measures may be subject to national legislation, but in anyway the FCM should comply with Article 3 of the Framework Regulation [4].

For compliance testing, determination of migration in actual foods is very impracticable. Therefore, mainly migration into simulants is determined. However, enforcement authorities may decide to take a packaged food sample from the market and analyze the food for migration of food contact substances. Testing with simulants under standardized conditions may show a deviation with migration in real food for many different reasons.

Global Legislation for Food Packaging Materials. Edited by Rinus Rijk and Rob Veraart

ISBN: 978-3-527-31912-1

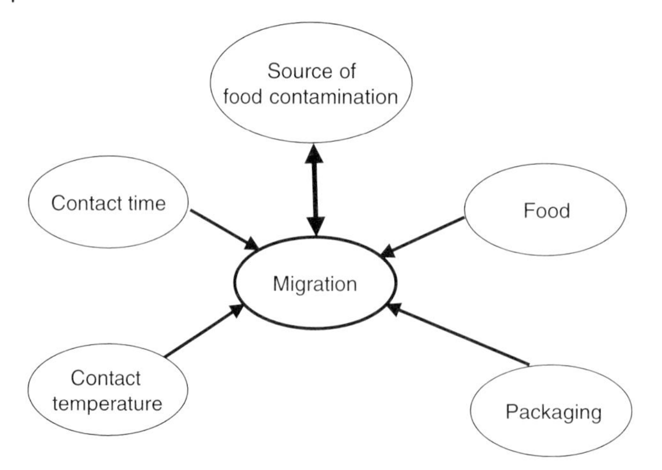

Figure 12.1 Migration related parameters.

Migration may be an important source of food contamination. For checking compliance, it should be realized that migration depends on the four pillars given in Figure 12.1. Each of the criteria has to be considered when designing migration experiments.

12.2
Composition of the FCM

EU directives and most national legislations are based on the positive list principle. This means that substances not listed are not authorized to be used in the FCM. The EU list for plastic additives is to be completed soon. But meanwhile national legislation may be supplementary. Also, substances excluded from EU directives may be subject to national legislation. For materials not yet regulated at the EU level, the relevant national legislation should always be considered. In the following sections, the general principle of the positive list will be elaborated on, and the experiments that may be needed.

12.2.1
How to Demonstrate Compliance of Composition

The most simple and convenient way to demonstrate compliance with the positive list has been achieved is when the composition is known. In many cases, the producer of the finished article may have limited factual information, but he may rely on the statements he should have been supplied. In such cases, experimental work may not be required. If for any reason no information is available, then the FCM should be analyzed using sophisticated analytical tools. The first part of the analysis should identify which substances have been used in the FCM using techniques such as gas chromatography with mass detection (GC-MS) and high-performance liquid chromatography with MS detection (HPLC-MS/MS), and other analytical techniques such as infrared nuclear magnetic resonance (NMR). Second, verify whether the substances identified are present on the relevant positive lists and what restrictions are established.

If a restriction on the residual content has been established, then this should be determined on the final product. For example, vinyl chloride monomer has a restriction of 1 mg/kg of FCM. If it concerns a multilayer material, then also the layers that are part of the FCM should be included in the determination. The final FCM should comply with this restriction. Many other examples could be indicated. A second type of restriction on the residual content is the maximum quantity expressed per area of FCM that is abbreviated as QMA. For plastic materials, the QMA is mainly established for monomers that are reactive with food or food simulants. Values are expressed in mg/6 dm^2 while 100% migration is assumed.

For regenerated cellulose most of the restrictions are based on a residual content or use values. If sufficient information from the manufacturing process is available, then compliance with the directive may be easy. Otherwise, the content of various substances should be determined and checked for compliance.

12.3 Selection of Contact Conditions

The migration, of course, depends not only on the sample composition but also on the test conditions during food contact. To allow FCM producers to demonstrate compliance with the relevant requirements, conventional conditions of contact time and temperature as well as food simulants have been defined. To demonstrate compliance, it is necessary to know the conditions of contact with the type of food. However, compliance can also be restricted for the FCM to certain types of food and under certain conditions of contact. In both cases, it needs a careful consideration to establish the relevant parameters to guarantee safety of the consumer.

12.3.1 Food Simulants

The migration depends on the food type. Components that are hydrophobic, for example, will migrate better to food with a high amount of fat on the surface. To avoid that testing has to be performed with all (potential) food types; the EU has defined some simulants that can be used. These food simulants are as follows:

- Simulant A: water, mimicking aqueous foodstuffs.
- Simulant B: 3% acetic acid, mimicking acidic foodstuffs with pH <4.5.
- Simulant C: 10% ethanol, mimicking alcoholic foodstuffs.;
- Simulant D: olive oil, or another approved oil, mimicking fatty foodstuffs.

- 50% ethanol: mimicking high alcohol containing products and milk and some dairy products.

For general purpose, the FCM migration into 3% acetic acid, 10% ethanol, and olive oil is considered to represent all food types. However, for defined foodstuffs or groups of foodstuffs, it is not necessary to use all simulants. The EU has prepared a table that can be used to select relevant simulants for the selected food. This table was published in Directive 85/572/EEC [5] and was amended in Directive 2007/19/EC [6]. In Table 12.1, an excerpt of Directive 85/572/EEC is given. It shows that for fried potatoes only simulant D is needed for testing. In addition, a simulant D correction factor (DRF), which is explained below, of 5 is assigned to the fried potatoes.

Both simulant B and C are mimicking the migration to the aqueous foodstuff, and therefore, if B or C is selected as a simulant, no testing is needed using simulant A.

In case the food type is not included in the table, an evaluation has to be made by the user himself.

To judge if a food can be regarded as a fatty foodstuff, CEN 14481 [7] can be used. This test method can be used to determine whether there is fatty contact and is applicable to all foods. Testing some foods can require modifications to the method. The method is applicable to contact situations ranging from −20 to 100 °C.

It is recommended to include simulant A if water is present in the food. If the pH of the food is 4.5 or lower, it is needed to include simulant B instead of simulant A to take the low pH effects into consideration.

If the ethanol amount is higher than 5%, simulant C should be used instead of stimulant A. If the ethanol amount is higher than 10%, the ethanol amount of

Table 12.1 Excerpt from Directive 85/572/EEC that needs to be used for the selection of the simulants.

Reference number	Description of foodstuffs	Simulants to be used			
		A	B	C	D
07.05	Rennet:				
	A. In liquid or viscous form	X(a)	X(a)		
	B. Powdered or dried				
08	Miscellaneous products				
08.01	Vinegar		X		
08.02	Fried or roasted foods:				
	A. Fried potatoes, fritters, and the like				X/5
	B. Of animal origin				X/4
8.03	Preparations for soups, brochs, in liquid, solid, or powder form (extracts, concentrates); homogenized composite food preparation, prepared dishes:				
	A. Powdered or dried:				
	I. With fancy substances on the surface				X/5
	II. Other				

simulant C should be adjusted. If a packaging is intended for wine with an alcohol content of 14%, the ethanol concentration of simulant C should be increased to 15%.

Simulants for dry food products have not been established and therefore migration into dry products such as flour, rice, and so on should be determined in the food itself or one of the above-mentioned simulants may be used as long as they are a worst-case simulant compared to the dry food.

If an alcohol-containing food has an ethanol amount higher than 10%, the ethanol amount in the simulant has to be adjusted accordingly (to the same level or higher than in the food).

In some cases, other simulants need to be chosen. This is the cases for some materials, for example, for ceramics, 4% acidic needs to be chosen as simulant, or under some member state legislation other simulants are sometimes assigned for nonharmonized legislation.

12.3.1.1 Substitute Fatty Food Simulants

Sometimes, it is not possible to use olive oil as a food simulant because the oil is absorbed too much by the sample or because the migrating substance reacts with the oil components. In such cases, the EU has assigned substitute simulants for simulant D. The simulants assigned are 95% ethanol and isooctane and in case of high temperatures (>100 °C) also modified polyphenylene oxide, better known as TENAX. The highest values of the tests performed should be used (e.g., the migration to 95% ethanol is 6 mg/dm^2 and to isooctane is 1 mg/dm^2, the value of 6 mg/dm^2 has to be used as the migration value for simulant D). In some cases, when the worst-case simulant can be assigned, only the testing using the worst-case simulant is sufficient (e.g., when a polyolefin has to be tested, the use of isooctane is usually sufficient). If, however, it can be demonstrated that a substitute simulant is not suitable of any reason, this simulant may be rejected.

12.3.1.2 Alternative Fatty Food Simulants

Because tests with olive oil are usually laborious and time consuming, it is allowed to use other solvent than those mentioned in the directive provided the results are similar or better than those with the official simulants. For example, migration to acetonitrile gives somewhat better results for the migration of some substances from some materials than migration into olive oil. It should be emphasized that comparative data must be available and provided when this choice is challenged. In many cases the simulants that are prescribed as substitute fatty foods simulants are used as alternative fatty food simulants.

12.3.1.3 Reduction Factors

DRF

For all types of food with a fatty character, olive oil or another type of oil needs to be selected as food simulant. However, olive oil has a much more fatty character than most foodstuffs. Therefore, the migration to olive oil may be much higher than in other foodstuffs. Because of this effect, the EU has inserted the simulant D reduction factor (DRF). This factor is established for all foodstuffs in the table in Directive 85/572/EEC. The higher the DRF the more worst-case olive is. For example, for fried potatoes a DRF of 5 is assigned. This means that the value obtained for the overall or specific migration to simulant D may be divided by 5 before it is compared with the overall or specific migration limit (SML).

The DRF may be applied to both specific and overall migration determined in olive oil or substitute simulants.

Example: the overall migration to olive oil of a plastic product is 40 mg/dm^2. This means that it is suitable for foodstuff with a DRF of 4 and 5. However, application of the DRF is conditional to the level of migration. If more than 80% of a substance is migrating to olive oil, then the DRF should not be applied.

FRF

One of the assumptions of the EU is that a person eats 1 kg of food, each day life long, and has a body weight of 60 kg. For many food types, it is possible to eat 1 kg each day, but for foodstuff with a very high amount of fat (more than 20%) it is not possible to maintain this consumption for a long time. Therefore, the EU has introduced the fat reduction factor (FRF) and assumed that 200 g of fat can be eaten each day. The FRF is calculated using the following equation:

$$\mathrm{FRF} = \frac{\mathrm{Fat}_{\%\ \mathrm{foodstuff}}}{20\%}.$$

The FRF may be applied only when the component meets the following restrictions:

- The component may not migrate to the nonfatty food because this would imply that the fat reduction is not justified. The EU authorities have developed a list of components for which an FRF may be used.
- The fat content of the food must be higher than 20%.
- The food must not be intended for babies and young children.
- The specific migration of a component is less than 80% of the amount of the same component available.
- The application of the FRF may not result in exceeding the overall migration limit.
- The relation between the area and the amount of food must be able to be estimated. If this is not possible, the FRF may not be used.

TRF

The total reduction factor (TRF) is calculated when both the FRF and DRF is applied. The TRF has to be calculated using the following formula:

$$\mathrm{TRF} = \mathrm{FRF} \times \mathrm{DRF}$$

If the calculated value of the TRF is larger than 5, the TRF is set to 5. If, for example, the DRF has a value of 2 and the FRF has a value of 3, the calculated value of the TRF is 6; however, it is reduced to 5 before it is applied in such a case.

12.3.1.4 Future Developments

No simulant has been established for dry foods without free fat on the surface. This does not mean that no migration to such a food will occur. Therefore, testing with the foodstuff itself may be needed to demonstrate compliance. Discussions are ongoing to assign the use of modified polyphenylene oxide (MPPO or Tenax) to simulate dry foods. This may be included in future legislation.

12.3.2 Selection of Time and Temperature Conditions

Directive 82/711/EEC [8] as last amended by Directive 97/48/EC [9] included provisions of time and temperature conditions that may be used in migration testing using food simulants. To mimic the actual contact of the foodstuff with FCM, conditions of time and temperature can be selected from the tables included in the Directive. One table transfers actual contact time into time periods that should be applied in simulating test. The second table transfers the contact temperature into temperatures to be used in migration testing. If a food is in contact with FCM for a combination of time and temperature conditions, the migration experiment should include the same contact conditions using the same test specimen. The tables below are copied from the Directive.

Selection of the test time from the actual contact time.

Contact time	Test time
$t \leq 5$ min	See below[a)]
5 min $< t \leq 0.5$ h	0.5 h
0.5 h $< t \leq 1$ h	1 h
1 h $< t \leq 2$ h	2 h
2 h $< t \leq 4$ h	4 h
4 h $< t \leq 24$ h	24 h
$t > 24$ h	10 d

a) In those instances where the conventional conditions for migration testing are not adequately covered by the test contact conditions mentioned in the table (for instance, contact temperatures greater than 175 °C or contact time less than 5 min), other contact conditions may be used that are more appropriate to the case under examination provided that the selected conditions may represent the worst foreseeable conditions of contact for the plastic materials or articles being studied.

Selection of the test temperature from the actual contact temperature.

Contact temperature	Test temperature
$T \leq 5$ °C	5 °C
5 °C $< T \leq 20$ °C	20 °C
20 °C $< T \leq 40$ °C	40 °C
40 °C $< T \leq 70$ °C	70 °C

(continued)

(Continued)

Contact temperature	Test temperature
70 °C < $T \leq$ 100 °C	100 °C or reflux temperature
100 °C < $T \leq$ 121 °C	121 °C[a]
121 °C < $T \leq$ 130 °C	130 °C[a]
130 °C < $T \leq$ 150 °C	150 °C[a]
T > 150 °C	175 °C[a]

a) This temperature shall be used only for simulant D. For simulants A, B, or C, the test may be replaced with a test at 100 °C or at reflux temperature for a duration of four times the time selected according to the general rules.

In the tables shown above, the test temperature and test time can be selected on the basis of the contact temperature and time.

- As an example, a food sterilized in the packaging for 40 min at 125 °C and then stored for several months at room temperature requires testing for 1 h at 130 °C followed by 10 days at 40 °C. For simulants A, B, or C, the test may be replaced with a test at 100 °C or at reflux temperature for a duration of four times the time selected. In the given example, this would result in testing for 4 h at 100 °C or reflux followed by 10 days at 40 °C.
- If a food contact material will be exposed to different food contact temperatures, this need to be taken into consideration also. If an ovenable tray is made to store food for 5 days up to 7 °C and the consumer can heat the food in the oven at 40 min at 160 °C, the proper contact conditions are 10 days at 20 °C followed by 60 min at 175 °C.

More information on the selection of time and temperature can be found in the Directive.

In case a microwave is used, it is not simple to specify the temperature from the table. The temperature reached depends on the food type (food types with a lot of free fat on the surface can become much warmer), power of the microwave applied, the time the power is applied, and the amount of food. The CEN has issued a method that can be helpful: CEN 14233 [10]. This European Standard specifies methods to measure the temperature reached by plastics materials and articles in contact with foodstuffs during microwave heating and conventional oven heating in order to select the appropriate temperature for migration testing. It is applicable to all plastic materials and articles.

There are some exceptions, and therefore, it is always important to see if the relevant EU or member state legislations dictate to select another time and/or temperature. For example, ceramics always needs to be tested using 22 °C for 24 h as contact conditions. For substitute simulants (see Section 12.3.1), other time and temperature conditions do apply.

Conversion table for substitute simulants for simulant D.

Test conditions			
With simulant D	With isooctane	With 95% ethanol	With MPPO
10 d 5 °C	0.5 d 5 °C	10 d 5 °C	—
10 d 20 °C	1 d 20 °C	10 d 20 °C	—
10 d 40 °C	2 d 20 °C	10 d 40 °C	—
2 h 70 °C	0.5 h 40 °C	2 h 60 °C	—
0.5 h 100 °C	0.5 h 60 °C	2.5 h 60 °C	0.5 h 100 °C
1 h 100 °C	1 h 60 °C	3 h 60 °C	1 h 100 °C
2 h 100 °C	1.5 h 60 °C	3.5 h 60 °C	2 h 100 °C
0.5 h 121 °C	1.5 h 60 °C	3.5 h 60 °C	0.5 h 121 °C
1 h 121 °C	2 h 60 °C	4 h 60 °C	1 h 121 °C
2 h 121 °C	2.5 h 60 °C	4.5 h 60 °C	2 h 121 °C
0.5 h 130 °C	2 h 60 °C	4 h 60 °C	0.5 h 130 °C
1 h 130 °C	2.5 h 60 °C	4.5 h 60 °C	1 h 130 °C
2 h 150 °C	3 h 60 °C	5 h 60 °C	2 h 150 °C
2 h 175 °C	4 h 60 °C	6 h 60 °C	2 h 175 °C

Some member state legislation may have some other contact time and temperatures for nonharmonized legislation.

At present, the longest contact period is 10 days. Amendment to the directives are under discussion and may include the requirement for longer contact periods when relevant. However, at this stage it is uncertain how long-term storage of foods will be simulated.

12.4 Migration Experiments

Migration from FCM covers a broad field of analytical chemistry and sometimes requires sophisticated equipment. However, the more basic conditions of contact and handling the test sample also need proper attention. CEN has drafted a couple of documents that provide in first instance good guidance about the initiation of migration experiments. It is useful to carefully study these documents. Guidelines for determination of overall migration [11] and specific migration of a restricted number of substances [12] are available.

12.4.1 Contact Methods

Now that the contact conditions are set as described above, the way how the food simulant is brought into contact with food must be selected. The choice depends not only on the composition of the material, the shape of the material to be tested, and the thickness of the sample but also on experience and instrumentation available.

Total Immersion

This is the simplest way of bringing the material in contact with food. During the contact time, the sample is immersed in the simulant. It is important that during the contact time different parts of the sample do not stick to each other. This method has some disadvantages or situations where immersion might not be the best choice:

- In case the food contact side and the nonfood contact side of a test sample are different, it is not possible to distinguish where the migrated component do originate from (with, for example, a multilayer). In this case, it must be assumed that all the components that migrate originate from the food contact site.
- In many cases, a test sample must be reduced in size to fit in the analytical equipment. Some polymers may "bleed" in the cutting edges.
- The migration will take place from a layer of maximum 0.25 mm thick. The migration will take place from two sides. If a sample has a thickness of 0.5 mm or larger, the area of both sides can be taken into account (in case both sides are identical). If the thickness is smaller than 0.5 mm, just one side of the test sample must be taken into consideration.

Single-sided contact by

Migration cell: When some materials are tested, only the food contact side should be in contact with the simulant, to ensure a representative migration behavior. An example of such cases is a can end or a multilayer film. The single-side contact can be obtained using a migration cell. Many types of migration cells are available. In Figures 12.1 and 12.2, typical cells are presented for testing multilayer films and can ends.

In cases of a multilayer material or sheet material, a migration cell must be able to bring a defined area in contact with a defined amount of simulant. The cells described are capable of achieving this. Depending on the simulant and the conditions of heating, they are suitable for testing in high temperature.

Can ends are placed on a glass cup with a flat and smoothed rim. Between the glass and the can end, a flexible ring is placed to assure a liquid-tight closure. During the contact time, the cell will be placed in up side down, to ensure that the test material is in contact with the simulant (Figure 12.3).

Pouch: Many multilayer materials do have a sealable layer on the food contact side, which enables the production of a bag from this material. A similar approach can be used in testing. A small bag is made by heat sealing and filled with the simulant.

Filling of an article: Some items, such as a bottle or a can body, can be filled with the simulant. The sample must be covered to reduce evaporation or oxidation of the oil at high temperature. In addition, care must be taken that no items falls into the simulant.

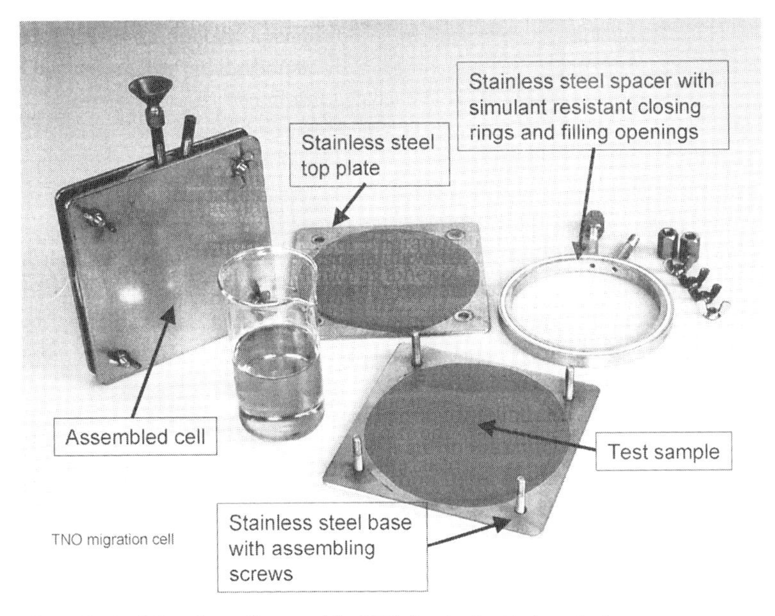

Figure 12.2 Migration cell as used by TNO for single-sided contact.

Reversed pouch: In case a multilayer material does not have a sealable layer on the food contact side, but has a sealable layer on the nonfood contact side, a reversed pouch can be made by heat-sealing two sheets at the nonfood contact side. The food contact side will be positioned on the outside of the bag. After the bag has been constructed, the bag is immersed in the simulant and the food contact side comes in contact with the food simulant.

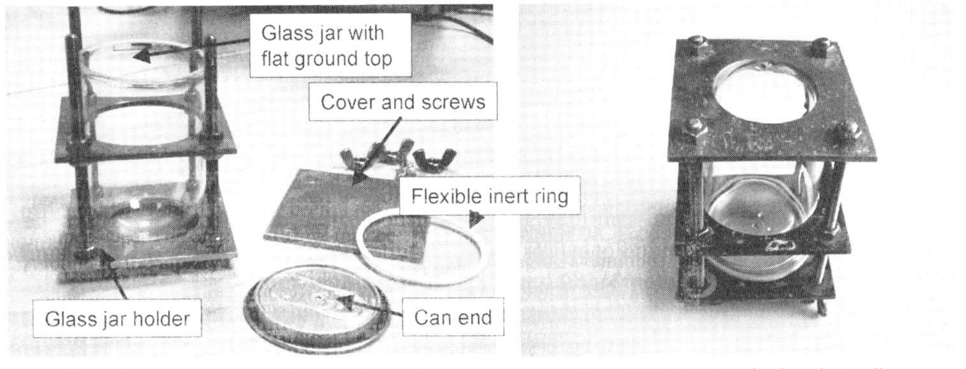

Figure 12.3 Can-end migration cell, as used by TNO.

12.4.2 Overall Migration

The overall migration is the total amount of components that migrate. The overall migration limit is a measure of the inertness of the material and prevents an unacceptable change in the composition of the foodstuffs, and, moreover, reduces the need for a large number of SMLs or other restrictions, thus giving effective control. Compliance with the overall migration limit is only a part of the safety evaluation of a food contact material.

The overall migration can be expressed per area; mg/dm^2 and per kg foodstuff; mg/kg. A limit for the overall migration of $10\,mg/dm^2$ or 60 mg/kg foodstuff is applicable. For plastics, this is described in Article 2 of Directive 2002/72/EC. At the national level, the overall migration may also be applied to nonplastic materials; different limits may apply.

12.4.2.1 Overall Migration into Aqueous Food Simulants

Methods of determining the overall migration into the aqueous simulants (A, B, and C) are as follows:

- Simulants, time, and temperature are selected using tables as shown in Sections 12.3.1 and 12.3.2, taking into account the expected food contact use.
- Depending on the sample, simulant, and other items as discussed in item 4, the contact method is selected.
- The contact area and amount of simulant is exactly determined.
- The food contact material is brought in contact with the simulant for a selected time and temperature. During testing, it is important that the time starts when the temperature selected has been reached in the simulant. If, for example, a temperature of 1 h at 100 °C must be selected according to the table mentioned below, and it takes 2 h before the simulant reaches 100 °C because the oven is not heating the solution very fast due to improper heat transfer, the sample must be in the oven for 3 h.
- The sample is separated from the simulant.
- The weight of an empty cup is determined.
- The simulant is evaporated and the residue is collected in the preweighted cup.
- The weight of the cup + residue is determined.
- The overall migration is calculated.

As a consequence, volatile chemicals that migrate are not included in the overall migration value for the aqueous simulants. A similar approach is chosen when a volatile substitute simulant is selected.

12.4.2.2 Overall Migration into Fatty Food Simulants

Because the fatty food simulant olive oil cannot be simply evaporated, the determination of the overall migration into olive oil is more complicated. The value of overall migration is measured by determining weight loss from the sample. But because the sample might have absorbed components of the fatty simulant during contact, the weight loss of the sample must be corrected for the amount of absorbed fat. The procedure of determining the migration into fat is as follows:

- The simulants, time, and temperature are selected using tables as shown in Sections 12.3.1 and 12.3.2, taking into account the expected food contact use.
- Depending on the sample, simulant, and other items as discussed in item 4, the contact method is selected.
- The contact area is determined exactly (A).
- The weight of the sample is determined before contact ($W1$).
- The food contact material is brought in contact with the simulant for a specific time and at a defined temperature.
- The sample is separated from the simulant and as much simulant as possible is removed from the sample with a tissue paper.
- The weight of the sample after contact ($W2$) is determined.
- The amount of the fat absorbed in the sample is extracted from the sample and then determined using a suitable gas chromatographic method (F).
- Calculate the migration (M_{mg/dm^2}) in mg/dm^2 using the formula $((W1 - W2 + F)/A)$. If migration shall be expressed in mg/kg food, then apply the conventional factor of 6 or apply the actual ratio of the amount of food in contact with a specified FCM area.

The determination of migration into simulant D is very complicated and errors can be made easily resulting in higher or lower values. The following critical issues need to be considered during the determination of overall migration into fatty foodstuffs:

- The weight before and after the contact time is determined. Therefore, it is important that in case the material is moisture sensitive, the amount of moisture is identical while determining $W1$ and $W2$. This could be achieved by conditioning the sample before determining $W1$ and $W2$, by means of vacuum drying or using constant relative humidity.
- During conditioning by vacuum drying, volatile components (not being water) are removed, resulting in a lower value for overall migration. Therefore, this phenomenon must be checked before vacuum conditioning is selected as a way of obtaining a constant amount of water in the sample.
- In determining the amount of fat absorbed by the sample (F), not all the fat is extracted, resulting in a value for F, which is too low.

Besides the above there, are many more sources of possible error. More detailed information about the determination of the overall migration can be found in CEN methods. The use of olive oil is preferred to the use of alternative simulants because in most cases 95% ethanol or isooctane is a more stringent simulant, which may result in a much higher value of overall migration than the value that would be obtained when olive oil is used. The use of other edible oils should have no significant effect on the level of migration.

12.4.2.3 Overall Migration into MPPO

At high temperatures, modified polyphenylene oxide (MPPO), better known under its trade name Tenax, can be used as a simulant. This simulant is a dry powder that has a strong adsorption capacity. To simulate migration in dry foods or to demonstrate compliance with the restrictions where olive oil is not suitable, the use of Tenax may be the only alternative. Particularly for determining migration from microwave susceptors under conditions of microwave heating, the Tenax may be the only substitute.

The method consists of the following steps:

- The simulants, time, and temperature are selected using tables as shown in Sections 12.3.1 and 12.3.2, taking into account the expected food contact use.
- The contact area is determined exactly and the amount of Tenax is adapted to it.
- A sample is placed in a petri dish with the food contact side upward.
- A known amount of purified MPPO is added on top of the sample at the selected time and temperature.
- The sample is separated from the simulant.
- The MPPO is extracted with an organic solvent.
- The weight of an empty cup is determined.
- The organic solvent is evaporated and the residue is collected in the preweighted cup.
- The weight of the cup + residue is determined.
- The overall migration is calculated.

12.4.3 Specific Migration

Determination of the specific migration of a component from an FCM may be performed in a foodstuff, but for compliance testing it is more convenient to use the authorized food simulants and test conditions as indicated above. The specific migration of a substance is determined in the food simulant using an appropriate analytical method. Usually, sophisticated analytical equipment and well-trained analysts are required to achieve the final goal. The number of substances with a specific migration limit for plastics is already huge. In general, the testing laboratory has to develop and validate its own analytical methods. Some guidance and validated methods are available from the CEN norms EN 13130. These norms provide information about treatment of the sample and the critical steps while performing the migration experiments. In addition, for some substances validation criteria have been established. These criteria may support the development and validation of in-house made methods.

If a method is selected or set up, it must be suitable for the component to be determined at the level required and the simulant used. Therefore, validation in the simulant is very important. It is important that the component is stable in the simulant during the migration period. If the component to be determined reacts with the simulant (e.g., primary amines react with aldehydes in olive oil), another simulant must be selected. The recovery or stability experiments, which include the storage conditions, must be performed at the level of migration or at the level of the SML.

Additional methods can be found on the web site of the Joint Research Center (JRC) (http://crl-fcm.jrc.it/) or in the public literature.

In principle, the test sample should be brought into contact with the relevant food simulant and stored for the selected time and temperature conditions. After the contact time, the simulant is analyzed for the migration of the substance of interest. Results are expressed in mg/kg simulant of in mg/dm^2 of FCM. Migration should not exceed the restrictions laid down in the directive or national legislation. Usually, an analytical tolerance is allowed. Where the directive mentions ND (not detectable), a value of maximum 0.01 mg/kg food is considered not detectable.

12.4.4 Mathematic Modeling

To avoid migration testing, it is allowed to demonstrate compliance with specific migration limits using mathematical modeling. The modeling is based on the fact that components want to migrate to places where the concentration of the same component is the lowest. This phenomenon is also referred to as diffusion. The calculation can be performed for a group of validated plastics for which adequate parameters are set.

The modeling can be done using one of the commercial or freeware software packages that are available on the market or by using calculations. Depending on the software, mathematical modeling can take into consideration the following aspects:

- The presence of multiple layers
- A cascade of time and temperature conditions
- The solubility of the component in the simulant
- Packaging area and amount of food/simulant

Mathematical modeling is a very useful tool because calculations can be done without doing real testing. It is even possible to perform calculations without having the sample. This enables you to see what the effects would be if a product is changed (e.g., the additive concentration is increased). However, mathematical modeling has some restrictions of where, when, and how it can be used, such as the following:

- Because the use of the modeling will result in an overestimation, it is possible to only confirm that a migration is lower than the limit. If the modeled result is higher than the specific migration limit, analytical testing is needed. Or, in other words, it is not possible to conclude that a material is not in compliance.
- The component must be homologically present in a layer. Some materials are not distributed homologous by definition (e.g., release agents).
- The components may not be charged like salts or heavy metals.
- The food contact must be a homogeneous monolayer plastic or a multilayer plastic. Migration from nonplastics cannot be estimated using mathematical modeling.
- Swelling of the polymer may not occur.

12.5 Residual Content

The residual content is the amount that is present in the finished product. The residual content can be expressed as QM in mg/kg FP or as QMA in mg/ 6dm^2 FP.

This means that, for example, if a component is used in one layer of a multilayer, the weight of all the layers must be taken into account when calculating the QM.

12.5.1 Worst-Case Calculations

In some cases, the amount of a material added is very low. If it can be demonstrated by calculation that even if all the substance added remains in the food contact material, the residual content cannot exceed, then the material may be considered in compliance without analytical determination.

However, in many cases the amount added will be quite high compared to the maximum residual amount. These cases could be when a very thin layer is used in a multilayer material.

12.5.2 Analytical Determination

The determination of a component in a material can be challenging. One way is to prepare a solution by extracting the food contact material with a suitable solvent. A suitable solvent is a very good solvent for the component to be determined and which causes some swelling of the FCM. To demonstrate that the solvent has extracted everything, successive extractions should be used. If a second extraction does not extract any additional quantity of the component, then the method can be regarded as sufficient.

Another way is first dissolve the polymer and the components to be determined. Hereafter, by adding a nonsolvent of the polymer, the polymer will precipitate while the component to be analyzed remains in the solution.

Again in both cases it is important that a good validation is performed to demonstrate that the method is suitable and the analytes can be retained from the food contact material.

In EN 13130, some methods are presented that deal with the determination of the residual content of specified substances (carbonyl chloride, ethylene oxide, propylene oxide, 1,3-butadiene, vinylidene chloride, isocyanates, and epichlorohydrin). These methods may give guidance for in-house development and validation of analytical methods.

Additional methods for the determination of residual content can be found on the web site of the JRC (http://crl-fcm.jrc.it/) or in the public literature.

12.6
Miscellaneous

12.6.1
Not Intentionally Added Substances

During the production or the use of food contact materials, reaction or decomposition products may be added or formed. Examples of these NIAS are impurities in monomer or additive used, oligomers formed during the polymerization process, and reaction products of an antioxidant that has reacted with oxygen. The NIAS must be evaluated before compliance with Article 3 of the Framework Directive (article must be safe) can be claimed.

Three connected questions must be answered during the evaluation:

- Which components are present?
- How much of these components can migrate to the food?
- Does this migrated amount pose a toxicological concern?

These questions are connected because depending on the toxicity of a component a smaller or larger migration can be regarded as safe. In principle, it is possible to analyze the sample down to a virtual zero concentration, identify everything, and evaluate all components that can migrate. This is seldom done because it is very laborious and extremely expensive.

In many cases, a tiered approach will be chosen. The result is based on the fact that the more information (and the more positive toxicological knowledge about a component is available), the higher the safe limit will be set.

Examples of a tiered approach were published by a group of European scientists [13], the United States [14], and ILSI [15]. The approaches published are all similar. There are different levels of exposure that are regarded as safe. The more nonadverse toxicological data are available, the higher the level is that can be regarded as safe.

Analytical equipment involved will be a variety including, but not limited to, liquid chromatography with mass detection, gas chromatography with mass detection, multidimensional analysis (GC/GC-MS or LC/GC-MS), and NMR.

In many cases, some components can be expected because the presence is known (e.g., an impurity present in an additive) or can be expected (e.g., breakdown product of a catalyst). Of course, many people will try to find these expected components, but it is also important to screen the sample for unexpected components using the so-called nontarget analysis.

In many cases, it may be helpful to start with experiments on the polymer because the higher concentration of the NIAS in the polymer simplifies the identification. A worst-case migration (assuming 100% migration or mathematic modeling) can be used to determine if the toxicological relevant level can be exceeded. Hereafter, the relevant components can be determined in the simulants if needed.

The combination of analytical and toxicological challenges makes the determination of the NAIS an expert job.

12.6.2 Organoleptic Testing

As is described in Article 3 Item 1c of the Framework Regulation (EC) No. 1935/2004, components that migrate to the food may not result in deterioration in the organoleptic characteristics.

This requirement applies to finished articles in contact with the food. This requirement makes the testing very difficult because the conclusion whether a food is organoleptically changed depends on the person (every person has a sense of odor and taste) and the food type (still water is much more sensitive than French cheese).

Of course, not all final applications can be tested, and therefore, some standardized tests can be used to rule out that the packaging is changing the organoleptic characteristics. These methods can be developed in-house (as long as they cover the final application) or some standard methods can be applied. It is very important that the panel that is performing the testing is well trained. Not every person is (always) suitable to take place in the test panel; if a test person is having a cold, for example, he cannot participate. In addition, attention must be paid to the presentation of the samples (must be anonymous and must be randomly coded), the room in which the testing is performed (must be free of odors), and much more.

Methods are published by the UK legislator for an odor test [16] and sensory test [17] by the Dutch [18] and the German legislators [19]. In addition, the EU has published a nonofficial document [20] that describes the following two unofficial methods.

1) **Testing for Microwave Application**:

   Sensory testing of food contact materials for use in microwave ovens is carried out with food simulants. For ready-to-eat-packages, the test can also be performed with the packed food. The object to be tested has to be cleaned as usual in the household or suggested by the producer. As food simulants, drinking water or neutral frying oil are used.

   For testing with drinking water (at temperatures below 100 °C), the testing object is filled to half of its volume; for testing with fat or oil (at temperatures between 100 and 150 °C), 100 g fat/dm^2 bottom area, which corresponds to a layer about 1 cm high.

   Covers and covering films are tested over appropriate vessels filled with water. Heating is carried out in a microwave oven under the following conditions:

   - Drinking water (max. 100 °C): 600 W, 2 min/100 ml.
   - Frying fat (max. 121 °C): 600 W, 3 min/100 g.
   - Frying fat (max. 150 °C): 600 W, 4 min/100 g.

   After heating, the food simulants are allowed to cool in the object tested to a temperature of 40–50 °C and tasted at that temperature against reference samples, which were treated the same way in glass.

   Otherwise, the test is carried out according to DIN 10955 (sensory testing of packaging materials and packages for food products).

If a food simulant is evaluated with the mark 3 (distinct deviation from the reference sample), it is assumed that real food is also reduced in its odor and/or taste quality, which means an offence against Article 2 of Directive 89/109/EEC.

2) **Testing at Oven Conditions:**
Sensory testing of baking papers and boards is carried out with a neutral sponge mixture, which is spread on the baking tray as suggested by the producer and covered with the mixture.

The standardized mixture contains
200 g wheat flour
150 g margarine
1 egg
2 tablespoons of water
and has to be cooled for at least 1 h to 4–8 °C.

After cooling, the mixture is rolled out on the baking paper or board covered by the testing material, over an area of about 12 dm^2 and about 0.5 cm thick. For smaller testing areas, correspondingly less dough is used.

The baking conditions are chosen corresponding to the instructions of the producer or selected from the following possibilities:

- 180 °C/30 min,
- 200 °C/25 min, or
- 220 °C/20 min.

For the reference sample, the mixture is baked on aluminum foil covered by testing material, under the same conditions.

After baking, the "cakes" are allowed to cool down to about 40 °C on the testing material. Then, the testing material is tested against the reference sample.

Alternatively, the test is carried out according to DIN 10955 (sensory testing of packaging materials and packages for food products).

If the testing material is evaluated with the mark 2.5 (remarkable deviation from the reference sample), it is assumed that a real food is also reduced in its odor and/or taste quality, which means an offence against Article 2 of Directive 89/109/EEC.

12.6.3 Other Tests

There may be many other tests laid down in the EU or Member State legislation. A short selection is shown here. In many cases, a test method is specified if a specific test is prescribed in the legislation. Therefore, always check the legislation to see according to which method the test must be performed.

12.6.3.1 Demonstrating the Absence of PAHs

Carbon black can be produced using many sources and purification procedures. Because many of the production processes include partial burning of organic materials, there is always a potential risk of the formation of polycyclic aromatic hydrocarbons (PAHs). These PAHs are known carcinogens and, therefore, it is required that they are not present. The legislator has set limits for the specifications for carbon black:

- Toluene extractables: maximum 0.1%, determined according to ISO method 6209.
- UV absorption of cyclohexane extract at 386 nm: <0.02 AU for a 1 cm cell or <0.1 AU for a 5 cm cell, determined according to a generally recognized method of analysis.
- Benzo(*a*)pyrene content: max. 0.25 mg/kg carbon black.

The three above-mentioned tests are all for the purpose of detecting the presence of PAHs. Benzo(*a*)pyrene, one of the most carcinogenic PAHs, must be individually determined.

A similar approach is present in some member state legislation (e.g., the Netherlands) for solid parafins and microcrystalline parafins. An extraction is performed. If PAHs are present, they will be extracted and determined using UV detection. Limits are set for some wavelength ranges.

12.6.3.2 Color Release

One of the undesired effects is that the food contact material transfers a color to the food. Both in the German and in the Dutch legislations, restrictions have been specified to verify the absence of the effect. In addition, a CEN method for paper has been published to investigate this phenomenon (CEN 646).

The method is based on wetted filter paper that is held for a certain time and at a certain temperature in contact with the test sample. A weight is positioned on the sample to ensure sufficient contact between the sample and the filter paper. After this time is over, it is verified if the filter paper is colored in respect to a wetted piece of filter paper that was not in contact with the test sample. In the following table, the differences are given between the Dutch and the CEN method:

	The Netherlands	CEN 646
Contact temperature	40 ± 2 °C	23 ± 2 °C
Contact time	5 h	10 min (short-contact applications) 24 h (long-contact applications)
Simulants	3% acetic acid Fatty simulant (sun flower oil, olive oil, etc.)	Water 3% acetic acid in water Sodium carbonate solution 5 g/l Isooctane Rectified olive oil

12.6.3.3 Colorants Purity

Some Member States, such as the Netherlands and Germany, do have purity requirements for colorants that do contain heavy metals. In principle, the colorants are insoluble. Pure colorants (not the colored finished products) are extracted with a strong hydrogen chloride. Hereafter, the hydrogen chloride solution is analyzed for the presence of some specific heavy metals using techniques such as atomic emission spectrometer (AES), atomic absorption spectrometer (AAS), or inductive coupled plasma (ICP) combined with mass selective detection (MS) or AES or optical emission spectrometer (OES). In addition the amount of primary aromatic amines that can be extracted from the colorants using hydrogen chloride is determined.

12.7 Declaration of Compliance and Supporting Documentation

A declaration of compliance is intended to transfer knowledge on the material from the supplier to buyer (of food contact materials or articles not already in contact with food) to ensure proper use of the materials for the purpose intended. It is also mentioned in the legislation that in view of potential liability, there is a need for the written declaration provided whenever professional use is made of plastic materials and articles, which are not by their nature clearly intended for food use.

Until the publication of the Framework Regulation (EC) 1935/2004, a material only had to comply with the legislation and a declaration of compliance was required. It was up to the Member State authorities to demonstrate that a material was not in compliance. This was problematic for the Member State authorities and also very laborious, as they had no information on the exact composition of the materials. Upon publication of the Framework Regulation (EC) No. 1935/2004 article 16 requires that the person who sells (ingredients for) food contact materials must have documentation supporting the declaration of compliance. This may include analytical reports and physical and chemical properties as well as compositional information. Enforcement authorities are authorized to check the supporting documentation.

This article is applicable only when a specific measure does require it. All EU directives and regulations include this requirement. Member States may require supporting documentation for nonharmonized materials as well!

The supporting documentation needs to contain documents that can prove the validity of the declaration of compliance. It is not required to test every type of material. In many cases, statements from suppliers can also be sufficient. In some cases, additional experiments have to be performed to demonstrate that the restrictions and requirements are met, such as overall migration, specific migration, and residual content. Data obtained by mathematical modeling, extrapolation of results obtained from comparable materials or based on compositional data may be valid. The supporting documentation, however, should clearly mention this in such a way that this is obvious when the data of a nontested material is reviewed on which data the compliance relies.

Therefore, it is recommended that for every food contact material, a dossier is made. It must be ensured that the rationale why the conclusion that a material meet the legal requirements was made, even after some years and changes of personnel.

12.7.1 Ceramics

For ceramics, provisions regarding supporting documentations are made in Article 2a. The article is very specific describing that the material is in compliance with the specific migration restrictions for lead and cadmium. In addition, it is noted that information regarding the test performed and the results obtained as well as the identity of the laboratory that performed the test should be made available to competent authorities.

Although compliance with Regulation (EC) No. 1935/2004 must be mentioned in the declaration of compliance, this is not a requirement for the supporting documentation.

Regarding the declaration of compliance, Directive 84/500/EEC does require at least the following items to be present:

- The identity and address of the company that manufactured the finished ceramic article and of the importer who imported it into the Community.
- The identity of the ceramic article.
- The date of the declaration.
- Confirmation that the ceramic article meets relevant requirements in this Directive and Regulation (EC) No. 1935/2004.

The written declaration shall permit an easy identification of the goods for which it is issued and shall be renewed when substantial changes in the production bring about changes in the migration of lead and cadmium.

12.7.2 Plastic Materials

In Article 9 of the Directive 2002/72/EC it is mentioned that plastics do need to be accompanied with a written declaration of compliance and that supporting documentation shall be available. In Annex Via of the directive a list of information to be included in the declaration of compliance is presented. In addition to some general information on the identity of the business operator and the substance or material, information on the use (time, temperature, food types) shall be included. In case a claim for a functional barrier is made then adequate information on the relevant substances should be provided. The declaration shall allow an easy identification of the material involved.

When recycled plastics are used as raw material, the declaration of compliance should specify that the recycled material is obtained by an authorized recycling process and in compliance with the specifications of the authorization.

12.7.3 Materials with BADGE

For materials containing BADGE, it is mentioned in Article 5 of Regulation (EC) No. 1895/2005 that a written declaration must be present. The exact reference mentioned in the legislation is "At the marketing stages other than the retail stages, materials and articles containing BADGE and its derivatives shall be accompanied by a written declaration in accordance with Article 16 of Regulation (EC) No. 1935/2004." In addition to this, it is also mentioned that appropriate documentation shall be available to demonstrate such compliance. That documentation shall be made available to the competent authorities on demand. The exact content of the supporting documentation is not specified.

12.7.4 Other Materials

For some items, the format of the declaration of compliance is not obligated. In this section, a suggestion for the content is given. A proper declaration of compliance should at least contain the following items:

- The identity and address of the company that manufactures the material and if applicable the importer who imports it into the Community.
- The identity of the article.
- The date of the declaration.
- The confirmation that the article meets relevant requirements in Regulation (EC) No. 1935/2004.
- The confirmation that the material meets relevant requirements in other specific directives and legislation in Member States.
- Mention if the material can be directly used or only behind a functional barrier (must be specified) or if a barrier approach was used.
- The intended use
  –Contact temperature restrictions with the food, if applicable
  –Contact time restrictions with the food, if applicable
  –Food-type restriction, if applicable (including DRF and FRF)
  –In case of the raw material, the maximum amount to be used in the final product
  –Restrictions on the food contact area and food content.

The supporting documentation must cover the items claimed in the declaration of compliance, as mentioned above.

List of Abbreviations

AAS	atomic absorption spectrometry
AES	atomic emission spectrometry
BADGE	bisphenol A Diglycidyl ether
CEN	European Committee for Standardization (Comitée Européen de Normalisation), see: www.cen.eu
DRF	simulant D reduction factor
FP	finished product
FRF	fat reduction factor
GC	gas chromatography
ICP	inductive coupled plasma
LC	liquid chromatography
MS	mass spectrometry
NMR	nuclear magnetic resolution
OES	optical emission spectrometry
OM	overall migration limit, expressed as mg/6dm^2 or mg/kg
PAH	polycyclic aromatic amines
QM	residual content, expressed as mg/kg FP
QMA	residual content, expressed as mg/6 dm^2 FP
SML	specific migration limit

References

1 Directive 2002/72/EC Commission Directive of 6 August 2002 on plastic materials and articles intended to come in contact with foodstuffs. (Plastics: Unofficial consolidated version including 2002/72/EC, 2004/1/EC, 2004/19/EC, 2005/79/EC, 2007/19/EC, 2008/39/EC.)

2 Directive 2007/42/EC Commission Directive of 29 June 2007 related to materials and articles made of regenerated cellulose film intended to come in contact with foodstuffs. (Codified version.)

3 Directive 2005/31/EC Commission Directive of 29 April 2005, amending Council Directive 84/500/EEC as regard to a declaration of compliance and performance criteria of the analytical method for ceramic articles intended to come into contact with foodstuffs.

4 Regulation (EC) No. 1935/2004 of the European Parliament and of the Council of 27 October 2004 on materials and articles intended to come in contact with food and repealing Directives 80/590/EEC and 89/109/EEC (L338/4).

5 Directive 85/572/EEC Council Directive of 19 December 1985 laying down the list of simulants to be used for testing migration of constituents of plastic materials and articles intended to come in contact with foodstuffs (Plastics: list of simulants for testing migration).

6 Directive 2007/19/EC Commission Directive of 2 April 2007 amending Directive 2002/72/EC related to plastic materials and articles intended to come into contact with food and Council Directive 85/572/EEC laying down the list of simulants to be used for testing migration of constituents of plastic materials and articles intended to come in contact with foodstuffs.

7 CEN EN 14233 Materials and articles in contact with foodstuffs – Plastics – Test methods for the determination of fatty contact. www.cen.eu.

8 Directive 82/711/EEC Council Directive of 18 October 1982 laying down the basic rules necessary for testing migration of the constituents of plastic materials and articles intended to come into contact with foodstuffs. (Plastics: Basic rules for testing migration – Unofficial consolidated version including 82/711/EC, 93/8/EEC and 97/48/EC).
9 Directive 97/48/EC Commission Directive of 29 July 1997 amending for the second time Council Directive 82/711/EEC laying down the basic rules necessary for testing migration of the constituents of plastics materials and articles intended to come in contact with foodstuffs (Plastics: Basic rules for testing migration, second amendment).
10 CEN EN 14233 *Temperature at the Plastic/Food Interface During Microwave and Conventional Oven Heating*, CEN EN, www.cen.eu.
11 CEN EN 1186 *Materials and Articles in Contact with Foodstuffs – Plastics, Part 1–15*, CEN EN, www.cen.eu.
12 CEN EN 13130 *Materials and Articles in Contact with Foodstuffs – Plastic Substances Subject to Limitation, Part 1–28*, CEN EN, www.cen.eu.
13 Kroes, R. *et al.* (2000) *Food and Chemical Toxicology*, **38** (2–3), 255–312.
14 Cheeseman, M.A. *et al.* (1999) *Food and Chemical Toxicology*, **37**, 387–412.
15 Kroes, R. *et al.* (2004) *Food and Chemical Toxicology*, **42**, 65–83.
16 S 3755 (1964) *Methods of Test for the Assessment of Odour from Packaging Materials Used for Foodstuff.*
17 ISO S 5929, part 3 (1984, ISO 4120-1983) *Sensory Analysis of Food*, ISO.
18 Sensory analysis; packaging and food utensils regulation, Staatscourant No. 88 of 12/05/98.
19 DIN 10955 (1983) *Sensory Testing of Packaging Materials and Packages for Food Products*, DIN.
20 Practical guide for food contact materials, 15 April 2003, SANCO D3/LR D.

13
Food Packaging Law in the United States

Joan Sylvain Baughan and Deborah Attwood

13.1
Introduction

There have been food purity concerns for centuries; the regulation of food in the United States can be traced all the way back to a 1646 Massachusetts Bay colony law regulating bakers. Regulation of the materials used to package or hold food, however, is a much more recent development. In 1913, Congress passed the Gould Amendment [1] to the Pure Food and Drug Act of 1906, which aimed at preventing deceptive packaging practices such as filling a large package only half full, but it was not until 1958, with the Food Additives Amendment to the Federal Food, Drug and Cosmetic Act (FDCA), that the US national government began to formally regulate the safety of the substances that comprised packaging materials. In the mid 1950s, consumer concerns about reports and allegations that chemical food additives were causing serious diseases, such as cancer, led to Congressional hearings headed by Representative James J. Delaney (D-NY). These hearings failed to show that packaging materials caused adulteration of food yet, despite vigorous industry opposition, the amendment passed [2]. The amendment recognized that substances used in packaging may, under some circumstances, migrate to and become part of food, and that it was necessary to determine whether exposure to those substances is safe [3].

American law uses the term "indirect food additives" when referring to these migrating substances; however, the USFood and Drug Administration (FDA), the main federal agency responsible for enforcing the laws and regulations dealing with food and food additives, identifies indirect food additives under the umbrella term "food contact substances" (FCS). Food contact substances are defined in the FDCA as "any substance intended for use as a component of materials used in manufacturing, packing, packaging, transporting, or holding food if such use is not intended to have any technical effect in such food." [4] Coatings, plastics, paper, and adhesives can all be food contact substances, as can colorants, antimicrobials, and antioxidants. Food contact substances, usually mixed with other substances, combine to make a "food contact material." Finally, a food contact material is processed to make a "food contact article." Examples of a food contact article are films, bottles, and trays [5].

Global Legislation for Food Packaging Materials. Edited by Rinus Rijk and Rob Veraart

ISBN: 978-3-527-31912-1

The Food and Drug Administration is the government body charged with primary enforcement of the FDCA and other related acts. Pursuant to the FDCA, FDA ensures that foods, cosmetics, medicines, medical devices, radiation-emitting products, and animal food and drugs, consumed or used by the US public are safe [6] and effective. "Safe" in the FDCA context does not mean that every substance is determined to be absolutely harmless, as FDA recognizes that this is almost impossible to meet this burden of proof when considering all potential circumstances; instead, it means that there is a "reasonable certainty in the minds of competent scientists that a substance is not harmful under the intended conditions of use." [7]

Within FDA, there are various product-oriented centers, each responsible for enforcing one part of the FDA's mandate. It is the Center for Food Safety and Applied Nutrition (CFSAN, pronounced "sif-san") that ensures the safety of food [8], with the Office of Food Additive Safety having the specific responsibility for food contact substances. The safety of food additives is determined through the Food Contact Notification (FCN) system, Food Additive Petitions (FAP), and Threshold of Regulation (TOR) requests, although manufacturers determine for themselves whether a product complies with the FDCA and therefore does not need to undergo a separate FDA safety determination [9]. In fact, avoiding government involvement is generally preferable, as a former FDA General Counsel has noted:

> [I]t is the primary and initial responsibility of the manufacturer of a product to determine the proper classification of this product, and to make certain that it meets all applicable legal requirements. It is in no instance necessary, and in most instances inadvisable, to ask the Food and Drug Administration for its opinion on the proper jurisdiction over the product [10].

13.2 FDA Rules and Regulations

The FDA requires clearance for food packaging or processing equipment materials only if those materials meet the definition of a "food additive" under the FDCA. If a material fits within the food additive definition, then it will be automatically considered unsafe unless it is subject to a food additive regulation or has received premarket authorization from FDA through submission of a Food Contact Notification [11]. The food additive regulations relevant to food contact substances are found in Title 21 of the Code of Federal Regulations, in Parts 174–186 (21 C.F.R. Parts 175–186), while Food Contact Notifications are listed on FDA's web site. Whether a material is a food additive or not is determined by the manufacturer, using the regulations, statutory definitions, and exemptions.

13.2.1 Definition of a Food Additive

The FDCA defines a "food additive," in relevant part, as a substance that is reasonably expected to become a component of food under the intended conditions of use [12].

With regard to food packaging materials, this definition looks to whether any component of the packaging that touches the food may be reasonably expected to migrate into the food. If a substance migrates and thus meets the food additive definition, it will be considered "unsafe" unless it is used in accordance with an applicable food additive regulation, an effective Food Contact Notification [13], or is statutorily exempt. Statutory exemptions are provided for substances that are "generally recognized as safe" (GRAS) [14], are the subject of a sanction or approval issued prior to the enactment of the Food Additives Amendment of 1958 [15], or are the subject of a Threshold of Regulation letter [16]. If the material is an unsafe food additive, that is, it is not being used in accordance with an applicable regulation, exemption, or notification, then the food it becomes part of is deemed adulterated [17]. If a substance is not reasonably expected to become a component of food under the intended conditions of use, as determined by the manufacturer, then it is *not* a food additive by definition and *does not require* any premarket clearance from the FDA.

13.2.2
Suitable Purity, the Delaney Clause, and FDA's Constituents Policy

The FDCA does not allow the presence of food additives that may render the food injurious to health [18]. If a component of a packaging material does migrate and become part of the food as an additive, FDA regulates the material to ensure the food is still safe for consumption. FDA's Good Manufacturing Practice (GMP) regulations require that the migrating components be of a "purity suitable for [their] intended use." [19] To meet this suitable purity standard, food contact substances must comply with any applicable food additive regulation and must not render food injurious to health or otherwise unfit for consumption.

This suitable purity requirement is particularly important with respect to carcinogenic constituents of food additives. Section 409(c)(3)(A) of the FDCA, also known as the "Delaney Clause" after Representative James Delaney, states that the FDA cannot deem any food additive safe if the additive is found "to induce cancer when ingested by man or animal, or if it is found, after tests which are appropriate for the evaluation of safety of [the additive], to induce cancer in man or animal" [20]. The FDA has acknowledged that "all chemical substances, including those used as additives, contain numerous impurities such as residual reactants, intermediates, manufacturing aids, and products of side reactions and degradation" [21]. Improving technology means that such impurities are detectable at increasingly minute levels, forcing the FDA to re-evaluate its policy for determining the safety of such impurities. As part of this re-evaluation, FDA began to distinguish between the additive as a whole and its individual "constituents." Under this approach, a food additive is regarded as the substance that is actually intended for use in food or for food contact, while all "nonfunctional chemicals" present in the additive are its constituents [22]. This so-called "constituents policy" allows FDA to find the use of food additives safe, even if they contain small amounts of carcinogenic substances as unintended contaminants, as long as the food additive, as a whole, is not carcinogenic [23]. Based on this policy, FDA interprets the Delaney Clause as prohibiting only a carcinogenic food additive, allowing FDA to permit the use of "safe" levels of carcinogenic constituents. FDA then uses risk assessment techniques to determine whether the level of a carcinogenic constituent in a food additive is low enough to be "safe," such that there is "a reasonable certainty that no harm will result from the proposed use of the additive" [24].

To determine the safety of the constituents, FDA estimates the potential daily exposure to a person through his or her diet. FDA uses animal carcinogenicity study data and quantitative extrapolation procedures to calculate the potential risk to a person from the estimated daily exposure to the carcinogenic constituent as a result of the intended use of the product. If the risk over a lifetime from *all* different uses of the product, as determined by the calculated upper bound, is less than 10^{-6} (one in 1 million), then the risk is considered negligible and the exposure to the constituent is considered to be safe [25]. This daily dose at a "1 in 1 million" level of risk is referred to by some as the "virtually safe dose" (VSD). Because it is likely that the constituent may enter the diet through more than one source or product, it is clear that each source should not contribute the entire VSD. Therefore, when considering multiple sources of exposure, one constituent should not contribute the entire VSD.

13.3 Exemptions

The FDCA contains a number of explicit exemptions regarding which substances constitute food additives. Other nonstatutory exceptions have developed as a result of legislative history or agency policy. These are all very important to food contact material manufacturers because, if a substance is not a food additive, then it does not have to be cleared by FDA before being marketed.

13.3.1 "No Migration"

The FDCA allows companies to determine for themselves whether a particular substance used in a food contact material will become a food additive. If a manufacturer determines that the substance is not reasonably expected to migrate to food when the food contact material is used as intended, then the substance is not a food additive under the food additive definition [26]. This "no migration" position may be supported by properly designed and conducted extraction studies. Alternatively, calculations may show that the potential level of migration is negligible [27]. However, these approaches still leave the question of whether there must be literally no migration, or whether an insignificant amount of migration is acceptable. The FDA has never provided definitive criteria for determining when a substance may reasonably be expected to become a component of food, but industry has come to rely on various sources of guidance, such as the "Ramsey Proposal" and the *Monsanto* v. *Kennedy* decision [28].

The "Ramsey Proposal," circulated by the FDA in 1969, would have permitted the use, without the prior promulgation of an applicable food additive regulation, of substances that migrate to food in quantities not greater than 50 parts per billion (ppb). Named after its author, Dr Lessel Ramsey, then Assistant Director of Regulatory Programs at FDA's Bureau of Science, this regulation would have applied to all substances except those known to pose some special toxicological concerns, for example, a heavy metal, a known carcinogen, or a substance that produces toxic reactions at levels of 40 parts per million (ppm) or less in the diet of man or animals. The proposal was never formally adopted by FDA, but the agency deemed the proposal's standards scientifically acceptable.

The *Monsanto* case involved an appeal from a decision by the FDA Commissioner that the substance acrylonitrile copolymer, used to manufacture unbreakable beverage containers, was an unsafe food additive. FDA argued that the migration of *any* amount of a substance is sufficient to make it a food additive, which would have allowed the FDA to require a food additive clearance for every food contact material, even without evidence that any part of it actually migrates to food. The United States Court of Appeals rejected FDA's argument, however, stating:

> Congress did not intend that the component requirement of "food additive" would be satisfied by . . . a mere finding of any contact whatever with food. . . . For the component element of the definition to be satisfied, Congress must have intended the Commissioner to determine with a fair degree of confidence that a substance migrates into food in more than insignificant amounts [29].

The Ramsey Proposal and Monsanto case suggest that substances that migrate to food in *de minimis* amounts are not food additives. A 1974 statement from FDA's Office of General Counsel supports this position, to the extent that the legal determination that a substance is a food additive must be based on more than evidence of a tiny amount of migration:

> Finally, if any court action is brought, we [FDA] have the burden of proving two things: first, that the ingredient may reasonably be expected to become a component of the food, and, second, that the amount of migration involved is not generally recognized as safe. We would need expert testimony on both issues. The fact that extreme conditions produced extraction would not be sufficient evidence in and of itself to justify a food additive conclusion. We would be required to put on evidence of experts showing that the extraction studies are reasonably related to actual use conditions and, thus, that the results can be extrapolated to normal use. We would also be required to show that the amount that might reasonably be expected to migrate is not generally recognized as safe and, thus, is a food additive [30].

For migrating substances that are of relatively high toxicity, such as heavy metals, or those that will have widespread use in food contact applications that could lead to a higher overall exposure, such as milk or soda bottles, or those that are intended for use in applications by a sensitive class, such as children and infants, the amount of migrating substance should be proportionally less. Substances with carcinogenic impurities must be evaluated on a case-by-case basis using risk assessment procedures.

13.3.2
Functional Barrier

A subset of the "no migration" exclusion is the functional barrier doctrine. This doctrine operates on the premise that if a substance is not part of the food contact surface of a package and is separated from the food by a barrier that prevents migration of the substance to food, then the substance cannot reasonably be expected to migrate to the food and is, therefore, not a food additive. Whether a true functional barrier exists often may be determined through a common sense consideration of the package structure and the exposure conditions anticipated for the package. Occasionally, there may be more complex applications such as cases involving interior layers of laminates, outer layers of packages, and external printing inks that require calculations or migration testing.

The functional barrier doctrine was a well-established, though unpublished, FDA position [31] before being acknowledged in the 1975 case of *Natick Paperboard* versus *Weinberger* [32]. In this case, the US Court of Appeals for the First Circuit looked into whether polychlorinated biphenyl (PCB) contaminants were impermissible food additives in paper and paperboard. The court found that FDA had the authority to seize such products as adulterated food, but distinguished food additives separated by a functional barrier, stating that if "the food placed in or to be placed in the paper container is or will be insulated from PCB migration by a barrier impermeable to such migration, so that contamination cannot reasonably be expected to occur, the paperboard would not be a food additive." [33]

13.3.3
Prior Sanction

The prior-sanction exemption is drawn directly from the FDCA. Prior to the Food Additives Amendment of 1958, the FDA and USDA received many inquiries from manufacturers regarding the suitability of using particular substances in food or as components of food contact materials to which the FDA typically responded via letter. When Congress enacted the 1958 Amendment, these informal acceptances attained the status of de facto regulatory exemptions because the FDCA specifically excludes from the definition of "food additive" any substance that is the subject of one of these "prior sanctions" [34].

Whether a substance is prior-sanctioned depends on whether a manufacturer possesses an appropriate pre-1958 letter. A prior sanction, however, does not mean that the FDA cannot control these substances. The agency has attempted to limit the scope of the exclusion by consistently construing prior sanctions as narrowly as possible, and can prohibit or set conditions on the use of any substance when FDA has proof that the substance is adulterating food [35].

13.3.4
GRAS

Generally recognized as safe or "GRAS" substances are explicitly exempted from the "food additive" definition in the FDCA [36]. For a substance to be GRAS, experts qualified by scientific training and experience must have evaluated the safety of the substance and determined that it is safe for food-related uses (or, for substances used in food prior to 1958, there must be support for the safe use in food based on common experience). Manufacturers who use a substance in a food contact material may independently determine that a particular intended use of the substance is GRAS without making any submission to the FDA. If, however, the FDA considers such a determination to be erroneous, the agency can take adverse regulatory action. FDA has issued regulations regarding the eligibility requirements for a substance to be considered GRAS: unless the substance was commonly used in food prior to 1958, the manufacturer must use scientific procedures, and the data used to support this position must be published and requires the same quantity and quality of scientific evidence as is required to obtain a food additive regulation [37].

The presence of substances at low levels in the diet may be considered an adequate basis to establish that the substances are GRAS in specific instances. For example, a Canadian Center for Toxicology (CCT) study that examined *de minimis* carcinogenic and noncarcinogenic risks resulting from low level exposure to food contact substances concluded that substances present in the diet at concentrations below 1.0 ppb can be considered safe even in the absence of toxicity testing provided that the structure of the substance does not indicate unusual toxicological properties [38]. Furthermore, FDA has effectively established what it considers to be a *de minimis* dietary level for food contact materials under the Threshold of Regulation policy, as discussed *infra*.

FDA regulations provide that a manufacturer who wishes to have FDA concurrence that the intended use of its product is GRAS can file a GRAS affirmation petition. However, under a rule-making proposal now treated as policy, FDA will no longer entertain or act on these petitions [39]. Instead, manufacturers submit a notification to FDA that a particular use of a substance is GRAS, and the agency reviews the claim and issues one of three types of letter: (i) FDA does not question the basis for the GRAS determination; (ii) the notice does not provide a sufficient basis for a GRAS determination; or (iii) agency has, at the notifier's request, ceased to evaluate the GRAS notification [40]. FDA maintains a database containing all GRAS notifications it has received and its responses [41].

13.3.5
Threshold of Regulation

FDA described the Threshold of Regulation rule, adopted in 1995, as the "process for determining when the likelihood or extent of migration to food of a substance used in a food contact article is so trivial as not to require regulation of the substance as a food additive" [42]. Under this rule, FDA may formally exempt noncarcinogenic food contact substances from the need for premarket clearance if a manufacturer shows that (i) the dietary concentration of the substance does not exceed 0.5 ppb or (ii) if the substance is cleared as a direct food additive, the dietary concentration that will result from the intended food contact use will not exceed 1% of the acceptable daily intake (ADI) for the substance [43]. While the TOR policy helps FDA to consider the suitability of food additives present in the diet at *de minimis* levels, it also provides guidance to the private sector's evaluation of food contact materials since this policy represents the FDA's tacit acknowledgment that the presence of a substance in the diet at extremely low levels is safe.

A study by three officials of FDA's CFSAN supports this position [44]. The study examined carcinogenicity data for 709 carcinogens [45] and concluded that the structure of an untested substance can be a strong indicator of whether it is likely to be a carcinogen [46]. Their research also indicates that results of short-term toxicity data and genotoxicity tests, that is, Ames assays and LD_{50} tests [47], can be strong indicators of carcinogenic potency as well. The study argued for an expanded dietary threshold for regulation of indirect additives: (i) a dietary threshold of 4 to 5 ppb for those substances lacking structural alerts, regardless of the results of an Ames assay; (ii) a dietary threshold of 4–5 ppb for substances other than *N*-nitroso and benzidine-like compounds testing negative in the Ames test, even if the substance has a structural alert; and (iii) a dietary threshold of 10–15 ppb for those substances testing negative in the Ames test *and* having an LD_{50} above 1000 mg/kg.

This study by no means represents new FDA threshold criteria; however, it provides further guidance for the private sector's self-evaluation of food contact materials, and may be considered to provide a suitable scientific basis for taking a self-determined GRAS position for qualifying substances with dietary exposures at levels higher than 0.5 ppb. The study also satisfies FDA regulations that a GRAS determination must be based upon published data. Thus, although the TOR clearance program has been largely superseded by the Food Contact Notification program, the regulation is still on the books and the FDA's analysis supporting the policy remains valid, making it difficult for the FDA to oppose a GRAS determination based on the described *de minimis* exposure criteria.

13.3.6
Housewares

A food contact substance that can be characterized as a "houseware" is exempt from the requirement for premarket clearance from the FDA because of its specific applications and the conditions of its intended use. A houseware typically is considered to be an article that is sold empty (not containing food), and is intended for use by consumers or commercial food service establishments (but not food processing facilities) to prepare, hold, or serve food. Cups, paper towels, and eating utensils for use in homes or restaurants are examples of products that are considered to be "housewares." These products are exempted from FDA premarket clearance because such products generally do not give rise to any public health concern based on their short duration of contact with food and, in some cases, their repeated use over time.

Although the "housewares exemption" was never formally written into law or regulations, the legislative history of the 1958 Food Additives Amendment explicitly indicates that Congress did not intend the FDA to have premarket clearance authority over housewares. The exemption resulted from concerns expressed during the legislative proceedings about the scope of the amendment and its broad definition that would have incorporated all components used in any product intended to hold food. To clarify the status of housewares, Congressman John Bell Williams, Chairman of the House Subcommittee on Health and Science and the floor manager of the bill, specifically stated that the legislation was "not intended to give the Food and Drug Administration authority to regulate the use of components in dinnerware or ordinary eating utensils" [48].

At a Food and Drug Law Institute Conference in 1958, shortly after the passage of the legislation, an FDA panel recognized the housewares exemption. The panel, which received written questions from the floor, responded to the following question as indicated below:

Question: Does the Food Additives Amendment of 1958 require pretesting of containers that are not intended for the commercial packaging of foods, but that are used for

(1) Dispensing?
(2) Preparing or serving?
(3) Temporary one-time use?

Examples are:

(1) Paper cups used in soft-drink or coffee dispensers.
(2) Baby bottles, cooking utensils, refrigerator bowl covers, plastic tableware.
(3) Plastic or paper plates and eating utensils intended for picnic use.

Answer: The amendment was not intended to cover the containers listed as examples as they are used in the home, the restaurant, or beverage dispensers. However, if such a container were used as a package for food being merchandised in a retail market, we think it could very well become subject to the amendment.

No further question was ever raised about the exemption's validity until 1974 when FDA surprisingly proposed a new rule to "revoke" the exemption. FDA obliquely recognized the exemption in the proposal's preamble, saying, "Since the enactment of the Food Additives Amendment . . . letters and oral opinions have at times been issued by [FDA] advising that ordinary houseware articles such as cutting boards, pots and pans, and eating utensils . . . are not subject to regulations under Section 409 of the [FDCA]" [49]. FDA then indicated that it was basing its proposed rule [50] on a US District Court case, *US* versus *Articles of Food Consisting of Pottery Labeled Cathy Rose* (The Cathy Rose Case) [51], which upheld the FDA's authority to seize adulterated food products under the Act. FDA's reliance on this case however, appeared tenuous at best; the Cathy Rose court specifically noted that "ordinary packaging or food holding devices from which there is no migration are not subject to the Act" [52] and did not mention or discuss the housewares exemption at all, let alone indicate judicial rejection of it. Similarly, in the Natick Paperboard case discussed *supra*, the First

Circuit upheld FDA's seizure of allegedly PCB-contaminated paper food packaging material on the basis of potential migration but did not discuss the housewares exemption [53]. In fact, there appears to be no case law discussing the housewares exemption, giving the FDA no specific judicial basis for its revocation. Ultimately, industry vigorously opposed the 1974 proposal and it was never promulgated nevertheless; the fact that FDA made such an attempt is good evidence that the exemption exists.

Indeed, the aberration of the 1974 proposal notwithstanding, the FDA has continued to acknowledge the exemption through publications, letters, and presentations. For example, a May 30, 1979 letter from Dr. Richard C. Kraska of FDA's Petition Control Branch summarized the FDA's stance as follows: "In April, 1974[,] the FDA published a proposal . . . to revoke the housewares exemption. This proposal has not yet been made final, and housewares remain exempt from the food additive regulations." In 1992, Dr. Patricia Schwartz, Regulatory Policy Strategic Manager at FDA's Center for Food Safety and Applied Nutrition, reaffirmed the housewares exemption, citing cookware, dishes, and cutlery as examples, in a speech comparing US and EU regulations on food contact materials [54]. In addition, in its booklet entitled "Requirements of Laws and Regulations Enforced by the US Food and Drug Administration," the FDA states that housewares are not subject to regulation as food additives [55].

FDA is willing to exempt housewares from the premarket clearance requirement because it recognizes that such products generally do not give rise to any public health concern; they undergo repeated use with large volumes of food and a short time of contact. Housewares must still comply with the adulteration provisions of the FDCA, however, and FDA will take action against housewares that may cause a health hazard or otherwise adulterate food. Therefore, houseware manufacturers must still take every reasonable action to ensure that their products are suitably pure for use in contact with food and will not create a health hazard under the intended conditions of use.

13.3.7 Basic Polymer/Resin Doctrine

Substances that fall within the scope of the "basic polymer" or "basic resin" doctrine are exempt from the FDA premarket clearance. As explained in 1966 by Dr. Joseph McLaughlin, then of the Bureau of Science's Division of Toxicological Evaluation, the FDA considers the basic polymer to be the total polymer, as it comes out of the polymerization process, without adjuvant-type ingredients such as plasticizers [56]. The basic polymer includes the cross-linking agents, catalysts, or other necessary substances that are essential components of the polymer and without which it is normally impossible or impractical to make the polymer [57]. Thus, these substances are cleared along with the basic polymer clearance and are exempt from separate preclearance requirements. Despite this exemption, however, the basic resin must still be made with substances of suitable purity for the intended use. If a particular catalyst renders the finished food contact material unsafe or unfit for use, that material would not be suitably pure. Also, the resin produced must comply with any applicable limitations in the regulation under consideration such as specific extraction requirements or physical properties.

The basic polymer or basic resin exemption is retained on the basis that these substances generally are used only in small quantities and either become a part of the resin during polymerization or are removed (e.g., washed) from the resin at the conclusion of polymerization. Therefore, their potential for migrating to food in more than insignificant amounts is virtually nonexistent, and without a reasonable expectation of migration there is no need for premarket clearance as a food additive. What this means for industry is that, as long as a polymer is listed in a regulation, is GRAS or prior-sanctioned, and is manufactured in accordance with good manufacturing practices, the polymer is covered even though different manufacturers may make it by different processes. The doctrine does not apply to substances extraneous to the polymerization reaction, however, meaning that stabilizers, antioxidants, pigments, lubricants, and other adjuvants added after polymerization generally require specific clearance like other food contact substances [58].

13.3.8 Mixture Doctrine

Although it is not an exemption, per se, from the need for a food contact clearance, the "mixture doctrine" provides for some additional flexibility in using cleared food contact substances. In producing food contact materials, manufacturers are permitted to physically blend different substances if all of the substances of interest are permitted for the intended application [59]. Under this "mixture doctrine," such blends require no further FDA clearance provided each substance in the mixture complies with any limitation applicable to that substance in its respective clearance. If any of the combined substances chemically react, as opposed to a physical mixing only, then the result of the reaction is considered to be a new substance requiring its own clearance, and the mixture doctrine does not apply. Any limitations, such as extractive limitations, related to the individual components of a mixture apply to those individual components rather than to the mixture; however, limitations on the end-use application of the mixture, such as food-type or temperature limitations, must be applied to the mixture as a whole. The mixture must meet the most restrictive end-use limitations applicable to any of the components.

13.4 The Food Contact Notification System: An Outgrowth of the Food Additive Regulation System for Food Contact Substances

13.4.1 Introduction

The Food and Drug Administration Modernization Act of 1997 established the FCN program to remove much of FDA's regulatory burden with regard to clearance of food additives [60]. Until this legislation, if a food contact substance was not already regulated as GRAS or prior-sanctioned, was not covered under one of the exemptions previously described, or if the manufacturer simply wished to have evidence of some official FDA acceptance of its product for corporate or customer assurance purposes, a food additive petition was required. Food additive petitions ultimately resulted in the promulgation of Food Additive Regulations for the food contact substances that were the subject of the petitions. These regulations are found in Title 21 of the Code of Federal Regulations, in parts 174–186 (21 C.F.R. Parts 174–186). They are organized in the following way: Part 174 contains general provisions applicable to all food contact materials, Part 175 contains regulations applicable to adhesives and coatings, Part 176 contains regulations applicable to paper and paperboard, Part 177 contains regulations applicable to polymers, Part 178 contains regulations applicable to adjuvants, production aids, and sanitizers, Part 179 contains regulations applicable to the use of irradiation in the production, processing, and handling of food, Part 180 contains regulations that permit the use of substances in food or in contact with food on an interim basis pending additional study, Part 181 contains regulations on certain prior-sanctioned substances (these regulations, however, do not provide an exhaustive list of prior-sanctioned substances), Part 182 contains regulations on certain substances that are listed by FDA as GRAS for direct addition to food, Part 184 contains regulations on certain substances that are affirmed by FDA as GRAS for direct addition to food (substances that are GRAS for direct addition to food are also considered to be GRAS for use in food contact applications), and Part 186 contains regulations on certain food contact substances that are affirmed by FDA as GRAS. (Note that the regulations in Parts 182, 184, and 186 do not provide an exhaustive list of all GRAS substances.) The regulations are generic in nature; that is, any manufacturer may market a substance that complies with a relevant food additive regulation applicable to its intended use, even if that manufacturer is not the party that was responsible for bringing about the regulation.

The food additive petition process, which is still used today for some substances, was very time consuming; FDA required significant amounts of data for the petition, and even if the petition received approval the manufacturer would have to wait for an average of 2–4 years for the agency to promulgate and publish a formal food additive regulation [61]. The FCN program significantly streamlines the petitioning process for food contact substances, although the documentation burden is similar to that of a FAP. The advantage of the FCN process is its speed, as a notification for safety of a substance becomes effective in 120 days by operation of law unless FDA issues formal substantive objections [62]. Another aspect of the FCN, as compared to an FAP, is that FCNs are proprietary in nature; that is, the substance for which the FCN was obtained may be marketed only by the manufacturer identified in the FCN.

It should be clearly understood that the FCN process has *not* removed the right to rely on the existing food additive regulations or any of the enumerated exemptions or exclusions. The FCN program simply adds another clearance option, albeit one that FDA has indicated that it favors for food contact substances. FDA still has the right to insist on the filing of a food additive petition in some situations, but almost all food contact substances are now cleared via the FCN process.

The FCN program allows a manufacturer or supplier of a food contact substance to submit a notification to the FDA regarding the identity and the use of the new substance, along with information supporting the conclusion that the substance is safe for the intended use [63]. The manufacturer identified in the FCN may begin marketing of the new food contact substance, or materials made with it, 120 days after filing the notification unless FDA determines that, based on the information and data submitted in connection with the notification, such use has not been shown to be safe, or the agency instead requires submission of a food additive petition to ensure public safety [64]. Within approximately 30 days after the FCN is filed, if the agency does not have any questions regarding the FCN, FDA will provide the submitter with a letter acknowledging the date of receipt of the submission, thereby setting the effective date (120 days later) of the notification if FDA does not object. The letter will also provide a description of the substance and any applicable limitations on the use of the substance, which the notifier should review closely as it provides an opportunity to ensure that FDA understands the terms of the notification and to request any necessary modifications [65]. If, after review, the FDA objects to the notification, it will issue a letter giving its reasoning and suggesting additional information that would be required to support the safety of the substance for its intended use. If FDA does not object, it will issue a letter and include the substance in an online list of effective notifications [66]. The letter and the web site listing should be sufficient to assure customers of the satisfactory status of the relevant product.

13.4.2
Prenotification Consultations

FDA encourages but does not require manufacturers to consult with the agency prior to submission of an FCN or FAP. This is called a "prenotification consultation," and is particularly advisable when current guidance does not completely apply to a given situation. There are three circumstances under which FDA recommends a prenotification consultation: (i) prior to the submission of an FAP, to ensure that a petition is truly required in lieu of an FCN and that an appropriate level of information is supplied in the petition; (ii) when there are uncertainties about how certain data may be interpreted and those uncertainties are of such magnitude that they may affect the outcome of the overall safety determination; and (iii) when different interpretations of available data would result in different conclusions regarding whether a notification or a petition should be submitted [67].

13.4.3
Requirements for an FCN

The requirements for an FCN are substantially similar to those for a food additive petition [68].

13.4.3.1 Comprehensive Summary

An FCN should include a summary and comprehensive discussion of the basis for the notifier's determination that the use of the FCS is safe. The summary should discuss cumulative dietary exposure to the FCS and any potential impurities, the results of toxicity studies, and any ADI derived from those studies, while addressing all safety data included in the notification. The notifier should discuss any data or information that appears inconsistent with the determination that the use of the FCS is safe.

13.4.3.2 Chemical Identity [69]

An FCN must include complete information regarding the identity and composition of the food contact substance and how it is produced. The substance's chemical name, structure, and molecular formula, including its Chemical Abstract Service (CAS) Registry Number, is required, and analytical data will be needed to identify each substance (e.g., a typical infrared spectra). The manufacturing specifications for the substance, related primarily to identity, purity, and safety, must be provided to show that adequate quality control procedures have been developed.

13.4.3.3 Intended Conditions of Use

FDA requires that FCN submitters describe the conditions under which an additive will contact food and provide data on the quantity of any substance likely to become a component of food under the intended conditions of use. Information on the conditions of intended use should include the concentration of the food contact substance in the final food contact material, temperature of use, types of food contacted, duration of contact, and whether the food contact material is intended for repeated or single-use application. The intended conditions of use requirement is perhaps the single most important factor with regard to the cost and degree of difficulty in bringing about a satisfactory regulatory status for a substance, as it will determine the anticipated exposure, and hence, will dictate the level of toxicology data necessary to establish its safety. In turn, the toxicology tests that are needed will tremendously impact on the cost of bringing a product to market and on the time it will take to make marketing a realistic possibility [70].

13.4.3.4 Intended Technical Effect

Notifiers should present data to show that the FCS will achieve the intended technical effect on the food article, rather than the food, and that the proposed use level is the minimum level required to accomplish the intended technical effect. In the case of a new polymer, notifiers should present data that demonstrate the specific properties of the polymer that make it useful for food contact applications [71].

13.4.3.5 Estimation of Dietary Intake

Another critical aspect of an FCN concerns the amount of the additive expected to enter a person's diet as a result of the intended use of the substance. The FCN must provide enough information for the FDA to determine the safety of the additive, based on a comparison of the estimated daily intake to available toxicity data. The FDA estimates probable exposure to the substance by combining migration data with information regarding typical uses of food contact articles that may contain the substance. From this daily exposure, the estimated daily intake (EDI) is calculated as the product of that concentration and the total food intake (solids and liquids), assumed to be 3 kg per person per day [72]. In addition to the EDI, notifiers must also calculate the cumulative estimated dietary intake (CEDI) of potentially migrating substances. The CEDI is estimated by looking at all sources of exposure, rather than just the EDI from the use that is the subject of its notification [73]. These calculations invariably represent a conservative estimate of the dietary intake of the additive because they are based on the assumption that the additive will always migrate at the maximum levels found in the extraction studies and that all food contact materials of a given type will be made using the subject substance.

13.4.3.6 Toxicity Information [74]

The type of toxicology data required to clear the proposed use of a given substance will depend on the nature of the material and the CEDI for the substance. At all exposure levels, carcinogenic constituents should have an estimate of the potential human cancer risk from the constituent due to the proposed use of the substance. For substances with CEDIs under 0.5 ppb, no toxicology data are required. For exposures between 0.5 and 50 ppb, two genotoxicity studies (a bacterial mutagenicity assay plus an *in vitro* cytogenetic damage or mouse lymphoma assay) are needed to provide an indication as to whether a given substance is likely to be a carcinogen. Where the intake exceeds 50 ppb but is below 1 ppm, a third genotoxicity study (in addition to the two noted above) in the form of an *in vivo* chromosomal aberration study is recommended, as well as two subchronic (90-day) studies (one in a rodent and one in a nonrodent animal). Where cumulative dietary exposure exceeds 1 ppm, FDA typically recommends the filing of a food additive petition instead of an FCN.

13.4.3.7 Environmental Information

The National Environmental Policy Act (NEPA) [75] requires any federal agency to assess the potential environmental impact of actions that it takes. FDA has long considered this requirement to apply to its promulgation of regulations in terms of clearing food additives, and believes that such information also needs to be submitted for notifications. As a result, an FCN must include either an environmental assessment (EA) or a claim for categorical exclusion from the requirement for an

EA [76]. FDA regulations contain various categorical exclusions from the EA requirement including exclusions for substances used only in coatings, substances used only in repeated use food contact applications, and substances that comprise less than 5% by weight of finished food packaging material and that remain in the packaging material through the use by the consumer [77].

13.5 The Food Additive Petition Process

The FCN program has substantially reduced the number of food additive petitions filed with FDA each year. Nevertheless, the FDCA states that the FCN process will be used for food contact substance authorizations except where FDA determines that an FAP is necessary to provide adequate assurance of safety [78]. FDA has published regulations for manufacturers to determine when an FAP may be required: (i) when the use of the food contact material will increase the CEDI of the substance from both food and food contact uses to a level equal to or greater than 1 ppm, assuming the substance is not a biocide; (ii) for a substance that is a biocide if the CEDI is increased to a level greater than 200 ppb; or (iii) when existing data for the substance include one or more bioassays that FDA has not reviewed and are not clearly negative for carcinogenicity [79]. Although FDA may require an FAP in these situations, it does not always do so.

Prior to filing an FAP, FDA recommends a prepetition consultation with the agency to either verify that an FAP is necessary or to ensure that the quality and quantity of information meets the minimum requirements for filing the petition [80]. FDA has published detailed regulations regarding the information similar to an FCN that an FAP should include: (i) the composition, specifications, and method of manufacture for a substance; (ii) its intended conditions of use; (iii) the quantity and identity of substances likely to become components of foods under the intended use conditions; (iv) an estimate of the concentration of the additive in the daily diet; (v) toxicology data demonstrating the safety of this intake level; and (vi) proposed tolerances (if necessary); and (vii) either an EA or a claim for categorical exclusion from the need to provide the EA [81].

Once FDA receives an FAP, it is assigned to a Consumer Safety Officer (CSO) who is the point of contact for all the consulting technical experts and the petitioner. The CSO distributes the relevant parts of the petition to the experts for evaluation, who prepare reports regarding the exposure and safety of the food contact substance. If the FAP is approved, then FDA publishes a final regulation in the *Federal Register* allowing, for food additives, the immediate use of that substance [82]. For a food additive, the period from petition submission to final rule averages about 2 years [83]. The advantage of an FAP is that it results in a food additive regulation, allowing anyone who complies with the conditions of use described in the regulation to use the additive. An FCN, as stated earlier, only allows the manufacturer identified in the FCN to use the product under the intended use conditions.

13.6 Conclusions

We hope that this discussion has provided a useful overview of the laws and regulations that apply to food packaging materials in the United States. It is possible to see the requirements as overwhelming, particularly when considering that this chapter only addresses food contact law, without looking at the direct food additives, drugs (human and animal), medical devices, cosmetics, or labeling issues that the FDCA also regulates. It may help, however, to think of the process as a series of questions. You must first determine whether the substance you want to include in your food packaging material is a food additive. To answer that question, you must determine what the substance is *not*; if it meets one of the exemptions or exceptions detailed above, then the substance is not a food additive and your only concern is whether the food packaging material as a whole is suitably pure. If the substance does not fit within one of the exemptions, then it is a food additive and you must file for premarket clearance with FDA. This premarket clearance can take one of the two forms: it can either be a food contact notification, which will apply for most substances, or a food additive petition, which will apply only under certain circumstances. The crux of the filing is to show that the substance you wish to use is safe for the intended use; that it will not adulterate the food and is suitably pure. If you successfully show that the substance is safe under its intended conditions of use, then the FDA clearance will become effective, and the substance will be permitted for use in the food contact application of interest. When analyzed this way, the FDCA does not seem quite so intimidating, and with the large amount of guidance published by FDA, manufacturers, either independently or with some outside assistance, should be able to work within the law to bring food contact substances to market in the United States.

References

1 37 Stat. 732 (1913). (This is the citation for the official federal session laws, referred to as the *Statutes at Large*, documenting the historical enactment of the Gould Amendment to the Pure Food and Drug Act of 1906, which has now been superseded. The *Statutes at Large* may be found in the Library of Congress in Washington, DC.)

2 Heckman, J. (2005) Food packaging regulation in the United States and the European Union. *Regulatory Toxicology & Pharmacology*, 42 (1), 96, 97.

3 See Federal Food, Drug, and Cosmetic Act (FDCA), §402(a), 21 United States Code (U.S.C.) §342(a). (The U.S.C. is the official federal code in which the federal laws of the United States are published.)

4 FDCA, §409(h)(6), 21 U.S.C. §348(h)(6). There are certain specific exemptions to this definition.

5 Center for Food Safety and Nutrition (CFSAN), FDA, Food Ingredients and Packaging Definitions (September 2007), Available at http://www.cfsan.fda.gov/~dms/opa-def.html (last visited June 25, 2008).

6 FDCA, §301, 21 U.S.C. §331.

7 21 C.F.R. §170.3(i).

8 CFSAN is also responsible for the safety of cosmetics.

9 FDCA, §201(s), 21 U.S.C. §321(s).

10 Hutt, P.B. (1969) Proper Classification of Products Under the Federal Food, Drug, and Cosmetic Act. Address at the Annual Convention of the Federal Bar Association, September 4, Miami, FL.
11 FDCA, §409(a), 21 U.S.C. §348(a). Note, however, that under its Threshold of Regulation rule, discussed in Section II.E., FDA may exempt certain food additives from the need for a regulation.
12 FDCA, §201(s), 21 U.S.C. §321(s).
13 Under this program, a manufacturer or supplier of a food contact material may submit a notification to FDA regarding the identity and use of the new food contact substance, along with information supporting the conclusion that the substance is safe for the intended use. If FDA does not object in writing to the notification within 120 days of its receipt, the submitter may market the product. Unlike food additive regulations, the notification may be relied upon only by the manufacturer and its customers.
14 FDCA, §201(s), 21 U.S.C. §321(s).
15 FDCA, §201(s)(4), 21 U.S.C. §321(s)(4).
16 21 C.F.R. §170.39.
17 FDCA, §402(a)(2)(C), 21 U.S.C. §342(a)(2)(C).
18 FDCA, §402(a), 21 U.S.C. §342(a), also known as the General Safety Clause.
19 21 C.F.R. §174.5(a)(2). Along the same lines, Title 21 of the Code of Federal Regulations, Section 110.80 (21 C.F.R. §110.80), containing FDA's good manufacturing practice regulations applicable to the manufacturing, packing, and holding of food products, states that appropriate measures should be taken to ensure that food is "suitable for human consumption" and that all materials for food containers are "safe and suitable" for their intended use.
20 FDCA, §409(c)(3)(A), 21 U.S.C. §348(c)(3)(A).
21 See 47 Federal Register (Fed. Reg.) 14,464 (April 2, 1982). (The *Federal Register* publishes federal rules and regulations before they are entered in the Code of Federal Regulations (C.F.R.).)
22 Ibid. at 14,466.
23 Ibid. at 14,466.
24 Ibid. at 14,468. In a November 26, 2004 *Federal Register* notice (69 Fed. Reg. 68831, 68836), FDA withdrew this advance notice of proposed rulemaking (ANPR) but stated that, "withdrawal of a proposal is not intended to affect whatever utility the preamble statements may currently have as indications of FDA's position on a matter at the time the proposal was published," and further that, "in some cases the preambles of these proposals may still reflect the current position of FDA on the matter addressed." The constituents policy as outlined in this preamble reflects current FDA thinking on the matter. Thus, despite the Agency's withdrawal of the ANPR, the constituents policy outlined in the April 2, 1982 *Federal Register* notice remains a valid policy by which to evaluate minor carcinogenic constituents of food additives.
25 This theoretical "1 in 1 million" risk level means that the possible risk to humans exposed to this dose of the chemical each day for a 70 year lifetime is the occurrence of one additional tumor in a population of 1 million. The actual increased incidence of cancer, however, is likely to be much less.
26 FDCA, §201(s), 21 U.S.C. §321(s).
27 Heckman, supra note 2, at 102.
28 *Monsanto vs. Kennedy*, 613 F.2d 947 (D.C. Cir. 1979). (This is the citation to the publication in the *Federal Reporter* of the *Monsanto* vs. *Kennedy* case, which was argued before the Court of Appeals for the District of Columbia.)
29 Ibid. at 955.
30 Memorandum from Peter Barton Hutt, General Counsel, FDA, to Sam D. Fine (October 31, 1974) (on file with author).
31 See, for example, letter from Frederick A. Cassidy, FDA, to Jerome Heckman (June 9, 1965) (on file with author).
32 *Natick Paperboard* vs. *Weinberger*, 525 F.2d 1103 (1st Cir. 1975), cert. denied, 429 U.S. 819 (1976).
33 Ibid. at 1107-08.
34 FDCA, §201(s)(4), 21 U.S.C. §321(s)(4). FDA regulations issued after the 1958 Amendment include lists of prior sanctioned substances, see 21 C.F.R.

§181.22, *et seq.*, although these listings are not exhaustive.

35 21 C.F.R. §181.1(b).

36 FDCA, §201(s), 21 U.S.C. §321(s).

37 21 C.F.R. §170.30(b).

38 See *Regulatory Toxicology & Pharmacology* (August 1990) containing the full report and the individual papers compiled as part of the CCT study; Munro, "Safety Assessment Procedures for Indirect Food Additives: An Overview," *Regulatory Toxicology & Pharmacology* 2 (August 1990).

39 62 Fed. Reg. 18937 (April 17, 1997).

40 Paulette Gaynor and Sebastian Cianci, "Regulatory Report: How FDA's GRAS Notification Program Works," *Food Safety Magazine* (December 2005–January 2006), available at www.fda.gov/food/FoodIngredients Packaging/GenerallyRecognizedasSafeGRAS/ucm083022.htm (last visited January 30, 2010).

41 Available at www.accessdata.fda.gov/scripts/fcn/funNavigation.cfm?rpt=grasListing (last visited January 30, 2010).

42 See 60 Fed. Reg. 36581, 36582 (July 17, 1995).

43 21 C.F.R. §170.39 (a) (2) (i).

44 See Cheeseman, M.A. *et al.* (1999) A tiered approach to threshold of regulation. *Food & Chemical Toxicology*, 37, 387.

45 Gold, L.S. and Zeiger, E. (eds) (1997) *Handbook of Carcinogenic Potency and Genotoxicity Databases*, CRC Press, Boca Raton, FL.

46 More specifically, the authors found that most compounds that are likely to be relatively potent carcinogens, based on structure, can be grouped into one of the following classes of substances: *n*-nitroso compounds, strained heteronuclear rings, alpha-nitro-furans, polycyclic amines, hydrazines/triazenes/azides/azoxy compounds, organophosphorous compounds, and heavy metal-containing compounds. The authors also indicated that the broad class of substances referred to in the paper as "endocrine disruptors" also includes many substances that are identified with carcinogenicity through a mechanism involving hormone modification.

47 The LD_{50} is defined as the dose that induces death in 50% of dosed animals.

48 104 *Congressional Record* (Cong. Rec.) 17,418 (August 13, 1958) (the *Congressional Record* publishes debates held in the United States Congress).

49 39 Fed. Reg. 13285 (April 12, 1974).

50 FDA withdrew this proposed rule as part of its November 26, 2004 purge, as discussed in footnote 21, supra.

51 370 Federal Supplement (F. Supp.) 371 (1974).

52 Ibid. at 373.

53 *Natick Paperboard Corp.* vs. *Weinberger*, 525 F.2d 1103 (1975).

54 P. Schwartz, "Regulation of Food Packaging in the EEC and the U.S.: A Comparative Analysis" (1992). In this speech, Dr. Schwartz stated: *FDA has traditionally not regulated "housewares" as food additives. Articles like cookware, dishes, and cutlery used exclusively in the home for food preparation and storage have not been subject to the pre-market safety evaluation to which commercial food packaging is subject. This does not mean that these "housewares" are not regulated at all by FDA. The Agency can still take action with respect to a "houseware" under the General Safety Clause (Sec. 402).*

55 Department of Health and Human Services, FDA (1989) *Requirements of Laws and Regulations Enforced by the U.S. Food and Drug Administration*, DHHS Publication No. (FDA) 89-1115, DHHS.

56 See Letter from Jerome H. Heckman, Keller & Heckman, to SPI Food Packaging Materials Committee (December 28, 1966) (on file with author).

57 Ibid.

58 See Heckman, supra note 2, at 105.

59 Letter from William F. Randolph, FDA (August 23, 1963); Address by W.F. Randolph, FDA, The Regulatory Control of Plastic Food-Packaging Materials in the United States, presented at the World Health Organization 5th Annual Seminar on Food and Drug Control for Central America and Panama, Managua, Nicaragua (May 1969); E.B. Detwiler, FDA, Synthetic Polymers and the Federal Food, Drug, and Cosmetic Act, *SPE Journal* 61 (January 1965).

60 Food and Drug Administration Modernization Act of 1997, Public Law Number (Pub. L. No.) 105-115, §309, 111 Stat. 2296, 2354-56.

61 FDCA, §409(b), 21 U.S.C. §348(b).

62 FDCA, §409(h)(2)(A), 21 U.S.C. §348(h)(2)(A).

63 FDCA, §409(h)(1), 21 U.S.C. §348(h)(1).

64 FDCA, §409(h)(2)(A)-(3)(A), 21 U.S.C. §348(h)(2)(A)-(3)(A).

65 CFSAN, FDA, Guidance for Industry, Preparation of Food-contact Notifications: Administrative (May 2002), available at www.fda.gov/Food/foodIngredients Packaging/default.htm > Guidance and Reference Information for Petitions and Notifications (last visited January 30, 2010).

66 Available at www.accessdata.fda.gov/scripts/fcn/fcnNavigation.cfm?rpt=fcs Listing (last visited January 30, 2010).

67 See CFSAN, supra note 66.

68 21 C.F.R. §170.101.

69 See CFSAN, FDA, Guidance for Industry, Preparation of Food-contact Notifications and Food Additive Petitions for Food-contact Substances: *Chemistry Recommendations* (December 2007), Available at www.fda.gov/food/Food IngredientsPackaging/default.htm > Guidance and Reference Information for Petitions and Notifications (last visited January 30, 2010), for more detailed information regarding requirements for the sections on Chemical Identity, Intended Conditions of Use, Intended Technical Effect, and Estimation of Intake.

70 Heckman, J. and Ziffer, D. (2001) Fathoming Food Packaging Regulation Revisited (March), www.packaginglaw.com/1804_.shtml.

71 See CFSAN, supra note 70.

72 Ibid.

73 Ibid.

74 See CFSAN, FDA, Guidance for Industry Preparation of Food-contact Notifications for Food-contact Substances: *Toxicology Recommendations* (April 2002), Available at www.fda.gov/Food/food IngredientsPackaging/default.htm > Guidance and Reference Information for Petitions and Notifications (last visited January 30, 2010).

75 42 U.S.C. §4321*et seq.*

76 CFSAN, supra note 66.

77 21 C.F.R. §25.32.

78 FDCA, §409(h)(3)(A), 21 U.S.C. §348(h)(3)(A).

79 21 C.F.R. §170.100(c).

80 See CFSAN, FDA, Guidance for Industry, Pre-petition Consultations for Food Additives and Color Additives (April 2005), available at www.fda.gov/Food/foodIngredientsPackaging/default.htm > Guidance and Reference Information for Petitions and Notifications (last visited January 30, 2010).

81 21 C.F.R. §171.1(c).

82 For color additives, the final rule becomes effective 31 days after the final rule is published.

83 See CFSAN, FDA, Guidance for Industry: Questions and Answers About the Petition Process (April 2006), available at www.fda.gov/Food/foodIngredients Packaging/default.htm > Guidance and Reference Information for Petitions and Notifications (last visited January 30, 2010).

14
Food Packaging Law in Canada – DRAFT

Anastase Rulibikiye and Catherine R. Nielsen

14.1
Introduction

Food protection laws in Canada had their genesis more than 130 years ago, with the passage of the Inland Revenue Act of 1875. Cited as "an act to impose licence duties on compounders of spirits; to amend the act respecting inland revenue; and to prevent the adulteration of food, drink, and drugs [1]," the Inland Revenue Act of 1875 was the Canadian Parliament's first attempt to regulate the safety of food and to provide a definition of adulterated food or drink and, indeed, of food and drink themselves. In 1878, 2 years after Inland Revenue analysts began to test food samples, 50.6% of food samples tested were deemed adulterated [2]. The law, which relied originally on just eight analysts appointed by the Commission of Inland Revenue to test foods, quickly had a deterrent effect; by 1883, 24.2% of samples tested were characterized as adulterated. However, enforcement suffered under the Inland Revenue Act due to a lack of definition and identified standards of quality [3]. Parliament remedied this problem in April 1884, by passing the Adulteration Act, "an act to amend and to consolidate as amended the several acts respecting the adulteration of food and drugs." The Adulteration Act defined "food" and "drugs," and paved the way for standards for individual foods, the first of which was issued in 1894 (prepared tea) [4].

The next major advancement in food protection in Canada came in 1919, when the responsibility for enforcement of the Adulteration Act passed to the Food and Drugs Division of the newly established Department of Health. Another change followed in 1920 when the Adulteration Act was amended and consolidated to become the Food and Drugs Act [5]. The new law not only addressed issues of adulteration in food but also, for the first time, addressed problems stemming from food packaging [6]. Since then, the Canadian government has significantly amended the Food and Drugs Act only twice, to incorporate cosmetics and medical devices. Today, the Food and Drugs Act stands much as it did in 1920, a law intended to protect Canadian consumers from hazards and frauds in the sale and use of food, drugs, cosmetics, and medical devices.

Global Legislation for Food Packaging Materials. Edited by Rinus Rijk and Rob Veraart

ISBN: 978-3-527-31912-1

14.2
The Legal Structure

14.2.1
The Agencies Involved

The regulation of food safety in Canada is divided among a variety of government actors. The regulation of the safety of food packaging, however, is managed by two federal agencies: Health Canada and the Canadian Food Inspection Agency (CFIA). The CFIA is a recently established agency for the purpose of consolidating food inspection-related resources at the federal level. The division of jurisdiction between the two agencies is quite complex; in simplest terms, Health Canada, which reports to the Minister of Health, is responsible for establishing policies and standards on the safety and nutritional quality of the food supply, while the CFIA, which reports to the Minister of Agriculture and Agri-Food, is responsible for enforcing these policies and standards and for food inspection activities.

Health Canada, established in 1993, evolved from the old Department of Health and Welfare. Within Health Canada, it is the Health Products and Food Branch (HPFB) that is responsible for establishing food policies and standards. The Food Directorate of HPFB is responsible, in cooperation with other agencies, for conducting scientific research, developing policies, standards, and guidelines, evaluating submissions from the food industry, and providing food and diet information to the Canadian public. As part of one of its key activities to evaluate submissions from industry, the Food Directorate reviews the safety of finished packaging materials and their ingredients. The Directorate also, under the authority of the Canadian Food Inspection Agency Act [7], assesses the effectiveness of the CFIA's food safety activities.

The CFIA was established in 1997 by the Canadian Food Inspection Agency Act [8], and is responsible for the administration and enforcement of several laws, including the Canada Agricultural Products Act [9], Meat Inspection Act [10], and Fish Inspection Act [11]. CFIA has the authority to inspect both CFIA-registered and non-CFIA-registered food processing facilities. The agency's Chemical Evaluation Program deals with the safe use of food packaging materials in federally registered food establishments and evaluates submissions for CFIA acceptance of these products for the use by food establishments. CFIA evaluation (which relies heavily on HPFB reviewers' input) entails a close examination of the product as a whole for its integrity to ensure that using the product will not present a food safety concern.

14.2.2
The Laws Involved

The Food and Drugs Act prohibits the sale of an article of food that (i) has in or upon it any poisonous or harmful substance; (ii) is adulterated; or (iii) is manufactured, prepared, packaged, or stored under unsanitary conditions [12]. "Unsanitary conditions" include circumstances that may render the article injurious to health [13]. The Food and Drugs Act further prohibits the sale of food that is not labeled or packaged as required by the regulations [14]. The mandates of the Food and Drugs Act are carried out, as to food packaging, by the Food and Drug Regulations, Division 23 of Part B, entitled "Food Packaging Materials" [15]. Division B.23.001 prohibits the sale of any food in a package that may yield to its contents any substance that may be injurious to the health of a person consuming the food [16], where "package" is defined as "any thing in which any food, drug, cosmetic, or device is wholly or partly contained, placed, or packed [17]." The term is read to include any article that a food contacts during processing or distribution for sale, including the food contact portions of food processing equipment, bulk food handling articles, and transportation vehicles [18].

Under the Meat Inspection Act, meat and poultry processing facilities must be registered with CFIA if the facility intends to sell meat and poultry products internationally, interprovincially, or to another CFIA-registered establishment or if the facility applies for a Canadian grade mark to the applicable meat or poultry product [19]. (The importance of this facility registration requirement, as to food packaging used therein, will be made clear later on.) Note that certain meat and poultry producers, however, are exempt from CFIA's registration requirements. Under Part I.4(1)(l) of the Meat Products Regulations [20], food products containing no more than 2% meat or poultry, calculated on the basis of the cooked weight of the food product, are exempt from the registration requirements of Sections 7, 8, and 9 of the Meat Inspection Act.

Under the Canada Agricultural Products Act, dairies are required to register with the CFIA if they intend to sell dairy products internationally, interprovincially, or to another registered establishment or if the facility applies for a Canadian grade mark applicable to the dairy product [21]. Dairy products are defined in Section 2 of the Dairy Products Regulations as "milk or a product thereof, whether alone or combined with another agricultural product, that contains no oil or fat other than that of milk." Other foodstuffs that have registration requirements under the Agricultural Products Act include eggs [22], fresh fruits and vegetables [23], honey [24], maple products (any product obtained exclusively by the concentration of maple sap or syrup) [25], processed egg (frozen egg, frozen egg mix, liquid egg, liquid egg mix, dried egg, dried egg mix, and egg product) [26], and grains.

Under the Fish Inspection Act, establishments that intend to sell fish internationally or interprovincially are required to register with the CFIA [27]. Fish is defined as "any fish, including shellfish and crustaceans, and marine animals, and any parts, products or by-products thereof [28]."

14.3 The Regulatory Scheme

14.3.1 Mandatory Requirements: Federally Registered Establishments

Finished articles, including equipment [29] and food packaging, which are intended for use in federally registered establishments (i.e., facilities that produce meat, poultry, seafood, dairy products, eggs, fresh and processed fruits and vegetables, grains, honey, and maple syrup), as opposed to products intended for the retail market, must be registered ("accepted") by the CFIA prior to use. This requirement is legislatively mandated. For example, Section 92(2)(b) of the Meat Inspection Regulations states that "No material used in packaging or labeling a meat product in a registered establishment shall come into contact with the meat product unless this material . . . (b) is suitable for the purpose for which it is to be used and is registered by the Director in a register kept for that purpose." The CFIA evaluates these articles as a whole to determine that their use will not present a food safety concern. Applications for registration/acceptance should include (i) a "no objection" letter from the HPFB [30]; (ii) a short description of the product's composition; (iii) any information regarding cleanability; and (iv) a sample of the material or schematic drawings [31]. Thus, while the HPFB maintains authority over food contact articles generally, including packaging, the CFIA inspects and regulates materials that are used in processing and *packaging* the products listed above, that is, meat, poultry, seafood, dairy products, eggs, fresh and processed fruits and vegetables, grains, honey, and maple syrup.

Thus, if the finished article is deemed safe, the CFIA issues a letter of acceptance to the applicant. These letters of acceptance have no expiration date. However, if the product's formulation or intended use changes, then the product must be reevaluated to regain acceptance status. The CFIA does not maintain a list of individual components or ingredients that are permitted to be present in the product, but does maintain a database of finished articles that have been accepted [32].

14.3.2 Not so Voluntary Requirements

As already mentioned, the Food and Drugs Act and concomitant Food and Drug Regulations are administered by the Health Products and Food Branch, and Division 23 of the regulations deals specifically with food packaging materials. Again, a package is defined under the Act to include "any thing in which any food, cosmetic, or device is wholly or partly contained, placed or packed," and it is interpreted broadly to include both food packaging materials for use in retail sales of food products and articles that come into contact with food during food processing and distribution [33]. However, the definition is not intended to extend to consumer products, such as kitchen utensils and household wrap, because Division 23 covers only packaging materials that are related to the sale of foods [34].

The Food and Drug Regulations specifically exclude food contact materials from the definition of a food additive [35], meaning that food contact materials are *not* generally subject to the regulatory premarket clearance requirements for food additives under section B16.002; again, though, registration and review is mandated for those food packaging materials used in CFIA federally inspected plants and for materials used to package infant formulas. The Article B25.046. (2) (j) of Food and Drugs Act requires the manufacturer of infant formula to include in its notification the description of the type of packaging to be used.

These registration and review processes are not legally mandated for food packages that are not subject to CFIA jurisdiction, that is, packaging for foods subject to HPFB jurisdiction.

However, broadly, Section B.23.001 of the Food and Drug Regulations states that "no person shall sell any food in a package that may yield to its contents any substance that may be injurious to the health of a consumer of the food." This language places the legal burden on the food seller to ensure the safety of any packaging material used in the sale of its food products. Because food may theoretically be contaminated by a food contact material, food manufacturers often seek assurance from their suppliers regarding the suitability of food contact materials for their intended use, regardless of whether food is a CFIA-regulated product or not. This assurance generally comes in the form of a "no objection" letter (NOL) regarding the food contact material from the HPFB. These letters can be obtained for any type of material, whether a finished product, a formulated product, or a single additive, and assure manufacturers that the products have been evaluated by the HPFB and deemed acceptable, from a chemical safety standpoint, for use in specified food packaging applications [36]. Ultimately, while no statutory mandates obligate a company to obtain an NOL for foods subject to HPFB's jurisdiction (as opposed to CFIA's), practically speaking many suppliers make it a practice to submit product formulations to the HPFB for evaluation. The assurances have become *de facto* necessary from a marketing and public relations standpoint.

Thus, the vast majority of food packaging materials are assessed for safety by the HPFB on a voluntary basis. Division 23 of the Food and Drug Regulations, which mostly sets general safety requirements, does explicitly ban or limit packages made of certain materials. For example, polyvinyl chloride containing any octyltin chemical in excess of certain specified levels is explicitly banned from food packaging [37]. Prohibitions also exist against packages that may yield (i.e., the substance could migrate from the packaging to the food contained inside it) *any* amount of vinyl chloride [38] and acrylonitrile [39].

14.3.2.1
The NOL Process

Premarket clearances for food packaging materials intended for the retail market may be obtained from the Food Directorate of the HPFB. The HPFB evaluates many types of products, ranging from finished articles, such as a coated films, plastic bottles, or cans, to single additives, such as antioxidants, colorants, or ultraviolet absorbers. The HPFB issues the clearance in the form of an NOL, which can be used to assure prospective customers of the product's safety, although it must be remembered that NOLs are not legal approvals and do not relieve sellers of their legal obligations of ensuring the safety and suitability of their packaging for holding food. As a practical matter, the intent of the letter is to make it highly unlikely that the specified, intended use of the product would lead to a regulatory violation [40].

As said before, the HPFB does not maintain a positive list for products with an NOL, but it recognizes various polymer resins as equivalent and interchangeable, thereby precluding the need for an NOL. The Food Directorate maintains a list of polymers for which NOLs have already been issued for food contact applications [41]. The purpose of the list is to allow the exchange of one polymer resin for another that is the subject of an NOL, as long as that is the only change made to a food packaging article, and the article already has received an NOL from the HPFB for food contact uses. However, it is necessary to advise HPFB of the change. Further, the agency can challenge the equivalency determination made by the manufacturer.

Normal Submission Requirements

The type and complexity of the information to be submitted to HPFB for evaluation of an NOL request absolutely depends on the type and complexity of the product under consideration. These can be broken down into two categories: (i) formulated products and finished articles, and (ii) specific constituents or single additives [42]. The former are usually submitted for the evaluation by converters, while the latter are usually submitted by raw material suppliers.

The first two elements of information required for the evaluation of a material – whether a single substance or a formulation – are the same: the identity and the proposed usage. However, the underlying information submitted in support of these elements differs depending on the type of product. For a formulated product, the identity serves to evaluate the chemical components that may be potentially extractable by foods contacting the material. The data to be submitted are the trade names, CAS numbers, structures, compositions (in the form of a quantitative listing of all components in which each one is identified by proper chemical name or trade name, grade, and supplier), specifications, and chemical and physical properties relative to the proposed use. For a single additive, the product identification consists of the chemical name, chemical formula (both empirical and structural), molecular weight, manufacturing process (including a detailed description and schematic diagram), purity specifications (such as the residual reactants or by-products), and chemical and physical properties.

The purpose of submitting information regarding the proposed use of the product is to permit the development of an estimated dietary intake of the particular packaging components involved and, where necessary, to establish appropriate test protocols. For a formulated product, the data submitted should include the form of the finished product (e.g., bottle, film, or casing, etc.), dimensions (volume, wall thickness, etc.), packaging ratio (the weight of food per unit area of packaging material in grams per square inch), conditions to which the article will be exposed during packaging, distribution, and use by consumers, and an estimate of projected market penetration. The product identity and proposed usage are the only required elements for formulated products and finished articles, although Material Safety Data Sheets, Technical Data Sheets, product literature, and samples may aid evaluation. For single additives, the proposed usage data requirements are the intended technical effect or purpose (e.g., antioxidant, stabilizer, etc.), types of substrates or polymers to which it will be added, maximum use level in each type of polymer, efficacy data to demonstrate that the additive will do what it is intended to do at the proposed use levels, types of foods involved, and conditions of use to which the packaging material will be exposed to during processing, transport, or consumer handling.

Single additive products require two additional elements of information for NOL submissions: (i) migration data and (ii) toxicity data. Migration data is developed through extraction studies and is used to identify and quantify the constituents that are likely to be extracted by foods and thus be potential migrants to food. The studies are usually conducted using food simulants under conditions that reflect, as close as possible, those of the proposed end-use applications. The HPFB evaluates this data in conjunction with the usage information to estimate the probable daily intake of extracted constituents in the average diet of the consumer. The final report submitted to the HPFB must include the parameters used in the studies, such as volumes of food simulants, surface area exposed, time, and temperature of testing, as well as full details of the analytical methodology used to measure the migration levels, including validation data and an estimate of the analytical method's limit of detection. In the absence of actual migration data, mathematical modeling or assumption of 100% migration could be used.

The amount and type of toxicity data that must be submitted to establish the safety of a single additive is based on the level of estimated dietary exposure. The toxicological data set, depending on the extent, can be used to establish a Tolerable Daily Intake for the migrating constituent in the average diet of a consumer. Theoretically, toxicological data is not required for constituents that have a dietary exposure less than 0.025 μg/kg bw, which is considered as the threshold of toxicological concern (TTC). The concept of defining a dose or dietary intake of a chemical where there is limited to no concern of potential toxic effects is based on extensive reviews of existing toxicology data for a variety of chemical structural classes, including carcinogens with identified structural alerts. A more detailed historical review of the TTC concept is provided in Kroes *et al.* [57]. While in theory, no toxicological data are required for those chemicals where exposure from food packaging uses is less than 0.025 μg/kg bw/day, an assessment of the chemical's carcinogenic potential is usually made, regardless of the exposure estimate. If the exposure level is above this threshold, then the toxicological data required to establish the safety of the constituent is determined on the basis of four levels of concern. Concern Level 1, or "very low," is between 0.025 and 0.1 μg/kg bw and requires data for structure–activity in accordance with the categories listed in the US Food and Drug Administration Red Book, or other methods. Concern Level 2, or "low," is between 0.1 and 2.5 μg/kg bw and requires the submission of both the Concern Level 1 information and two short-term genotoxicity studies (usually an Ames test and a chromosomal aberration assay, either *in vitro* or *in vivo*, in mammalian cells) and a 28-day feeding study in rodents. Concern Level 3, or "medium," is between 2.5 and 25 μg/kg bw and requires Concern Level 1 data, the two genotoxicity studies from Concern Level 2, plus a 90-day feeding study (instead of the 28-day study), a multigeneration study, and a teratology study, all in rodents. Finally, Concern Level 4, or "high," which is for a dietary exposure greater than 25 μg/kg bw, requires all the studies listed for Concern Level 3, plus a 1-year feeding study in nonrodents and a chronic toxicity/oncogenicity study in two rodent species. In general, these data requirements should be viewed as suggested guidelines or minimal requirements, with toxicological interpretation ultimately being used to establish safety or risk.

Recycled Products

With the heightened environmental consciousness of informed consumers, the use of recycled products in food contact has grown as well. Recycled plastic food contact products are required to comply with the same laws and regulations as virgin products, in that the packaging must not impart harmful substances to the contents. Owing to the permeable nature of plastics, it is possible for chemical contaminants resulting from consumer misuse or abuse to remain in the recycled product and thus migrate to food. Concerns also arise about structural integrity and microbial contamination of recycled food packaging, although these issues are not as significant as they could be because manufacturers of recycled plastic food packaging materials must test the physical properties of the product in the same manner as for virgin materials. Furthermore, microbial organisms are usually eliminated as a result of the high temperatures, cleaning agents, and sanitizers used in the recycling process. Health Canada has issued more specific guidance on the use of recycled plastic and additional data that manufacturers should submit to obtain an NOL [43].

Manufacturers of recycled plastic materials who wish to obtain an NOL from HPFB to market their products must demonstrate that the product complies with Division 23 of the Food and Drug Regulations. This entails demonstrating that any contaminants are removed, neutralized, or reduced to safe levels by the recycling process. Because of the difficulty of accurately characterizing the chemical composition of a recycled material, compared to virgin material, the HPFB has other special information submission requirements. First, the submission letter should identify the material by its manufacturer, trade name, and code number, clearly indicate the types of food to be packaged in the container and the conditions of its intended use, describe the structure and dimensions of the product, and describe the chemical composition of the product by listing all the ingredients by chemical name, trade name, and supplier. Next, the manufacturer should describe the recycling process from source collection to final fabrication, and describe the quality control program set up to eliminate or neutralize chemical and microbial contaminants. Finally, the manufacturer may have to demonstrate the efficacy of the recycling process at removing or reducing chemical contaminants, to ensure that the resulting packaging will not adulterate food.

The Health Canada guidance presents several considerations to assist manufacturers with developing an efficient recycling process that will ensure the resulting packaging does not pose a risk of chemical contamination of food:

1) **Use of a Functional Barrier:** If a manufacturer chooses to use a functional barrier, made of virgin material or aluminum, for example, between the recycled plastic and the package contents, then in most situations the HPFB will not be concerned about migration from the recycled material. The effectiveness of the barrier depends on the chemical nature and thickness of the barrier, as well as the conditions of use of the package. Under more severe conditions of use, the effectiveness of the barrier to prevent migration should be supported with scientific data.

2) **Source Control:** Plastic recyclers should develop a comprehensive source control program. This could consist of measures to ensure that the sources of the recycled materials are recorded and documented, to limit the source of collection to food contact plastics, to promote "food contact use only" collection sites, and initial sorting to detect and reject containers that may contain potentially hazardous or toxic chemicals.
3) **Use Limitations:** Manufacturers and consumers can limit potential migration by restricting the types of food contacting the package, limiting the use conditions of the package to reduce the likelihood of regulation (e.g., restricting the package to room-temperature exposures or below, where migration is less likely to occur) or restricting the use of the recycled material to food packaging applications where there is no or unlikely possibility of migration of contaminants to food, such as in secondary plastic wrapping for juice boxes or in egg cartons.
4) **Process Efficacy:** This refers to the ability of the recycling process to remove contaminants from the recycled materials. Recyclers can show the efficacy of the recycling process at removing potential contaminants through a protocol that exposes the plastic packaging (either in container form or as flaked or pelletized resin) to selected surrogate contaminants, and then subjects the material to the recycling process. A subsequent analysis of the resulting material should show whether the surrogate contaminants are removed by the recycling process. The surrogate contaminants should represent commercial contaminants available to consumers, such as automotive fuels and oils, solvents, pesticides, household cleaners, and so on. If the recycling process involves the chemical depolymerization of the source plastic, followed by regeneration and purification of the resulting monomers, the contaminant-removed efficacy of the process can be shown by "spiking" the material with known levels of surrogate contaminants, subjecting the material to the depolymerization process, and analyzing the resulting monomers for residual contaminants.

14.3.2.2 The Result

An opinion letter from the HPFB typically identifies the product of interest and states whether the agency has any objection to the use of the material as proposed. A proviso generally is included in the letter indicating that the material must be technically suitable for the intended end use. As already noted, HPFB letters are not legally binding; therefore, an HPFB objection to a particular product does not literally preclude the marketing of that product in Canada. Practically speaking, however, most food manufacturers will respect the opinion of the HPFB and decline to use a material that is the subject of an unfavorable opinion, particularly where that opinion might be used in a court of law as expert support in an adulteration case.

Unlike the CFIA, the HPFB does not publish a "positive list" of materials that are considered to be safe for use in food packaging applications or that have been the subject of a favorable safety review. Products are reviewed individually and on a case-by-case basis, except that some suppliers, such as plastic resin manufacturers, have established master listings with the HPFB to cover products that may be used in multiple food packaging products [44]. An NOL does not expire and is considered valid as long as the composition and intended use of the product remain as described in the original submission. If these specifications do change, a new request should be submitted to the HPFB. In some cases, this request can be in the form of a simple letter to the authorities identifying the change [45].

14.4 Enforcement

All enforcement activities are carried out by the CFIA. The Food and Drugs Act authorizes inspectors to examine and sample products at any establishment manufacturing, preparing, preserving, packaging, or storing articles regulated under the Act [46]. The inspector also has the power to seize and detain any article that the inspector reasonably believes violates the Act or its regulations [47]. Articles that are found to violate the Act or its regulations may be subject to forfeiture and destruction under three circumstances: (i) if the person who possessed the article at the time of seizure consents; (ii) if the person who caused the violation is convicted; or (iii) if a judge of a superior court orders so upon application of an inspector [48]. Other acts enforced by the CFIA both provide inspection authority and lay down procedures, for regaining possession of seized articles [49].

14.5 Conclusions

The Canadian food packaging regulatory system is premised on high safety standards and is similar to the US system in the data requirements and details required in support of an NOL. The crucial difference – the food packaging constituents that migrate are not considered "food additives" that legally mandate preclearance by the HPFB – results in a somewhat simplified regulatory analysis. With the exception of CFIA-registered foods and their packaging, legally, a manufacturer of food packaging may market its products in Canada provided the product is determined by the manufacturer to be safe and suitable. But, as suggested above, though not legally required to do so, most manufactures prefer the added assurance and comfort an HPFB NOL provides to them in their marketing efforts.

In the future, manufacturers may be subject to additional regulatory requirements and stricter enforcement standards. On April 8, 2008, Bill C-51, "An Act to Amend the Food and Drugs Act and to make consequential amendments to other Acts [50]," was tabled in the Canadian House of Commons. These amendments would introduce additional regulatory requirements, including a prohibition on the importation of food packaged under unsanitary conditions [51], new inspection and enforcement powers [52], license and registration requirements for food importers and interprovincial traders [53], documentation requirements [54], and increased penalties [55], and would change the rules of disclosure for confidential business information [56].

While Bill C-51 did not become law before the 39th Parliament ended on 7 September 2008, its introduction presages heightened scruting by the Canadian government of the food industry and its suppliers. While the bill did not pass in 2008, its more introduction suggests that the regulatory status of food packaging law and regulations in Canada must be followed closely for the foreseeable future.

References

1 Pelletier, M.A. Food Packaging Regulations in Canada 1995 (on file with author).
2 The Role of Food and Drug Regulation in the Evolution of Health Canada, 1, http://www.hc-sc.gc.ca/dhp-mps/alt_formats/hpfb-dgpsa/pdf/homologation-licensing/evolution-fra.pdf (last visited 6 August 2008) [hereinafter *Evolution*].
3 Ibid.
4 Gnirss, G. (2008) A History of Food Law in Canada, Food in Canada, 38.
5 Evolution, *supra* note 2, at 5.
6 Gnirss, supra note 4, at 1.
7 Canadian Food Inspection Agency Act, R.S.C. c. 6 (1997) (Can).
8 CFIA Act, 1997, 2d Sess., 35th Parliament, 45-46 Eliz. II.
9 Canada Agricultural Products Act, R.S.C., c. 20 (4th Supp.) (1985) (Can).
10 Meat Inspection Act, R.S.C., c. 25 (1st Supp.) (1985) (Can).
11 Fish Inspection Act, R.S.C., c. F-12 (1985) (Can).
12 Food and Drugs Act, R.S.C, c. F-27, §4 (2001) (Can).
13 Ibid. at §2.
14 Ibid. at §5(2).
15 Ibid. at §30.
16 Food and Drug Regulations, C.R.C. §B.23.001 (2001) (Can).
17 Food and Drugs Act at §2.
18 See Pelletier, M. (1995) Health Protection Branch, "Food Packaging Regulations in Canada" presented to ICI/PIRA (June). On the other hand, consumer products, such as kitchen utensils, household wrap, and cookware for use in the home are not included. Ibid.
19 Meat Inspection Act at §3(1).
20 Meat Inspection Regulations, SOR/1990-288 (Can).
21 Canada Agricultural Products Act at §13.
22 Egg Regulations, C.R.C., c. 284 (2008) (Can).
23 Fresh Fruit and Vegetable Regulations, C.R.C., c. 285 (2008) (Can).
24 Honey Regulations, C.R.C., c. 287 (2008) (Can).
25 Maple Product Regulations, C.R.C., c. 289 (2008) (Can).
26 Processed Egg Regulations, C.R.C., c. 290 (2008) (Can).
27 Fish Inspection Act at §16.
28 Fish Inspection Act at §2.
29 The statute does not require suppliers of equipment parts to obtain clearance from the CFIA prior to the sale of the parts. Rather, it is the food-processing equipment in its entirety that must be approved by the CFIA. However, manufacturers of finished equipment and processing facilities may request assurances from suppliers regarding the suitability of the component materials in the form of "no objection" letters from the HPFB.
30 For example, if an article is to be used in a registered meat establishment, or to package meat, a "no objection letter" from the HPFB must be sent to the CFIA, Meat and Poultry Products Division, along with other required information discussed above.
31 A company is permitted to send a single submission to the HPFB, with a copy to the CFIA, requesting both agencies' "non-objection" to the use of the food package at issue in food contact applications. The CFIA relies on the HPFB's safety evaluation; it does not conduct an independent review. If the HPFB is satisfied regarding the suitability of the use of the food package at issue, it notifies the CFIA of this fact. "No objection" letters are then issued regarding the package from both agencies.
32 Materials that are approved for packaging in food facilities under CFIA's authority

are listed on CFIA's Reference List of Accepted Construction Materials, Packaging Materials and Non-Food Products, available at http://www.inspection.gc.ca/english/fssa/reference/refere.shtml (last visited 5 August 2008).

33 Baughan, J.S. (2001) Understanding Food-Packaging Regulation in Canada: A Different Approach, http://www.khlaw.com/showpublication.aspx?Show=951.

34 Ibid.

35 Under Section B01.001 of the Canadian Food and Drug Regulations, a food additive is defined, in part, as follows: "a substance, including any source of radiation, the use of which results, or may reasonably be expected to result in it or its byproducts becoming part of or affecting the characteristics of the food, but does not include . . . *food packaging materials and components thereof*" (emphasis supplied).

36 The principles and criteria used by the HPFB and the US Food and Drug Administration in assessing the safety of food contact materials have been described by Michel Pelletier, an HPFB official, as "closely similar." M. Pelletier, Health Protection Branch, "Food Packaging Regulations in Canada" presented to ICI/PIRA (June, 1995). Mr. Pelletier continued by stating that ". . . generally speaking, packaging materials permitted for use in the U.S. under FDA regulations would also likely be deemed acceptable, in principle, for use in Canada." Ibid. at 7. Notably, the HPFB toxicologists have agreed with the principles underlying the FDA's Threshold of Regulation approach. Ibid. at 17. Compliance of a product with FDA regulations alone, however, is not considered by the HPFB to be sufficient evidence of the acceptability of the product for use in Canada. Ibid. at 7.

37 Food and Drug Regulations at §B.23.002 and B.23.003. The latter section elaborates, stating that "a person may sell food, other than milk, skim milk, partly skimmed milk, sterilized milk, malt beverages and carbonated non-alcoholic beverage products in a package that has been manufactured from a polyvinyl chloride formulation containing any or all of the octyltin chemicals, namely, di(*n*-octyl)tin s,s′-bis (isooctylmercaptoacetate), di(*n*-octyl)tin maleate polymer and (*n*-octyl)tin s,s′,s″-tris(isooctylmercaptoacetate) if the proportion of such chemicals, either singly or in combination, does not exceed a total of three percent of the resin, and the food in contact with the package contains not more than one part per million total octyltin." Furthermore, B.23.004, B.23.005, and B.23.006 chemically identify the various octyltins.

38 Food and Drug Regulations at §B.23.007.

39 Food and Drug Regulations at §B.23.008.

40 Pelletier, supra note 1 at 6.

41 Available at http://www.hc-sc.gc.ca/fn-an/legislation/guide-ld/polymers_tc-polymere_tm-eng.php (last visited 5 August 2008).

42 As described on the Health Canada web site, Available at http://www.hc-sc.gc.ca/fn-an/legislation/guide-ld/guide_packaging-emballage01-eng.php (last visited 5 August 2008).

43 Available at http://www.hc-sc.gc.ca/fn-an/legislation/guide-ld/recycled_guidelines-directives_recycle01-eng.php (last visited 5 August 2008).

44 Baughan, supra note 33.

45 Ibid.

46 Food and Drugs Act at §23(1).

47 Food and Drugs Act at §23(1)(d).

48 Food and Drugs Act at §27.

49 See, for example, Meat Inspection Act at §16.

50 Current bill, C-51, Available at http://www2.parl.gc.ca/HousePublications/Publication.aspx?DocId=3398126&Language=e&Mode=1&File=36#4 (last visited 6 August 2008).

51 C-51 at §4 (Prohibitions).

52 C-51 at §10 (Inspection).

53 C-51 at §6 (Foods).

54 C-51 at §10 (Documents).

55 C-51 at §14 (Offences).

56 C-51 at §8 (Confidential Business Information).

57 Kroes, R. *et al.* (2004) Structure-based thresholds of toxicological concern (TTC): guidance for application to substances present at low levels in the diet. *Food and Chemical Toxicology*, 42, 65–83.

15
Food Packaging Legislation in South and Central America

Marisa Padula

15.1
South America

15.1.1
MERCOSUR

The creation of the common market bore the need to harmonize national legislations of the member states, including the rules and regulations related to packaging materials intended to come in contact with foodstuffs. The process of harmonizing laws and regulations was initiated in March 1992, coordinated by the Grupo Mercado Comun (GMC) (Common Market Group), the executive organ of the MERCOSUR.

The GMC has among its functions to coordinate and to guide the activities of the subgroups and to give legal force to the recommendations of these subgroups by approving them as GMC resolutions, which are the harmonized supranational MERCOSUR laws valid in the territory of all the member states.

Harmonization of the national legislation governing food contact packaging materials was discussed within the Food Committee of the Sub-workgroup 3 (SGT-3) – MERCOSUR Technical Regulations and Conformity Evaluation.

Prior to the formation of the common market, Brazil and Argentina had their own national rules and regulations in place regarding the use of materials intended to come in contact with foodstuffs in the form of resolutions and regulations based on the positive list principle, which means that only the substances and compounds listed may be used in formulating packaging materials within the limits set for overall migration and overall composition. In 1992, at the beginning of the harmonization process, the decision was taken that the legislation common to the four member states should follow the same model. At that time, the first MERCOSUR Resolution related to packaging materials – Resolution GMC 03/92 – "General Criteria and Classification of Packaging Materials and Articles Intended to Come into Contact with Foodstuffs and Included in the Annex to this Resolution" was adopted.

Global Legislation for Food Packaging Materials. Edited by Rinus Rijk and Rob Veraart

ISBN: 978-3-527-31912-1

Resolution GMC 03/92 continues to be in force and applies to packages, packaging equipment, and articles intended to be placed in direct contact with foodstuffs during the manufacture, production, portioning, handling, storage, distribution, commercialization, and consumption of foods.

The criteria defined by Resolution GMC 03/92 stipulate that any and all substances used in packaging and packaging materials intended to come in contact with foodstuffs must be included in the positive list and comply with overall migration limits, specific migration limits (when applicable), and composition limits. It is further required that the components used in the packaging should be of a degree of purity appropriate for the intended purpose [1].

The general criteria stipulate that packages must be manufactured in compliance with good manufacturing practices, must not cause unacceptable changes either to the composition of foods or to their sensory characteristics and must not pose any risk to human health.

An annex to this resolution classifies the packaging materials into different groups, each of which is subject to specific resolutions. According to this classification, packaging materials fall into one of the following categories:

- Plastics materials, including varnishes and coatings.
- Regenerated cellulose.
- Elastomers and rubber.
- Glass.
- Metals and their alloys.
- Cellulosic materials.
- Wood, including cork.
- Textile products, waxes, paraffins, and others.

At present, 37 GMC Resolutions on packages/packaging materials intended to come in contact with foodstuffs are in force: 20 related to plastics materials, 7 governing the use of cellulosic materials, 2 with regard to elastomeric materials, 2 regulating the use of regenerated cellulose, while paraffins, adhesives, glass, and metallic packages are the specific object of 1 GMC Resolution each, in addition to general Resolution GMC 03/92 and the Resolution on Reference Methods of Analysis. Table 15.1 shows a list of the MERCOSUR GMC Resolutions on food contact materials in force in March 2009.

15.1.1.1 Plastic Materials

The General Provisions (Resolution GMC 56/92) state that this technical regulation applies to all packages and articles (including those of domestic use), accessories, and coatings made of plastic and intended to come in contact with foods, raw materials, and mineral water. The technical regulation applies to compound materials made exclusively of plastics, to materials composed of two layers, each one of which consisting exclusively of plastics, to materials composed of two or more layers, one or more of which may not consist exclusively of plastics, provided that the layer in direct contact with food is exclusively of plastic [3].

Table 15.1 MERCOSUR GMC Resolutions on packaging materials for food contact use [2].

Materials	Subject matter	MERCOSUR GMC Resolution No.
General	General Criteria for Food Contact Packages and Articles (Framework Resolution)	3/92
	Reference Methods of Analysis for the control of food contact packages and articles	32/99
Plastic packages and articles	General criteria	56/92
	Classification of foods and food simulants	30/92, 32/97
	Overall migration	36/92, 10/95, 33/97
	Positive list of resins and polymers	24/04
	Positive list of additives	32/07
	Colorants and pigments	56/92, 28/93
	Residual vinyl chloride (CL)	47/93, 13/97
	Residual styrene (CL)	86/93, 14/97
	Mono- and diethyleneglycol (SML)	11/95, 15/97
	Fluorinated polyethylene	56/98
	Polymer- and/or resin-based film-forming preparations intended for food coatings	55/99
	Returnable PET packages for nonalcoholic carbonated beverages	16/93
	Multilayer PET packages with an intermediate layer containing recycled materials for nonalcoholic carbonated beverages	25/99
	Packages made of food-grade postconsumer recycled PET and intended for food contact	30/07
Metallic packages and articles	General provisions	46/06
Glass and ceramic packages and articles	General provisions	55/92
Cellulosic packages and articles	General provisions	19/94, 35/97, 20/00
	Overall migration	12/95
	Positive list of components	56/97
	Filter papers for hot filtration and cooking	47/98
	Recycled cellulosic materials	52/99
Regenerated cellulose	Regenerated cellulose films	55/97
	Regenerated cellulose casings	68/00

(continued)

Table 15.1 (Continued)

Materials	Subject matter	MERCOSUR GMC Resolution No.
Elastomeric packages and articles	General provisions	54/97
	Positive lists	28/99
Adhesives for the manufacture of packages	General provisions	27/99
Food contact paraffins	Technical regulation	67/00

CL, composition limit and SML, specific migration limit.

The General Provisions further state that only substances included in the positive lists of compounds (resins, polymers, additives, etc.) may be used in the manufacture of food contact plastic packages and articles and that such compounds should be of a degree of purity compatible with the intended use. In addition, such compounds should conform to specifically indicated limits, restrictions, and tolerance of use.

Plastic materials and articles should not, under foreseeable conditions of use, transfer any undesirable, toxic, or contaminating substances that pose a risk to human health in quantities above the overall and specific migration limits or that may bring about unacceptable changes in the composition of foods or their sensory characteristics.

The overall migration limit set for these materials is 50 mg residue/kg food simulant for (i) packages and articles with a capacity greater than 250 ml, (ii) packages and articles the contact surface of which cannot be estimated or easily determined; (iii) packages and articles containing sealing components or devices with a small surface area. An overall migration limit of 8 mg residue/dm^2 of plastic surface area is set for: (i) packages and articles with a capacity smaller than 250 ml; (ii) plastic materials in general.

The General Provisions also require that only virgin materials be used in the manufacturing of food contact plastics, and prohibit the use of plastic materials made from recycled packages, fragments of recycled plastic articles, and recycled plastics or that come from materials previously used for food contact applications. However, this prohibition does not apply to materials that have been reprocessed as raw material using the same transformation process by which it was originally produced. Specific technological processes for the manufacture of resins obtained from recycled materials must be evaluated by the competent authority.

Positive List

The positive list of additives permitted for use in the manufacture of food contact plastics was updated in 2007 and published in December 2007 as Resolution GMC 32/07 – "Positive List of Additives for Plastics Intended for Use in the Manufacture of Food-Contact Packages and Articles." This list includes substances that are added to plastic materials to obtain a desired thermal effect, antioxidants, antistatic agents, foaming agents, antifoam agents, lubricants, plasticizers, and so on and substances used to produce an appropriate polymerization media, such as wetting agents, surfactants, pH-regulating agents, and solvents. However, excluded from the list are impurities, intermediate products, degradation products, and inhibitors, accelerating agents, catalysts, catalyst modifiers and deactivators, molecular weight regulators, polymerization inhibitors, and redox agents. The list further sets purity criteria compatible with the intended use of each of the substances listed and permits the use of food additives as long as the restrictions established and applicable to their use in foods are respected [4].

This update incorporates the CAS number (CAS *Register Number – American Chemical Society*) into the basic framework of the positive list, as well as the Spanish and Brazilian Portuguese version of the list of additives, including the limits and restrictions of use.

Verification of compliance with the specific migration limits shall be conducted in accordance with the methods prescribed in the corresponding MERCOSUR resolutions.

The positive list of polymers and resins permitted for use in the manufacture of food contact plastics was updated in 2004 and published as Resolution GMC 24/04 – "Positive List of Polymers and Resins Intended for Use in the Manufacture of Food-Contact Packages and Articles" [5].

Food Simulants and Classification of Foodstuffs

The food simulants adopted by the MERCOSUR are distilled water, acetic acid 3% (w/v) in aqueous solution, and ethanol 15% (v/v) in aqueous solution – or at a concentration that reproduces, as closely as possible, the actual concentration – for aqueous, acidic, and alcoholic foods, respectively. Rectified olive oil and *n*-heptane are to be used as simulants for fatty foods. Rectified olive oil must be used as simulant in tests conducted to check compliance with any new material (polymers or additives) or whenever there is incompatibility of the material with *n*-heptane or when the intended application involves high temperatures [6].

The MERCOSUR legislation classifies foods into six categories (Types I, II, III, IV, V, and VI):

- **Type I** – aqueous foods (with pH above 5.0).
- **Type II** – acidic foods (having a pH equal to or lower than 5.0).
- **Type III-a** – aqueous foods that contain oil or fat.
- **Type III-b** – aqueous acidic foods that contain oil or fat.
- **Type IV** – oily or fatty foods.
- **Type V** – alcoholic foods.
- **Type VI** – dry solid foods or foods on which the extractive action of food simulants is of little significance.

An Annex to Resolution GMC 32/97 defines the simulants that are to be used for each type or group of foods. The Annex describes several foods or groups of foods, followed by the specific food simulant that should be used to conduct migration tests [7].

The test conditions employed for the extraction of migrating substances are standard conditions of time and temperature that simulate the actual use of the packaging material being tested. For example, for contact temperatures of up to 40 °C and contact times exceeding 24 h, the migration test conditions are set to 10 days at 40 °C. The test conditions described in Resolution GMC 36/92 apply to plastics and other materials, with the exception of cellulosic materials, which require differentiated test conditions [8].

Colorants and Pigments

According to the MERCOSUR legislation, colorants and pigments must not migrate into foods. Resolution GMC 56/92 establishes criteria of purity by setting limits for metals (described below) and limits on aromatic amine levels, which should not exceed 0.05% (m/m).

- Arsenic (soluble in NaOH 0.1 N) 0.005% (m/m)
- Barium (soluble in HCl 0.1 N) 0.01% (m/m)
- Cadmium (soluble in HCl 0.1 N) 0.01% (m/m)
- Zinc (soluble in HCl 0.1 N) 0.20% (m/m)
- Mercury (soluble in HCl 0.1 N) 0.005% (m/m)
- Lead (soluble in HNO_3 1 N) 0.01% (m/m)
- Selenium (soluble in HCl 0.1 N) 0.01% (m/m)

Resolution GMC 28/93 establishes the analytical test methods to be used to determine the level of metals in pigments. In addition, Resolution GMC 32/99 lays down the method of analysis for the determination of aromatic amines in pigments [9, 10]. The first resolution also stipulates additional requirements for black carbon pigment.

Determination of specific migration of metals contained in extracts obtained in overall migration tests may also be required for the following metals when it is not possible to analyze the pigment: antimony (Sb), lead (Pb), fluorine (F), arsenic (As), copper (Cu), mercury (Hg), barium (Ba), chromium (Cr), silver (Ag), boron (B), tin (Sn), zinc (Zn), and cadmium (Cd). The specific migration limits for these compounds are those established in the technical regulation concerning the presence of contaminants in foods.

The update of Resolution GMC 28/93 was published as Public Consultation in the MERCOSUR countries in August of 2008. Nowadays, it is in the last step toward approval at MERCOSUR. This resolution is being modified to be as close to the Resolution AP (89) 1 from Council of Europe as possible. The new resolution is establishing the purity criteria by setting limits for nine metals, arsenic (As), barium (Ba), cadmium (Cd), zinc (Zn), lead (Pb), mercury (Hg), selenium (Se), antimony (Sb), and chromium (Cr), and limits for unsulfonated and sulfonated aromatic amines and specific limits for benzidine, β-naphthylamine, and 4-aminebiphenyl (<10 ppm). For carbon black a maximum limit of 2.5% is being established along with limits for toluene extractables, UV absorption of cyclohexane extract and limit for benzo(*a*)pyrene. For specific migration of metals, limits are being established in this resolution based on international limits.

Postconsumer Recycled Materials

Resolution GMC 25/99 "Disposable multilayer PET packages intended to hold nonalcoholic carbonated beverages" permits the use of an intermediate layer of postconsumer recycled PET and stipulates that the layer in direct contact with the beverage be exclusively of virgin PET with a minimum thickness of 25 μm, in addition to stating that control analyses should be performed to guarantee the quality of such packages. Each manufacturer should apply for and receive authorization from the competent authority [11].

The use of postconsumer recycled PET obtained by the *bottle-to-bottle* process and used in the manufacture of packages intended for direct food contact was approved by Resolution GMC 30/07 in December 2007 [12].

This technical regulation on the use of postconsumer recycled food-grade polyethylene terephtalate (PET) (food-grade PCR-PET) in food contact packaging applications establishes the general requirements and evaluation criteria, approval/authorization, and registration procedures for PET packages made with varying degrees of virgin PET (food grade) and super-clean postconsumer recycled (food-grade) PET intended for food contact. The provisions of this technical regulation apply to finished products (food-grade PCR-PET packages), precursor articles of such packages, and the raw materials used in their manufacture (food-grade PCR-PET resins).

The proportion of food-grade PCR-PET to be used in the manufacture of food-grade PCR-PET packages is subject to the restrictions specified in the special authorization of use, such as letters of nonobjection and/or approvals issued by an internationally recognized regulatory authority or institution.

The basic criteria for approval of food-grade PCR-PET packages state that these packages must not transfer any substance extraneous to the food in quantities that could endanger human health or bring about changes in the sensory characteristics of the packaged foods.

Food-grade PCR-PET packages must comply with all applicable sanitary requirements laid down in the MERCOSUR legislation governing plastic packages and must be compatible with the foods they are intended to contain.

Food-grade PCR-PET packages, and/or their precursor articles, must be approved/authorized and registered with the Competent National Sanitary Authority, following the established procedures.

The responsibility for the quality of the PCR-PET material is divided between the manufacturer of the PCR-PET resin, the manufacturer of the PCR-PET packages, and the food manufacturer. The resin manufacturer must employ an approved/authorized process or technology for physical or chemical recycling. Each specific case must be duly registered with the Competent National Sanitary Authority based on a documented description of the technology involved, validation information (*challenge test*), authorizations, and the results of sanitary tests carried out to attest the conformity to sanitary requirements of the product and/or process. The resin manufacturer must also provide the package manufacturer with all necessary information concerning the PCR-PET resin (including the information for which foods and conditions of use the PCR-PET resin has been approved), keep an updated record of both the origin of the postconsumer PET and the final destination of the finished resin, in addition to implementing ongoing programs of quality assessment, analytical monitoring, and sensory analysis.

The manufacturing company of the packaging made of PCR-PET resin must be duly qualified and registered with the Competent National Sanitary Authority. Likewise, the package produced by such qualified/registered packaging manufacturing company must be approved/registered by the competent authority. The producer of the package must keep an updated and documented record of the origin and characterization of the PCR-PET and virgin PET resins used in the manufacture of the package, in addition to implementing a traceability system and guaranteeing the quality of the product that must be manufactured in conformity with good manufacturing practices. Furthermore, the packaging manufacturer must also provide the end user of the package with all the necessary information regarding the package.

Food producers who decide to use packages made from food-grade PCR-PET, or precursor articles of PCR-PET, must exclusively use materials approved/authorized and registered by the Competent National Sanitary Authority (following the established procedures) and ensure that such packages will be solely used to contain the foods specified by and under the conditions specified in the respective approval/authorization and registration documents.

The label of the product should contain the following: identification of the manufacturer (name and address), the lot number or code that allows traceability of the product, and the expression "PCR-PET."

15.1.1.2 **Elastomeric Materials**

The technical regulation on Elastomeric Packages and Articles for Food Contact applies to elastomeric (natural or artificial) packages and articles intended to be placed in contact with foodstuffs during their production, manufacture, handling, shipping, distribution, and storage, including multilayer packages and articles that also contain nonelastomeric layers provided that the layer in direct contact with food is exclusively elastomeric.

The General Provisions (Resolution GMC 54/97) establish that elastomeric packages and articles must be produced in accordance with good manufacturing practices and must not, under foreseeable conditions of use, transfer to foods any undesirable, toxic, or contaminating substances, in quantities exceeding the overall and specific migration limits, or endanger human health or bring about unacceptable changes in the composition of foods or their sensory characteristics [13].

The overall migration limit for constituents of elastomeric materials is the same as that for constituents of plastic materials and articles. Likewise, the requirements for colorants and pigments are also identical to those set for plastic materials.

Only those substances included in the positive list published as Resolution GMC 28/99 may be used in the composition and manufacturing of elastomeric food contact materials and articles [14].

The positive list of compounds permitted for use in the manufacture of elastomeric food contact materials is divided into four parts: polymers, reticulation or chain extension agents, additives, and pressure-sensitive adhesives, along with their respective specific migration limits, composition limits, and restrictions of use. In addition to the additives described in this resolution, additives included in the Positive List of Additives for Plastics Materials may also be used in the manufacture of elastomeric food contact materials and articles.

The following specific migration limits apply to the substances listed below when used in the manufacture of elastomeric food contact materials and articles:

- *N*-Nitrosamines: 1.0 mg/dm^2.
- Primary aromatic amines, calculated as aniline hydrochloride: 50 mg/kg food simulant.
- *N*-Alkyl-arylamines, calculated as *N*-ethylphenylamine: 1 mg/kg food simulant.
- Secondary aliphatic or cycloaliphatic amines: 5 mg/dm^2.

This technical regulation also applies to elastomeric materials intended for mouth contact.

15.1.1.3 Adhesives

Adhesives for Direct Food Contact

Adhesives permitted for direct food contact are those that comply with the requirements laid down in Part IV – Pressure-sensitive Adhesives – of Resolution GMC 28/99 – Positive List for Elastomeric Packages and Articles Intended to Come into Contact with Foodstuffs.

This list is divided into two items:

- Substances permitted for use in the composition of pressure-sensitive adhesives and which may be used on the contact surface of labels or self-adhesives for poultry, dry foods, and processed, frozen, dried, or partially dehydrated fruits and vegetables.
- And substances permitted for use in the manufacture of pressure-sensitive adhesives and which may be used on the contact surface of labels or self-adhesives for fresh fruits and vegetables and "*in natura*" eggs.

Substances and colorants permitted for use in foods may also be used provided they comply with all corresponding requirements set forth in food legislation.

Adhesives for Indirect Food Contact

Adhesives for indirect food contact are described in Resolution GMC 27/99 concerning adhesives used in the manufacture of packages and articles intended to be placed in contact with foodstuffs [15]. This resolution states that such adhesives may be made from one or more of the substances included in the Positive List of Polymers, Resins and Additives for Plastic Food Contact Packages and Articles, the Positive List for Cellulosic Food Contact Packages and Articles, and the Positive List for Elastomeric Food Contact Packages and Articles.

The substances used must meet the purity criteria established and the quantity of adhesive that contacts packaged foods at the edge of laminates must be minimal and must not exceed the limits of good manufacturing practices. Under normal conditions of use, the laminated packaging material must remain firmly bonded without visible separation.

15.1.1.4 Waxes and Paraffins

Published in 2000, Resolution GMC 20/00 – technical regulation on food contact paraffin waxes – applies to synthetic food contact paraffins, petroleum waxes (paraffinic and microcrystalline), and polyethylene waxes and to the products based on these substances and used in coatings for packages and articles intended to come in contact with foodstuffs and in cheese coatings. Item 3 contains a positive list of the substances permitted for use in the manufacture of these products, along with their respective purity criteria requirements [16].

The provisions of this technical regulation require the control of metals and easily carbonizable substances and establish specific requirements for synthetic paraffin (oil content, absorptivity, and freezing point) and petroleum waxes (absorptivity).

15.1.1.5 Cellulosic Materials

The resolutions governing cellulosic packages and articles for food contact are divided into the following headings: general provisions, overall migration, positive list, filter papers for hot filtration and cooking, and recycled cellulosic materials for food contact.

Resolution GMC 19/94 and complementary Resolutions GMC 35/97 and 20/00 establish general provisions for food contact cellulosic materials and their scope of application [17–19].

These resolutions apply to cellulosic packages and articles intended to be placed in contact with foodstuffs, including materials that are coated or surface-treated with paraffins, polymeric resins, and other substances. These resolutions also apply to multilayer packaging materials composed of different types of materials provided that the layer in direct contact with food is exclusively cellulosic.

Excluded from the scope of these resolutions are cellulosic packages and articles intended to come in contact with foodstuffs that require peeling prior to consumption (e.g, citrus fruits, nuts in their shell, coconuts, pineapples, melons, etc.).

These resolutions do not apply to secondary packages made of paper, paperboard, and cardboard, whenever it is ensured that they will not come in direct contact with foods.

As with all other resolutions related to food packaging, the resolutions governing cellulosic packages stipulate that cellulosic packages and packaging materials shall not transfer any substances to foodstuffs in quantities exceeding the established limits and must not bring about changes in either the composition or the sensory characteristics of foods. In addition, cellulosic packaging materials must comply with microbiological standards compatible with the foods they are intended to come in contact with.

The General Provisions also establish maximum levels for polychlorinated biphenyls (5 mg/kg) and pentachlorophenol (0.1 mg/kg). Although the resolutions state that the level of metals is to be controlled, they do not establish specific migration limits for cadmium (Cd), mercury (Hg), arsenic (As), lead (Pb), and chromium (Cr) and, when necessary, for antimony (Sb), boron (B), barium (Ba), copper (Cu), tin (Sn), fluoride (F), and zinc (Zn). In this case, it is stated that maximum amounts of these metals should not exceed the limits established for contaminants in food.

Resolution GMC 56/97 provides a Positive List of Components for Cellulosic Packages and Articles for Food Contact. The list of permissible substances for use in food contact cellulosic packaging materials is divided into fibrous raw materials (virgin cellulose fibers – bleached or unbleached – virgin synthetic fibers, cellulose fibers from materials recycled as part of normal industrial production procedures), and nonfibrous raw materials, processing aids, special processing aids for papers, and restrictions [20].

The analytical test procedures and overall migration limits are laid down in Resolution GMC 12/95 – Overall Migration Tests for Cellulosic Packages and Articles Intended to Come into Contact with Foodstuffs. The food simulants used are the same as those that are used in compliance tests conducted with plastic materials, with the difference that the contact conditions (time and temperature) set are specific to cellulosic materials. The overall migration limit is 8 mg/dm^2 [21].

Filter papers for cooking and hot filtration are regulated by Resolution GMC 47/98 [22]. This resolution applies only to papers with a thickness less than 500 g/m^2 and intended to come in contact with aqueous foods but not with fatty foods. The resolution contains a specific positive list for these materials and states special requirements for hot water extracts, such as the absence of formaldehyde or glyoxal and of metals such as cadmium, arsenic, chromium, mercury, and lead.

Resolution GMC 52/99 – Technical Regulation for Recycled Cellulosic Materials [23] – applies to packages manufactured from papers, entirely or partially composed of secondary fibers, and intended to pack solid dry foods or foods on which the extractive action of food is of little or no significance – type VI foods (Resolution GMC No. 30/92).

Secondary fibers should not be manufactured using recycled fibers obtained from indiscriminately collected waste materials and the cleaning and sanitation process must guarantee that the quality of the papers is compatible with their use in food contact applications.

The overall migration limit applicable in this case is also set at 8 mg/dm^2 and all other requirements are identical to those that apply to virgin materials.

15.1.1.6 Regenerated Cellulose Films

Resolution GMC 55/97, Technical Regulation on Regenerated Cellulose Films for Food Contact, applies to regenerated cellulose films intended to come in contact with foodstuffs and to multilayer packages composed of several types of materials and provided that the layer in direct contact with food is exclusively made of regenerated cellulose. This resolution neither does apply to synthetic casings of regenerated cellulose, which are subject to a specific technical regulation, nor does it apply to coated regenerated cellulose film the surface of which is intended to come in contact with food is coated with a resin or polymer layer having a thickness more than 50 mg/dm^2 (5 g/m^2) [24]. In this latter case, the Technical Regulation on Plastics Packages and Articles for Food Contact applies.

The positive list of constituents of regenerated cellulose films is divided into two parts: the first part concerns the substances used in the manufacture of uncoated regenerated cellulose film, while the second part lists the permissible substances for use in the manufacture of coated regenerated cellulose film.

The colorants and pigments used to color regenerated cellulose films must comply with the requirements established for plastic materials.

15.1.1.7 Regenerated Cellulose Casings

Resolution GMC 68/00 – Technical Regulation on Synthetic Casings of Regenerated Cellulose for Food Contact – applies to regenerated cellulose-based casings intended to come in contact with foodstuffs [25].

Only those substances on the positive list of this resolution may be used in the composition and manufacture of the casing. The positive list encompasses constituents of the base sheet, opacity and gliding agents, surface finishing agents, preservatives, and coatings.

Synthetic regenerated cellulose-based casings must comply with microbiological standards compatible with the food they are to be placed in contact with and should not transfer off-flavors or bring about any unacceptable changes in the sensory characteristics (e.g., taste and odor) of the food they are intended to come in contact with.

The regenerated cellulose casings must comply with all applicable overall migration limits.

15.1.1.8 Metallic Materials

The legislation for metallic materials was updated in 2006 and published as Resolution GMC 46/06 – Technical Regulation on General Provisions for Metallic Packages, Coatings, Utensils, Lids, and Articles for Food Contact [26]. This resolution applies to packages, coatings, utensils, lids, and articles manufactured from coated or noncoated metallic materials and intended to come in contact with foodstuffs during their production, processing, shipping, distribution, and storage. As mentioned before with regard to other packaging materials, metallic packages, materials, and articles must not transfer undesirable, toxic, or contaminating substances to foodstuffs or bring about unacceptable changes in the composition or in the sensory characteristics of foods.

This Resolution lays down a positive list of metallic raw materials, such as tinplate, chromium-coated steel, aluminum, and its alloys, among other materials. These packages and articles may have their contact surface protected by metallic coatings such as tin or by polymeric coatings. The constituents of polymeric coatings authorized for food contact applications are included in the positive list of resins, polymers, and additives for plastic materials and should comply with the established requirements regarding restrictions of use, composition limits, and specific migration limits.

The substances permitted for use in the manufacture of sealing compounds used on metallic lids to ensure total package integrity or hermeticity are described in the positive lists of elastomeric food contact materials and must comply with the restrictions of use, composition limits, and specific migration limits established.

Thermoplastic cements may be used for package molding provided they comply with the requirements set out in the regulations governing plastic and/or elastomeric food contact materials.

The updated and revised version of this regulation incorporated processing aids such as surface lubricants commonly used to facilitate stuffing, stretching, hot/cold stamping, and molding of metallic objects made from stored rolls or stacks of sheet metal, or used to form rolls of metallic laminates or store metallic laminates. The substances permitted for use in the manufacture of these lubricants are listed and described in item 3.5 of this resolution and they are divided into two categories based on the maximum permitted limit of the substance on the metallic surface.

Coated metallic materials must also fulfill overall migration requirements. Verification of compliance with migration limits is to be carried out using the same food simulants used for plastics materials; however, whenever the resulting data exceed the overall migration limit specified, the test results should be corrected for the migration of metals and for the migration of zinc if the varnish contains zinc oxide. The overall migration limit is set at 50 mg/kg or 8 mg/dm^2.

Specific migration of metals into foods must be closely monitored. Metals must not migrate in quantities exceeding the limits established in the MERCOSUR resolution governing the presence of contaminants in foods.

15.1.1.9 Glass

The Technical Regulation on Glass and Ceramic Packages Intended to Come into Contact with Foodstuffs dates back to 1992 and applies to packages, articles, and domestic utensils made of glass or ceramic materials that come in contact with foods, either for prolonged periods of time or during brief and repeated contacts of limited duration production, manufacture, shipping, distribution, storage, and commercialization of the food [27].

The types of glass that may be used for food contact purposes are borosilicate glass – for any contact situation, including sterilization and cooking in industrial or domestic ovens – and sodium-calcium glass – for any contact situation, including pasteurization and industrial sterilization. Crystal glass is permitted only for domestic use in brief and repeated contacts of short duration.

Glass packages may be returnable, provided that they are appropriately cleaned and sanitized, and may be recycled without any restriction.

Porous ceramic materials must not be used for food contact applications. For ceramic, vitrified and enamel-coated glass and metal packages and articles, overall migration testing (maximum limit of 50 mg/kg food simulant or 8 mg/dm^2) and specific migration testing of cadmium and lead are required, with the limits established in accordance with the holding capacity and size of the containers or packages.

15.1.1.10 **Legislation Update**

The updating process of the MERCOSUR legislation stagnated from 1999 to 2005. Only in 2006 discussions were reinitiated with the aim of updating the regulations governing food contact materials in force in the MERCOSUR member states.

In short, the criteria established for the inclusion of new substances in the positive lists require the presentation of a solid documented justification of the technological need of its use, in addition to references to the lists of approved substances laid down in Directives of the European Union and/or the Code of Federal Regulations of the United States. Exceptionally, a substance documented in other internationally recognized legislations may be approved. The petition or proposal for the inclusion of a new substance in the positive lists as part of the MERCOSUR legislation must include the following items: identification of the substance, concentration of impurities, and the percentage of impurities along with the analytical data that characterize the substance, its physical and chemical properties, its potential use and technological purpose or function, the maximum percentage of the substance to be used, and the minimum percentage needed to achieve the desired effect, in addition to information regarding the processing of the finished product and the conditions under which the substance is suitable for food contact. Furthermore, a report containing the toxicological data of the substance must be presented or attached to the proposal for inclusion in the MERCOSUR positive lists [28].

New packaging systems or new technologies also need to be approved by the competent authority. For that purpose, the proposal or petition for approval must be accompanied by complete technical studies to be submitted for evaluation and approval within the MERCOSUR.

15.1.1.11 **Implementation of GMC Resolutions in the MERCOSUR Member States' National Legislations**

In order to become effective in the MERCOSUR member states, the GMC Resolutions must be incorporated into the respective national legislations. In Argentina, the GMC Resolutions were incorporated by adopting them in the form of resolutions issued by the Ministry of Health and inclusion in the Código Alimentário Argentino (Argentine Food Code) [28]. In Brazil, GMC Resolutions were incorporated in the form of "Portarias" (ministerial decree) and resolutions issued for each type of packaging material by the Agência Nacional de Vigilância Sanitária – ANVISA (the National Agency for Sanitary Vigilance) of the Brazilian Ministry of Health. For example, Resolution No. 105 published on May 19, 1999 by ANVISA consolidated all

MERCOSUR resolutions on plastic materials approved up to that date [29]. In Uruguay, as well as in Paraguay, the GMC Resolutions came into force after publication by the Ministry of Public Health and the Ministry of Public Health and Social Welfare, respectively [28].

Although most harmonized resolutions have already been implemented into the MERCOSUR member states' national laws, there are still some differences with respect to the approval procedures adopted by the competent national sanitary authorities to allow the trade of locally produced or imported materials within their respective national territories.

In Argentina, manufacturers of plastic packages and articles must obtain regulatory approval (i.e., registration) from the competent authorities prior to marketing their products, in accordance with the rules laid down in MERCOSUR resolutions. The same applies to imported packages and articles. Packages or packaging materials imported from other MERCOSUR member states do not need to be registered in Argentina if they are registered in the country of origin [30].

In Brazil, with the publication on March 15, 2000 of ANVISA Resolution No. 23 – "Basic Procedures for the Registration and Exemption of Registration of Food-Related Products," food packages became exempt from registration [31]. This resolution is part of ANVISA's multiple strategy to (i) improve and modernize the control of industrially processed foods; (ii) reduce the bureaucracy dealing with the registration and approval of foods and packages; (iii) concentrate its efforts on updating technical regulations and standards; and (iv) adopt a more effective and exacting approach to its sanitary inspection and control activities [32]. With this decision, ANVISA assigned exclusively to the manufacturer of the packaging the responsibility to ensure the quality and safety of the products he produces.

Resolution No. 23/2000 further establishes that packages made from postconsumer recycled materials must be registered and that the interested manufacturing companies may submit to ANVISA a proposal of approval of the technology used to produce packages and packaging materials and articles from postconsumer recycled materials for direct food contact.

For imported products, ANVISA published on March 15, 2000 Resolution No. 22 – "Basic Procedures for the Registration and Exemption of Registration of Imported Food-Related Products" [33], which states that imported products are subject to the same requirements as those applicable to packages produced in Brazil.

The MERCOSUR resolutions have already been incorporated into Paraguayan legislation by the Paraguayan Ministry of Public Health and Social Welfare via two ministerial decrees [28].

Uruguay has adopted the MERCOSUR resolutions into its own laws via a decree published by the Ministry of Public Health (MSP). Food packages must be registered with departments of the Bromatology Division [28].

15.1.2
Venezuela

Venezuela signed the Protocol of Adhesion to the MERCOSUR in December 2005.

Article 3 of this protocol states that Venezuela will adopt the harmonized MERCOSUR rules and regulations currently in force in the respective member states over a period of 4 years beginning on July 4, 2006 (6 months after the assignment of the Protocol of Adhesion) [2].

The General Regulation on Foods (*Reglamento General de Alimentos*), Decree No. 525 of 1959, establishes in its Article 18 that raw materials and/or additives used in the manufacture of packages for food contact and packages intended to come in direct contact with foods require authorization of the competent authority [34]. The Complementary Standards of the General Regulation on Foods, Resolución SG-081-96 (*Normas Complementares do Reglamento General de Alimentos*) of March 15, 1996, published in the *Official Gazette* No. 35 921 by the Ministry of Health and Social. Development, describe in Articles 30–33 the requirements for registration and obtaining sanitary authorization [35]. In summary, the main requirements for packaging materials or packages – either locally produced or imported – refer to the following: qualitative and quantitative composition, description of the manufacturing process and the system of quality control (method of analysis and reference standard), proposed use and conditions of use, sanitary certificate of the country of origin (only for imported materials), and a duly signed declaration stating that the submitter understands and accepts the standards and specifications inherent to the request.

The sanitary authorization for raw materials and/or additives and packages is issued by the Board for Food Hygiene (*Dirección de Higiene de los Alimentos*) through the Department of Building, Equipment, and Packaging Control (*Departamento de Edificaciones, Equipos y Envases*) of the Ministry of Health. Directives of the European Union or the US Food and Drug Administration are consulted whenever a certain item is not or insufficiently covered by specific Venezuelan technical standards.

On May 4, 2007, the Venezuelan Ministry of Health published Resolution No. 82 on Good Manufacturing Practices, Storage and Shipping of Packages and/or Articles Intended for Food Contact. The articles refer to buildings and installations, water supply, manufacturing equipment, waste management, education and training, hygiene and sanitary practices, raw materials, quality, storage, and registration of manufacturing and distribution procedures [36].

Venezuela has put a number of Venezuelan Standards – COVENIN in place, worked out by the Foundation for Standardization and Quality Certification (FONDONORMA – *Fundo para la Normalización y Certificación da Calidad*) related to packages, including definitions, sizes, specifications, terminology, test methods, and so on. At present, 25 COVENIN standards have been approved related to packaging materials for food contact, covering items such as overall migration (plastics materials, cellulosic materials, and sealants), determination of metals in dyes and pigments, determination of the level of free vinyl chloride, determination of specific migration of mono- and diethyleneglycol, the maximum extractable fraction in *n*-hexane and the maximum soluble fraction in xylene, and specifications regarding the use of isocyanate-based adhesives [37].

15.1.3 Chile

Chile has so far no specific regulations or legislation in place governing packaging materials for direct food contact. Some requirements applicable to food contact packages are included in the Decree No. 977/1996 and its subsequently revised and updated versions, Sanitary Food Regulation (*Reglamentos Sanitario de los Alimentos*) Title II – Foods (*Título II – De los Alimentos*), Paragraph III – Packages and *Articles* (*Parráfo III – De los Envases y utensilios*) published by the Ministry of Health. Articles 123, 125, 126, 128, and 129 cover packages intended for foods.

In summary, Article 123 states that packages must not transfer toxic substances to foods or bring about a change in their sensory or nutritional characteristics. Article 125 refers to metallic materials and stipulates that these materials must not contain more than 1% impurities such as lead, antimony, zinc, copper, chromium, iron, and tin, in addition to establishing a maximum level of 0.01% arsenic. Article 126 states plastic packagings must not contain residual monomer levels more than 0.25% of styrene, 1 ppm of vinyl chloride, and 11 ppm of acrylonitrile. The same article further determines that plastic packages must not transfer to foods more than 0.05 ppm vinyl chloride or acrylonitrile or any other substance that may endanger human health. Article 128 permits the use of returnable packages provided these packages permit adequate cleaning and sanitation, while Article 129 establishes that packages that have been in contact with nonfood products or which are incompatible with foods must not be used to hold foodstuffs, or in other words, this Article prohibits filling of food into packages used for or approved for purposes other than holding food [38].

Technical standards related to the quality of packages are published by the Chilean National Institute of Standardization (INN – Instituto Nacional de Normalização).

15.1.4 Andean Community

The Andean Community is made up of four countries, Bolivia, Peru, Ecuador, and Colombia. The main objective of the Andean Community is to achieve more accelerated, balanced, autonomous, and sustainable development via Andean, South American, and Latin American integration.

The Andean Commission, created on May 26, 1969 is the executive body of the Andean Integration System and responsible – along with the Andean Council of Ministers of Foreign Affairs – for adopting decisions related to a number of different subject matters under discussion. At the request of one or more member countries or of the General Secretariat, the Commission's Chairman may call upon the Commission to meet as an Enlarged Commission, in order to address sectorial issues, to consider regulations for the coordination of development plans, and to harmonize the economic policies of the member countries, as well as to hear and resolve all other matters of common interest.

After a period of stagnation lasting up to 2007, the process of approximation and harmonization of the laws of the Andean Community member states (Bolivia, Peru, Ecuador, and Colombia) regarding packaging materials and articles intended to come in contact with food was taken up again. At present, a standard on the determination of overall migration is in the stage of approval. In this particular case, the MERCOSUR legislation was taken as the basis.

Bolivia

The Bolivian National Service of Livestock Health and Food Safety (*Servicio Nacional de Sanidad Agropecuaria y Inocuidad Alimentaria* – SENASAG) – a division of the Bolivian Ministry of Rural Development, Livestock and Environment (*Ministerio de Desarrollo Rural, Agropecuario y Meio Ambiente*) – is the government body responsible for food safety in Bolivia. This executive regulatory agency published on March 12, 2003 Administrative Resolution No. 019/2003 laying down an updated version of the Sanitary Requirements for the Manufacture, Storage, Shipping and Portioning of Foods and Beverages for Human Consumption. Article 44 of Chapter VIII – Portioning and Packaging – of this regulation determines that packaging materials must provide adequate protection to foods to avoid contamination and damage, in addition to allowing appropriate product labeling. This regulation further states that all packaging materials intended for food contact must be nontoxic and permits the reuse of food packaging materials provided they are adequately cleaned and disinfected between use.

The Bolivian Institute for Standardization and Quality – IBNORCA issued a series of standards covering packages for food contact (glass, metallic, paper and paperboard, and plastic food containers); however, none of these standards define how packages intended for food contact should be evaluated [39].

Ecuador

The Food Regulation (Reglamento de Alimentos), implemented via Executive Decree (*Decreto Ejecutivo*) 4114, of July 22, 1988, Title V (*Titulo V*), Chapter III (*Capitulo III*) – Filling and Packaging (*Del Envasado y Embalaje*) lays down general provisions and requirements for food contact packages. Articles 136–143 state that the package must ensure protection to and preservation of the food throughout the entire storage and distribution chain, in addition to being appropriate to processing and being made of virgin materials and comply with all applicable laws, regulations, and standards in force. The regulation permits the reuse of returnable glass containers provided they allow adequate and correct sanitization [40].

Ecuador has so far no specific technical regulations in place on packaging materials intended to come in contact with foods and in the absence of these adopts the rules and requirements established by the Food and Drug Administration of the United States of America.

Technical standards concerning the quality of packagings are published by the Ecuadorian Institute of Standardization (INEN – Instituto Equatoriano de Normalização) [41].

Peru

In Peru, the requirements for the control and surveillance of foods and beverages were established by the government and published as ministerial decree (*Decreto Supremo*) 007/1998 Regulation on the Sanitary Surveillance and Control of Foods and Beverages (*Reglamento sobre Vigilancia y Control Sanitario de Alimentos y Bebidas*). The requirements are stated under Title IV (*Titulo IV*), Chapter VII (*Capitulo VII*), Article 64, which permits the use of returnable packages, and under Title VIII (*Titulo VIII*), Chapter III (*Capitulo III*), Articles 118 and 119, which state that packages must not transfer toxic substances to foods or bring about changes in their sensory characteristics and that the packages must not contain metallic impurities such as lead (Pb), antimony (Sb), zinc (Zn), copper (Cu), chromium (Cr), iron (Fe), tin (Sn), mercury (Hg), cadmium (Cd), arsenic (As), and styrene, acrylonitrile, and vinyl chloride residues in quantities exceeding the established safe limits. The final provisions of Article 119 stipulate that the maximum levels for impurities shall be set forth in specific legislation, although those have not yet been established [42].

The Peruvian National Institute for the Defense of Competition and Protection of Intellectual Property – INDECOPI (*Instituto Nacional de Defensa de la Competencia y de la Protección de la Propiedad Intelectual*), through the Technical Committee for Standardization (*Comité Técnico de Normalización*) – CTN 014 – Flexible Packages for Foods has, since 2004, published a series of standards covering plastic packages for food contact, the observation of which remains voluntary until further notice. These standards include general provisions, a classification of foods and food simulants, positive lists of monomers and polymers, additives, colorants, and pigments (purity criteria), and procedures for the determination of overall migration and specific migration of selected monomers. These standards were worked out on the basis of the MERCOSUR resolutions governing plastics materials [43].

Colombia

According to the Colombian National Institute for Surveillance of Foods and Drugs – INVIMA (*Instituto Nacional de Vigilancia de Medicamentos y Alimentos*) of the Colombian Ministry of Social Protection (*Ministerio de la Protección Social*), so far Colombia has no specific technical regulation in place governing packages or packaging materials intended to be placed in direct contact with foodstuffs. However, Ministerial Decree No. 3075, 1997, of the Ministry of Health, lays down requirements for the processing, filling, shipping, and distribution of foods. According to Article 18, the packages must be manufactured using materials appropriate for food contact and comply with the requirements and regulations of the Ministry of Health, although those have not yet been established. Packaging materials must be adequate and provide appropriate protection against contamination, must not have been previously used for different purposes that might cause contamination of the food, and must be inspected prior to their use to ensure that they are in good condition, clean and disinfected [44].

In Colombia, technical standards for and certification of products are coordinated by the Colombian Institute of Technical Standards and Certification – ICONTEC (*Instituto Colombiano de Normas Tecnicas y Certificación*). There are two Colombian voluntary technical standards in place with regard to materials for food contact – NTC 5022 "Plastic Materials and Articles for Contact with Foods and Beverages – Determination of Overall Migration" and NTC 5023 "Plastics Materials and Articles for Contact with Foods." The first standard refers to the method to be followed for the determination of overall migration, including types of food simulants and contact conditions, and is based on the methods used in the European Union. The second standard specifies the materials and good manufacturing practices for plastic compounds and articles for contact with foods and beverages in such a way that, under normal conditions of use, such materials do not transfer any of their constituents to the food/beverage in quantities that pose a risk to human health or bring about an unacceptable change in the composition of the food/beverage or its sensory characteristics. This standard also provides positive lists, purity criteria, specific migration limits, in addition to describing the method of analysis to verify compliance with purity criteria and specific migration limits and is based on the Directives of the European Union and the regulations published in the Code of the Federal Regulations of the Food and Drug Administration [45, 46].

15.2 Mexico

The Mexican Ministry of Health assigned to the Federal Commission for the Protection against Sanitary Hazards – COFEPRIS (*Comisión Federal para la Protección contra Riesgos Sanitários*) – an organ of the Ministry of Health – the task of regulating and controlling sanitary hazards with the objective of reducing exposure of the population to chemical, physical, and biological hazards, which include foods, additives, and packages, among others [47].

In Mexico, there is no specific legislation in place on packages and packaging materials. The Regulation for the Sanitary Control of Products and Services (*Reglamento de Control Sanitario de Productos y Servicivios*) – published on August 9, 1999 – provides, under Title 24 (*Título Vigésimo Quarto*) Packages and Packaged Products (*Envases y envasado de productos*) – Sole Chapter (*Capítulo Único*), Articles 209–214, a description of some characteristics of packages for foods. In summary, the aforementioned articles state the following [48]:

- **Articles 209–211**: Establish that a classification of packages and the physical, chemical, and toxicity characteristics of each type of packaging material will be specified in corresponding standards.

- **Article 212**: Reused or recycled materials may be used in the manufacture of packages provided that packages obtained from such materials present a sanitary quality appropriate for and compatible with food contact.
- **Article 213**: The integrity of the package must be demonstrated to avoid health hazards and chemical and microbiological contamination of the product.
- **Article 214**: Establishes that packages may be reused to hold nonalcoholic carbonated beverages.

All technical standards and certifications are coordinated by the Mexican Organization for Standardization and Certification – Normex (*Organismo Nacional de Normalización e Certificación*). Most technical standards applicable to food packages were put in place in the 1990s. According to the Mexican Ministry of Economy, the new standards will be based on the regulations established by the Food and Drug Administration of the United States of America [47].

15.3 Central America

The seven countries that make up Central America, Costa Rica, Panama, Nicaragua, El Salvador, Guatemala, Honduras, and Belize do not have any specific legislation in place with respect to packaging materials intended for food contact. The General Health Laws or Health Codes of the respective countries contain some general requirements for direct food contact packages.

Packaging requirements aiming at guaranteeing the quality of foods and the appropriateness of packages to the processing and filling of foods are set forth in Resolution 176-2006 published by the Council of Ministers of Economic Integration, in accordance with the provisions of the General Treaty of Economic Integration signed by Costa Rica, El Salvador, Guatemala, Honduras, and Nicaragua. Resolution 176-2006 is composed of four Central American Technical Regulations – RTCA (*Reglamentos Técnicos Centroamericanos*):

- **RTCA 67.01.30:06**: Processed foods. Procedure for granting a Sanitary License to Manufacturing Plants and *Bodegas*.
- **RTCA 67.01.31:06**: Processed foods. Procedure for granting a Sanitary License and Sanitary Registry or Sanitary Inscription.
- **RTCA 67.01.32:06**: Requirements for the Importation of Processed Foods for Tasting and Display at Trade Shows and Exhibitions.
- **RTCA 67.01.33:06**: Food and beverage processing industries. Good Manufacturing Practices. General Principles.

Article 8 of the later regulation, Process and Production Control, describes, in item 8.3 – Packaging (*Envasado*) – requirements for packages as follows:

> "Any and all materials used for holding foods must ensure the integrity of the food product in foreseeable conditions of storage, should be inspected prior to

their use to ensure that they are in good state, clean and disinfected. In case such packages are to be reused, they must be adequately cleaned and sanitized".

15.3.1 Costa Rica

Some general requirements applicable to food packaging are set forth in the General Health Law – Ley no. 5395 of October 30, 1973 and its updated versions [49].

Decree No. 33 724, published in the *Gazeta* No. 82 of April 30, 2007 puts in force Resolution 176-2006 of the Council of Ministers of Economic Integration (General Treaty of Central American Economic Integration) [50].

15.3.2 El Salvador

Decree No. 955 of 28 April – Health Code – published by the Ministry of Public Health and Social Assistance – stipulates, in Section 12 Foods and Beverages, requirements for foods and commercial establishments selling foods. However, this piece of legislation does not contain any specific reference to packages intended to be placed in direct contact with foods or beverages.

The El Salvador National Council for Science and Technology – CONACYT (*Consejo Nacional de Ciencia e Tecnologia*), which is the government body responsible for working out mandatory technical standards in El Salvador has published several standards covering food packaging concerning, predominantly, specifications and physical–mechanical tests [51].

15.3.3 Guatemala

References to food packages can be found in the Health Code of the country.

The Guatemalan Health Code was published as Decree 90–97 on October 2, 1997 by the Congress of the Republic of Guatemala [52].

Book II (*Libro II*) Health Actions (*De las acciones de salud*), Title I (*Título I*) Health Promotion and Disease Prevention (*De las acciones de promoción y prevención*), Chapter V (Capítulo V) related to foods, establishments and retail commerce of foods (*alimentos, establecimientos y expendios de alimentos*), Section I (*Seccion I*) Health Protection Related to Foods, establishes in its Article 137 – Packaging Materials and Packages – that only the use of packaging materials or packages that are compatible with the foods they are intended to hold will be permitted and that such packaging materials or packages must not bring about changes in the composition and other characteristics of the foods as a result of package/food interaction.

Book III (*Libro III*) Health Infractions and Sanctions (*Infracciones contra la salud y sus sanciones*), Sole Title (*Título Único*), Chapter II (*Capítulo II*), Special Part (*Parte especial*), Section I (*Seccion I*) Infractions Punishable with Fines (*De las infracciones *sancionadas con multa*), establishes in Article 226 Special Cases (*Casos especiales*), Item 44, that the use of packages that negatively change the quality of food products constitutes what is considered a "special case of health infraction."

The Guatemalan Committee for Standardization (COGUANOR – Comissão Guatemalteca de Normas) is the government body responsible for working out standards, which are voluntary. A series of standards was published between 1985 and 1994 concerning metallic, plastic, and paperboard packages for foods. However, most of these standards are related to specifications and physical tests [53].

15.3.4 Honduras

Decree No. 65 Health Code, issued on May 28, 1991 and published in the *Gazeta* No. 26 509 on August 6, 1991, establishes in Book II (Libro II), related to Health Promotion and Disease Prevention (*Promoccion y Proteccion de la Salud*), Title II (*Título II*) Foods and Beverages (*De los Alimentos y de las Bebidas*), Article 0084 that surfaces that come in contact with foods or beverages must not bring about unacceptable changes in the sensory, physical–chemical, and biological characteristics of the food product and must be free of any contamination [54].

Article 0086 permits the reuse of containers provided they are adequately cleaned and sanitized and do not pose a contamination hazard to foods or beverages.

15.3.5 Belize

At present, in Belize, packaging materials are subject only to the provisions and requirements of the Food and Drug Act, Chapter 291, revised in 2000, which prohibits the addition to foods of substances that may affect health and make foods unfit for consumption [55].

15.4 Cuba

There is no specific legislation in place for food packaging in Cuba.

The requirements for packages for food contact are established by the Cuban obligatory standards: NC 452:2006 Packaging and Auxiliaries, general sanitary requirement and NC 456:2006 Equipment and articles for food contact, general sanitary requirements [56].

15.5 Conclusions

MERCOSUR members are the most organized countries in Latin America regarding food contact legislation.

In other countries, the food package requirements are described in the General Food Law or in the Health Code or even in the technical standards published by National Standards Association.

References

1 GRUPO MERCADO COMUN. MERCOSUR/GMC/RES no. 03/92. Criterios generales de envases y equipamientos alimentarios en contacto con alimentos. Buenos Aires, 01 de abril de 1992. Available at http://www.mercosur.int/msweb/Normas/normas_web/Resoluciones/ES/1992/RES_003_1992_ES_RnvEqu_Aliment.doc. Accessed December 2007.

2 Mercosur GMC resolutions on packaging materials for food contact use. Available at http://www.mercosur.int/msweb/portal%20intermediario/pt/index.htm. Accessed December 2007.

3 GRUPO MERCADO COMUN. MERCOSUR/GMC/RES no. 56/92. Disposiciones generales para envases y equipamientos plasticos em contacto con alimentos. Available at http://www.inmetro.gov.br/barreirastecnicas/PDF/GMC_RES_1992-056.pdf. Accessed December 2007.

4 GRUPO MERCADO COMUN. MERCOSUR/GMC/RES no. 32/07. Regulamento técnico Mercosur sobre "lista positiva de aditivos para materiais plásticos destinados à elaboração de embalagens e equipamentos em contato com Alimentos" (revogação das res. GMC no. 95/94 e 50/01). Montevidéu, 11 de dezembro de 2007. Available at http://www.mercosur.int/msweb/Normas/normas_web/Resoluciones/PT/2007/RES_032-2007_PT_Lista%20Positiva%20Aditivos-Anexos%20incluidos.doc. Accessed December 2007.

5 GRUPO MERCADO COMUN. MERCOSUR/GMC/RES no. 24/04. Reglamento técnico Mercosur sobre la lista positiva de. Polímeros y resinas para envases y equipamientos plásticos. Available at http://www.mercosur.int/msweb/Normas/normas_web/Resoluciones/ES/GMC_2004-10-08_NOR-RES_24_%20ES_RTM%20Lista%20Positiva%20de%20Pol%C3%ADmeros.PDF. Accessed December 2007.

6 GRUPO MERCADO COMUN. MERCOSUR/GMC/RES no. 30/92. Embalagens e equipamentos plasticos destinados a entrar em contato com alimentos: classificação dos alimentos e simulantes. Available at http://www.inmetro.gov.br/barreirastecnicas/PDF/GMC_RES_1992-030.pdf. Accessed December 2007.

7 GRUPO MERCADO COMUN. MERCOSUR/GMC/RES no. 32/97. Reglamento tecnico Mercosur sobre la incorporacion de la tabla no. 1: clasificacion de alimentos simulantes, como anexo de la Res GMC no. 30/92 "envases y equipamientos plasticos en contacto con alimentos:clasificacion de alimentos y simulantes". Montevideo, 05 de setembro de 1997. Available at http://www.cancilleria.gov.ar/comercio/mercosur/normativa/resolucion/1997/res3297.html. Accessed December 2007.

8 GRUPO MERCADO COMUN. MERCOSUR/GMC/RES no. 36/92. Ensaios de migração total de embalagens e equipamentos plásticos em contato com alimentos. Rio de Janeiro, 18 de janeiro de 1992. Available at http://www.cancilleria.gov.ar/comercio/mercosur/normativa/resolucion/1992/res3692.html. Accessed December 2007.

9 GRUPO MERCADO COMUN. MERCOSUR/GMC/RES no. 28/93. Disposiciones sobre colorantes y pigmentos en envases y equipamientos plásticos en contacto con alimentos. Available at http://www.cancilleria.gov.ar/comercio/mercosur/normativa/resolucion/1993/res2893.html. Accessed December 2007.

10 GRUPO MERCADO COMUN. MERCOSUR/GMC/RES no. 32/99. Regulamento técnico Mercosul sobre metodologias analíticas de referencia para controle de embalagens e equipamentos em contato com alimentos. Asunción, 10 de junho de 1999. Available at http://www.inmetro.gov.br/barreirastecnicas/PDF/GMC_RES_1999-032.pdf. Accessed December 2007.

11 GRUPO MERCADO COMUN. MERCOSUR/GMC/RES no. 25/99. Regulamento técnico Mercosul sobre embalagens descartáveis de PET multicamada destinadas ao acondicionamento de bebidas não-alcoólicas carbonatadas. Asunción, 10 de junho de1999. Available at http://www.inmetro.gov.br/barreirastecnicas/PDF/GMC_RES_1999-025.pdf. Accessed December 2007.

12 GRUPO MERCADO COMUN. MERCOSUR/GMC/RES no. 30/07. Reglamento técnico Mercosur sobre envases de Polietilentereftalato (PET) postconsumo reciclado grado alimentario (PET-PCR grado alimentario) destinados a estar en contacto con alimentos. Available at http://www.mercosur.int/msweb/Normas/normas_web/Resoluciones/ES/2007/RES_030-2007_ES_PET-PCR.doc. Accessed December 2007.

13 GRUPO MERCADO COMUN. MERCOSUR/GMC/RES no. 54/97. Reglamento tecnico mercosur sobre envases y equipamientos elastoméricos destinados a entrar en contacto con alimentos. Montevideo, 13 de dezembro de 1997. Available at http://www.cancilleria.gov.ar/comercio/mercosur/normativa/resolucion/1997/res5497.html. Accessed December 2007.

14 GRUPO MERCADO COMUN. MERCOSUR/GMC/RES no. 28/99. Reglamento técnico Mercosur sobre lista positiva para envases y equipamentos elastoméricos em contacto com alimentos. Available at http://www.cancilleria.gov.ar/comercio/mercosur/normativa/resolucion/1999/res2899 html. Accessed December 2007.

15 GRUPO MERCADO COMUN. MERCOSUR/GMC/RES no. 27/99. Reglamento técnico Mercosur sobre adhesivos utilizados en la fabricación de envases y equipamientos destinados a entrar en contacto con alimentos. Asunción, 10 de junho de 1999. Available at http://www.mercosur.int/msweb/Normas/normas_web/Resoluciones/ES/Res_027_099_RTM_Lista%20Pos-Env-Equip_Elastom_Alimentos_Acta%202_99.PDF. Accessed December 2007.

16 GRUPO MERCADO COMUN. MERCOSUR/GMC/RES no. 67/00. Reglamento técnico Mercosur sobre parafinas em contato com alimentos. Brasilia, 07 de dezembro de 2000. Available at http://www.mercosur.int/msweb/Normas/normas_web/Resoluciones/ES/Res_067_000_RTM_Parafinas_Contac-Alim_Acta%204_00.PDF. Accessed December 2007.

17 GRUPO MERCADO COMUN. MERCOSUR/GMC/RES no. 19/94. Envases y equipamientos celulósicos en contacto con alimentos. Available at http://www.mercosur.int/msweb/Normas/normas_web/Resoluciones/res94es/9419.pdf. Accessed December 2007.

18 GRUPO MERCADO COMUN. MERCOSUR/GMC/RES no. 35/97. Modificación de la RES GMC no. 19/94 disposiciones generales sobre envases y equipamientos celulósicos en contacto con alimentos. Montevideo, 05 de setembro de 1997. Available at http://www.inmetro.gov.br/barreirastecnicas/PDF/GMC_RES_1997-035.pdf. Accessed December 2007.

19 GRUPO MERCADO COMUN. MERCOSUR/GMC/RES no. 20/00. Modificação do regulamento técnico Mercosur "embalagens e equipamentos celulósicos em contato com alimentos" (resolução GMC no. 19/94). Buenos Aires, 28 de junho de 2000. Available at http://www.inmetro.gov.br/barreirastecnicas/PDF/GMC_RES_2000-020.pdf. Accessed December 2007.

20 GRUPO MERCADO COMUN. MERCOSUR/GMC/RES no. 56/97. Reglamento técnico mercosur sobre lista positiva para envases y equipamientos

celulósicos en contacto con alimentos. Montevideo, 13 de dezembro de 1997. Available at http://www.mercosur.int/msweb/Normas/normas_web/Resoluciones/ES/Res_056_097_.PDF. Accessed December 2007.

21 GRUPO MERCADO COMUN. MERCOSUR/GMC/RES no. 12/95. Ensayo de migración total de envases y equipamientos celulósicos. Disponivel em: http://www.mercosur.int/msweb/Normas/normas_web/Resoluciones/ES/GMC_1995_RES_012_ES_EnvEqui_Celulo.PDF. Accessed December 2007.

22 GRUPO MERCADO COMUN. MERCOSUR/GMC/RES no. 47/98. Reglamento técnico Mercosur sobre papeles de filtro para cocción y filtración en caliente. Rio de Janeiro, 8 de dezembro de1998. Available at http://www.mercosur.int/msweb/Normas/normas_web/Resoluciones/ES/Res_047_098_RTM_Pap-Filtro_Cocci%C3%B3n_Filt-Caliente_Acta%204_98.PDF. Accessed December 2007.

23 GRUPO MERCADO COMUN. MERCOSUR/GMC/RES no. 52/99. Reglamento técnico Mercosur sobre material celulósico reciclado. Montevideo, 29 de dezembro de 1999. Available at http://www.mercosur.int/msweb/Normas/normas_web/Resoluciones/ES/Res_052_099_RTM_Material%20Celul%C3%B3sico_Reciclado_Acta%203_99.PDF. Accessed December 2007.

24 GRUPO MERCADO COMUN. MERCOSUR/GMC/RES no. 55/97. Reglamento técnico Mercosur para plículas de celulosa regenerada destinadas a entrar em contacto com alimentos. Montevidéo, 13 dezembro de 1997. Available at http://www.inmetro.gov.br/barreirastecnicas/PDF/GMC_RES_1997-055.pdf. Accessed December 2007.

25 GRUPO MERCADO COMUN. MERCOSUR/GMC/RES no. 68/00. Reglamento técnico mercosur sobre tripas sintéticas de celulosa regenerada en contacto con alimentos. Brasília, 07 de dezembro de 2000. Available at http://www.mercosur.int/msweb/Normas/normas_web/Resoluciones/ES/Res_068_000_RTM_Tripas%20Sint%C3%A9ticas_Alim_Acta%204_00.PDF. Accessed December 2007.

26 GRUPO MERCADO COMUN. MERCOSUR/GMC/RES no. 46/06. Reglamento técnico mercosur sobre disposiciones para envases, revestimientos, utensilios, tapas y equipamientos metálicos en contacto con alimentos (derogación de las RES. GMC no. 27/93, 48/93 y 30/99). Brasilia, 24 de novembro de 2006. Available at http://www.mercosur.int/msweb/Normas/normas_web/Resoluciones/ES/RES%20046-006_ES_%20Envases%20metalicos.PDF. Accessed December 2007.

27 GRUPO MERCADO COMUN. MERCOSUR/GMC/RES no. 55/92. Reglamento técnico Mercosur sobre envases y equipamientos de vidrio y cerámica destiuadas a entrar em contaco com alimeutos. Available at http://mercosur.org.ey/show? contetid=610. Accessed December 2007.

28 Padula, M., Garcia, E.E.C., and Cuêrvo, M. (2006) Food contact legislation: Mercosur and South América. *World Food Regulation Review*, **15** (12).

29 BRASIL. Agência Nacional de Vigilância Sanitária - ANVISA. Resolução 105, de 19 de maio de 1999. Aprova regulamentos técnicos sobre disposições gerais para embalagens e equipamentos plásticos para contato com alimentos. Diária Optical de República do Brasil Brasila. DE, 20 de maio de 1999 Seçâo 1.

30 Padula, M. and Anosti, A., Leglislación MERCOSUR sobre la aptitud sanitária de los envases para alimentos, in: R. Catalá and R. Gavara (Eds) *Migracióon de componentes y residuos de envases em contacto com alimentos*. Valencia: Instituto de Agroquímica y Tecnología de Alimentos. CSIC/CYTED/IATA, 2002, p. 45–83.

31 BRASIL. Agência Nacional de Vigilância Sanitária. Resolução no 23, de 15 de março de 2000. Dispõe sobre o manual de procedimentos básicos para registro e dispensa da obrigatoriedade de registro de produtos pertinentes à área de alimentos. Diário Oficial [da] República Federativa do Brasil, Poder Executivo, Brasília, DF, 16 de março de 2000. Available at http://e-legis.

anvisa.gov.br/leisref/public/showAct.php?id=22680&word=. Accessed December 2007.

32 Garcia, E.E. (2000) Dispensa do registro de embalagens. Informativo CETEA, Campinas, 12 (2), 9–11, Available at http://www.cetea.ital.sp.gov.br/infCETEA.htm. Accessed December 2007.

33 BRASIL. Agência Nacional de Vigilância Sanitária. Resolução no. 22, de 15 de março de 2000. Dispõe sobre os procedimentos básicos de registro e dispensa da obrigatoriedade de registro de produtos importados pertinentes à área de alimentos. Diário Oficial [da] República Federativa do Brasil, Poder Executivo, Brasília, DF, 16 de março de 2000. Available at http://e-legis.anvisa.gov.br/leisref/public/showAct.php?id=136&word=. Accessed December 2007.

34 LA JUNTA DE GOBIERNO DE LA REPUBLICA DE VENEZUELA. Decreto no. 525, de 12 de enero de 1959. Reglamento general de alimentos. Gazeta Oficial, Caracas, 16 de enero de 1959. n. 25.864, 14 p. Available at http://www.gobiernoenlinea.ve/docMgr/sharedfiles/reglamentogeneralalimentos.pdf. Accessed December 2007.

35 REPÚBLICA BOLIVARIANA DE VENEZUELA. Ministério del Poder Popular para la Salud. Resolución SG-081-96. Normas complementarias del reglamento general de alimentos. Gaceta Oficial, Caracas, 15 de marzo de 1996. n. 35.921.

36 REPÚBLICA BOLIVARIANA DE VENEZUELA. Ministério del Poder Popular para la Salud. Normas sobre prácticas para la fabricación, almacenamiento y transporte de envases, empaques y/o artículos destinados a estar en contacto con alimentos. Gazeta Oficial, Caracas, 4 de mayo de 2007. n. 38.678, 13 p. Available at http://www.avipla.org/uploads/documentos/Resnro_82_envases.pdf. Accessed at December 2007.

37 FONDO PARA LA NORMALIZACIÓN Y CERTIFICACIÓN DE CALIDAD – FONDONORMA. Available at http://www.fondonorma.org.ve.

38 REPUBLICA DE CHILE. Ministério de Salud. Depto. Asesoria Jurídica. Decreto no. 977/96, de 6 de agosto de 1996. Reglamento sanitário de los alimentos. Diário Oficial de la República de Chile, Santiago, 13 de maio de 1997. Available at http://www.sernac.cl/leyes/compendio/DS/ds_977-96_reglamento_alimentos.pdf. Accessed at December 2007.

39 INSTITUTO BOLIVIANO DE NORMALIZACIÓN Y CALIDAD – IBNORCA. Disponible em: http://www.ibnorca.org. Accessed at December 2007.

40 EQUADOR. Decreto Ejecutivo 4114. Reglamento de alimentos. Registro Oficial, de 22 de julio de 1988. n. 984, 26 p.

41 INSTITUTO ECUATORIANO DE NORMALIZACIÓN – INEN. Available at http://www.inen.org.ec.

42 PERU. El Presidente de La República. Decreto Supremo no. 007-98-AS, de 24 de setembro de 1998. Aprueban reglamento sobre vigilancia y control sanitario de alimentos y bebidas. Diário Oficial El Peruano, Lima, año 16, n. 6666, 25 de setiembre de 1998. pp. 164319–164334. Available at http://www.minsa.gob.pe/leyes/ds00798sa_rvcsab/ds.htm. Accessed December 2007.

43 INSTITUTO NACIONAL DE DEFENSA DE LA COMPETENCIA Y DE LA PROTECCIÓN DE LA PROPIEDAD INTELECTUAL – INDECOPI. Available at http://www.indecopi.gob.pe/. Accessed December 2007.

44 COLOMBIA. El Presidente de la República. Decreto no. 3075, de 1997. Por el cual se reglamenta parcialmente la Ley 09 de y se dictan otras disposiciones. Available at http://www.suratep.com/index.php?option=com_content&task=view&id=118&Itemid=136. Accessed December 2007.

45 INSTITUTO COLOMBIANO DE NORMAS TÉCNICAS Y CERTIFICACCIÓN – ICONTEC. NTC 5022: materiales y artículos plásticos destinados a estar en contacto con alimentos y bebidas. Determinación de migración global, 2003, 15 p.

46 INSTITUTO COLOMBIANO DE NORMAS TÉCNICAS Y CERTIFICACCIÓN – ICONTEC NTC

5023: materiales, compuestos y artículos plásticos para uso en contato con alimentos y bebidas, 2001, 16 p.

47 Soto-Valdez, H. (2006) *Latin American food contact legislations, in Food Contact Legislation for Packaging in Emerging Markets, 2006, San Francisco*, PIRA, Leatherhead, p. 19.

48 MEXICO. Secretaria de Salud. Reglamento de control sanitario de productos y servicios. Diario Oficial dos Estados Unidos Mexicanos, 09 agosto 1999. 98 p. Available at http://www.cibiogem.gob.mx/Normatividad/normatividad_SSA/071005RCSPS.pdf. Accessed December 2007.

49 COSTA RICA. La Asamblea Legislativa de la República. Ley no. 5395, de 30 de octubre de 1973. Ley general de salud. Gaceta No. 222, San José, Costa Rica, 24 nov. 1973. 97 p. Available at http://www.ministeriodesalud.go.cr/leyes/leygeneraldesalud.pdf. Accessed December 2007.

50 COSTA RICA. El Consejo de Ministros de Integración Económica. Decreto no. 33724. Pone vigencia Resolución 176-2006 (COMIECOXXXVIII)> alimentos procesados proced. licencia sanitaria, proced. otorgar registro sanitario y inscripción sanitaria, requisitos importación alimentos procesados, industria alimentos bebidas procesados. Gaceta no. 82, San José, Costa Rica, 30 abr. 2007, 20 p.

51 EL SALVADOR. Consejo Nacional de Ciencia e Tecnología – CONACYT. Available at http://www.conacyt.gob.sv.

52 GUATEMALA. Congreso de la República de Guatemala. Decreto no. 90–97. Código de salud. Available at http://www.disaster-info.net/PED-Sudamerica/leyes/leyes/centroamerica/guatemala/salud/Codigo_de_Salud.pdf. Accessed December 2007.

53 COMISIÓN GUATEMALTECA DE NORMAS – COGUANOR. Available at http://www.coguanor.org/.

54 HONDURAS. Ministerio de Salud Publica. Decreto no. 65–91, de 28 de maio de 1991. Codigo de salud. Gaceta no. 26509, 06 ago. 1991. Available at http://www.ministeriodesalud.go.cr/leyes/leygeneraldesalud.pdf. Accessed December 2007.

55 BELIZE. Law Revision Commissioner. Food and Drugs Act. Chapter 291, 69 p.

56 CUBA. Oficina Nacional de Normalización. Available at www.mconline.cu. Accessed December 2007.

16
Israel's Legislation for Food Contact Materials: Set for the Global Markets

Haim H. Alcalay

Abstract

Present Israeli regulations for food contact materials (SI-5113) define the *updated* EU food contact materials Directives *or* the FDA regulations as acceptable. This approach of accepting both regulations is appropriate for countries, such as Israel, which have large exports and wish to be integrated into the global economy.

The legislative process – led by SII (Standards Institution of Israel) – is described. It is based on a dual Technical and Expert's committee. The provisions and documentations of the SI-5113 (for local and imported packaging) are enumerated and approved Israeli tests laboratories are listed.

Issues related to actual implementation of the food contact material standards are discussed.

The "guiding principles" for food contact materials are reviewed again in order to place the FC regulations in the proper perspective with regard to "risk assessment."

Finally, some thoughts on "undertakings for the future" are presented, including creation of a "global database for food contact regulations," listing of approved test labs worldwide, and creation of a global task force for a continued support on these matters.

To enable the global market to grow, within internationally safe and acceptable food contact regulations, it appears that there should be a "mutual respect" among the various legislative bodies and "reciprocal arrangements" should be agreed upon!

16.1
Introduction

This chapter intends to cover several aspects of food contact legislation in Israel.

The primary purpose is to describe the present regulations as adopted in Israel under Israel Standard SI-5113 "Plastics Materials and Plastic Articles in Contact with Food and Beverages" [1].

Furthermore, the actual "legislation process" will be reviewed, describing the parties and considerations involved. Some comments will be made with respect to

Global Legislation for Food Packaging Materials. Edited by Rinus Rijk and Rob Veraart

ISBN: 978-3-527-31912-1

"implementation issues." The importance of framing regulations that are appropriate in the current "global market" economy will be described since it reflects the needs of a small country such as Israel that depends on exports to all world markets. Finally, some thoughts on undertakings for the future will be enumerated, based on the guiding principles of the global food contact legislation.

16.2
The Standards: Legislative Process in Israel

The Standards Institution of Israel (SII) is the only statutory body in Israel that develops and establishes standards. SII was created by the Knesset (Israeli Parliament) in 1953 and was mandated with the responsibility of the preparation and publication of technical specifications and standards for products and services that are produced locally or are imported.

SII encompasses standardization, testing, inspection services, and conformity assessment management system certification (ISO). SII members' have participated in several international certification bodies.

The overall policy and specific approvals of action are subject to the "Commissioner of Standards" under the Ministry of Industry, Trade, and Labor.

16.3
Technical and Expert Forums at SII

The specific route for SI-5113 is described in Figure 16.1.

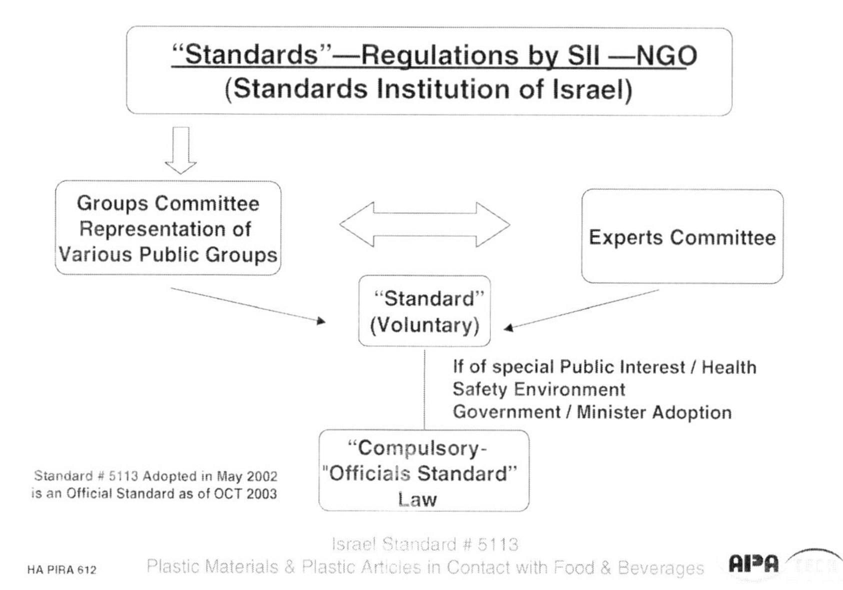

Figure 16.1 Israel food contact legislation process.

Once the "General Plastic Committee at SII" approves a subject matter for certification, a specific subcommittee made up of several (predetermined) representatives of *interest groups* is convened to discuss the framework and scope of the proposed standard. The representatives include industry and consumer groups, trade organizations, university members, and SII representatives, as shown below.

- Technical Committee # 701 Polymeric Resins and Analytical Methods
  - Association of Plastics and Rubber Industry
  - Association of Kibbutz Industries
  - Chamber of Commerce
  - Israel Plastics and Rubber Society
  - Technion – Israel Institute of Technology – Material Dept.
  - Israel Industries Association
  - Workers Union – Consumers
  - The Standards Institution of Israel (SII)
  - Israel Plastics and Rubber Center
  - IDF – Israel Defence Forces

Actual details of proposed standards are referred to the Committee of Experts that consists of the following people:

- Committee of Experts
  - Dr. H. A. (Chair) – Israel Plastics and Rubber Industries
  - Dr. R. A. – SII – Standards Institution of Israel
  - Dr. S. H. – Israel Manufacturers Association
  - Mr. J. R. – Technical Director – Nestle–Osem – Israel
  - Dr. A. S. – Israel Ministry of Health
  - Mr. N. S. – Technical Director Kafrit Industries
- Also, contributing to the preparation of standards:
  - Mr. E. F. – Technical Director – Agresco
  - Prof. J. M. – Head Food Lab – Technion Israel
  - Mr. R. M. – Global R.S – Lamination Co.
  - Ms. C. S. – Ministry of Commerce and Trade – Israel
- Government

Experts are drawn from various disciplines, including industry, laboratory, and packaging production experts, health experts, universities (food lab), and users of packaging in export markets. The integrated experience of the experts is translated into recommendations for the actual proposed standard that are sent back to the technical subcommittee for final approval to "publish" the new standard for public review and comments.

16.4 Voluntary Standards and "Compulsory – Official Standards"

Upon completing a proposed new standard, the SII distributes the draft to numerous "potential interested parties," and moreover, the draft standard is published on the

Figure 16.2 SII marks.

SII web site for public review and comments. After public review, the proposed new standard has to be approved by the Commissioner of Standards before it becomes official. Although standards are "voluntary," yet they set the requirements for product specification. Companies that decide to abide by a standard can submit their product and production process to SII review and certification, thus enabling them to get compliance SII mark (Figure 16.2).

If a given standard is deemed to have special public interest because of health, safety, or environmental implications, the Minister in charge of the given field can adopt the standard and designate it an "official standard," thus becoming a compulsory law. SI-5113 was published as a standard in 2002 to become later on an "official standard," the law of the land, which all must follow.

16.5
Food Contact SI-5113 Provisions

After intense discussions, it was decided that Israel's Food Contact Regulation will be based on existing food contact regulation of other countries, rather than developing a new standard.

The actual decision was to adopt two standards meeting food contact regulations, the *European Directives* and the *FDA-USA CFR-21 Regulations*. It is important to stress that it is essential to meet the requirements of the *updated* regulation, thus taking into account the changes that take place all the time in the scope of regulations.

The decision to adopt two of the most common world standards relates in no small part to the fact that Israel exports a lot of its packaging goods – thus, knowing and meeting the destination countries' norms is essential for high-quality exports. Actually, over 70% of Israeli packaging is exported to the United States and Europe (about equal share), thus there was no reason for Israel to choose one over the other. Despite the differences of approach between the European and the US regulations, it was not deemed necessary to rate one legislation over the other. It was believed that any legislation meeting either standard will have a safe package.

In the absence of compliance to EU Directives or FDA guidelines, requirements as defined by the Ministry of Health will apply. In addition, Standard-5113 Provisions have requirements for recycled materials.

16.6 Documentation Requirements

SI-5113 requires that a statement of compliance be made by the producer to which US or EU standards they conform. The documentation requirements do include the following items:

1) The producer of " finished polymeric article" will provide the verification body a statement of compliance stating that all materials used meet Standard-5113 requirements.
2) The producer of "finished article" will keep in his records all compliance statements (DOC) and support documents (SDs) provided by all parties in the production chain – ingredient supplier, resin manufacturer, and converter – to be submitted to the " verification body" upon demand.
3) Upon the demand of the Ministry of Health, compliance statements will be accompanied by test results from an accredited laboratory (for specific tests required) and by MSDS for all materials.
4) The producer of "finished article" will keep all DOC and SDs for at least 2 years after the said article is no longer sold on the market.

Furthermore, the packer must keep *compliance documents*/MSDS from all suppliers in "production chain." The Ministry of Health can require test results to be shown from an accredited lab (at present, some of the "demand functions" have been "transferred in practice" to the SII). It is well known, of course, that several international companies have their own criteria and may impose additional requirements in order to approve their supplier.

16.7 Approved Test Laboratories in Israel

There are three laboratories in Israel that are "approved" to carry out the various food contact tests (migration) and provide "compliance certificates."

- Packaging and Food Laboratories, Technion, Israel Institute of Technology, Haifa.

- IPRC, Israel Plastics and Rubber Center, Shenkar College, Ramat Gan
- SII (Standards Institution of Israel), Chemistry Labs, Tel-Aviv

With respect to testing imported packaging material, the SII is the sole agency established by law to carry out conformation tests.

16.8
Introducing the New Standard in Israeli

A new standard must be properly introduced to those who must conform to it. Several steps have been taken to familiarize and explain SI-5113 provisions and requirements to producers of FC plastic packages in Israel. These steps are as follows:

1) Publish the draft of the standard “open to public” for comments.
2) Formally present it in government publications.
3) Publish articles in local trade journals (*Packaging News, Plastics Magazine, Export Institute News,* and so on).
4) Giving a lecture before the board members of the food and packaging industry organization.
5) Holding seminar and panel discussion with technical directors, quality managers, laboratory personnel, and industry and university R&D people.
6) In-house seminars in key food and packaging companies.

The standard is published in official gazette and trade magazines, and lectures and seminars are organized for lab QC personnel, for the management personnel of package producers (flexible and rigid), and users (food companies).

16.9
Imports of Packaging Materials into Israel

The Ministry of Industry and Trade has set up Procedure No. 401 to deal with official standards. It is divided into four groups that depend on the level of hazards involved in food packaging materials.

- **Group 1**: goods with the highest hazard level, in order to deliver from customs, must have a certificate of compliance from the SII.
- **Group 2**: goods with medium hazard, the importer must provide a representative sample of the product and must have a declaration of compliance with the entire product records available for 7 years from the first importation.
- **Group 3**: goods with low hazard, the importer must provide SII acceptance based on simply the declaration of compliance. Product records must be kept for 7 years from first importation.
- **Group 4**: goods for industrial use only and not to be directly used by consumers, there is no restriction on import. Details on above are available on the web site of the Ministry of Industry and Trade [2].

The general information required for registration of plastic packaging materials for food contact with food and beverages includes the following:

1) List of raw materials
2) Declaration of food approval for new raw materials, description of the use of final article (specifying the type of food, the time, and the temperature) that will be held in the packaging material.

For imported articles, an import form must also be attached. The labs of SII will decide if and which test will have to be performed. Usually, the permission granted by the standards institute will be valid for 12 months with a possibility of renewal 1 month before expiration of permission.

16.10 Global Israeli Approach to Food Contact Legislation

In order to enable to have a greater participation in global markets, Israel has accepted the use of either the EU Directives [3] or the US FDA regulations [4]. Furthermore, internationally recognized testing procedures (DIN, ASTM) have been adopted and certification bodies allowed. There is acceptance of internationally accredited labs, but this must be based on reciprocal agreements. Finally, there is an extensive participation of Israel in international standards organizations such as ISO and IEC.

The participation of Israel in the global economy is also demonstrated by the fact that several international companies operate within Israel.

With all these measures, it is expected that Israeli products will gain acceptance as quality and safe products on the world markets.

16.11 Kosher Regulations

It may be worth mentioning that kosher regulations are not part of the official food contact regulations of Israel. However, kosher food is of great importance in Israel and around the world. To be "Kosher" one must meet a very stringent set of regulations. Several "authorizing bodies exist" and appropriate markings on the "packed food" such as "OU," "U," "K," and so on attest to norms it meets.

"Kosher" may be one of the earliest known "regulatory systems." It is basically derived from "religious principles" that possibly relate to dietary laws that in fact ensure health and safety of those who abide by them (just the same as food contact regulations are to provide). Demonstrating the large spread of "Kosher" brands are "OU" (Orthodox Union) regulations that certify more than 400 000 products. Packaging must meet these standards since some "additives" made from, say, animal fat may not be acceptable.

16.12 Implementation Issues (FAQ)

Although the food contact regulations, the EU Directives or US FDA regulations are quite detailed, packaging material producers often ask/need clarifications. Frequently asked questions (FAQ) are listed below.

- Validity of certificates: 1, 5 years or indefinite, as long as all parameters are the same.
- Which packages or package families must be submitted for actual tests (migration – total/specific) – for example, for articles made from the same resins and components but different sizes – is a repeat test necessary?
- Similarly, the same article made from "generic" resins or components, but from different manufacturers (all having MSDS and certificates attesting that all components are approved for food contact), need separate tests?
- Are US certificates acceptable in Europe and vice versa?
- Which labs are recognized/accredited, where to find them worldwide?
- How to choose "representative samples" from a large array of articles of rather similar construction (e.g., laminated films)?

It may be worthwhile to include an appendix or a listing on the official regulatory agency's web site to deal with them.

16.13 Guiding Principles

In order to place a global food contact legislation in appropriate priority among "global issues," it is worthwhile to reemphasize the "guiding principles" for it to exist:

- Mitigate risk.
- Balance and prioritize requirements with proper perspective to take into account health risks and social and environmental impacts.
- Apply market-based mechanisms of sustainability, mandatory disclosure, and incentives.

With regard to priorities, it is important to view the data available with respect to "hazards" and portion related to packaging. Professor Vincent Aegarty lists food-related EU notifications, by hazard types [5]. Of all hazard types – about 2500 – on the top of the list are mycotoxins (881) and chemical contamination (636), while *defective/ incorrect packaging* is at the bottom of the list with 18 notifications (Italy 8, Denmark 4, Malta and France 2 each, Belgium and UK 1 each).

References

1 SII, The Standards Institution of Israel www.sii.org.il.

2 The Ministry of Trade and Industry www.tamas.gov.il.

3 The EU web site http://www.ec.europa.eu/food/food/chemicalsafety/foodcontact/documents_en.htm.

4 The FDA web site http://www.cfsan.fda.gov/~lrd/foodadd.html.

5 Alcalay, H., "Current Developments in Israel Regulations" at Global Food Contact Legislation Conference Madrid 3–4 July 2006.

17
Rules on Food Contact Materials and Articles in Japan

Yasuji Mori

17.1
Introduction

In Japan, there are two national laws for food safety. One is "the Food Sanitation Law" and other is "the Food Safety Basic Law." They were enacted under "The Constitution of Japan" Article 25 [1] that pledges: "All people shall have the right to maintain the minimum standards of wholesome and cultured living. In all spheres of life, the State shall use its endeavors for the promotion and extension of social welfare and security, and of public health."

In 1947, the Food Sanitation Law [2] was enforced. This law regulated the safety of all foods, food additives, apparatus, and packages/containers without drugs and quasidrugs regulated under "the Pharmaceutical Affairs Law" [4], and this law was placed as a food safety management regulation, recently. In 2003, the food safety basic law [3] was enforced. This law was promulgated after the BSE (bovine spongiform encephalopathy), or mad cow disease, outbreak, which was a turning point, and was placed as a food safety assessment regulation recently.

Under the food sanitation law, the Minister of Health, Labor, and Welfare [9] may establish specifications for apparatus and packages/containers of foods from the viewpoint of public health.

"Specifications and standards for foods, food additives, and other materials" [7] were established by the Ministry of Health and Welfare in 1959. This is known as the Notification No. 370 in Japan. In Japan, milk and milk products are distinguished as special foods for baby, children, and health-poor people. The Ordinance of Specifications and Standards for Milk and Milk Products [8] was promulgated by the Ministry of Health and Welfare in 1951; this is called "the ordinance of milk" or "the Ordinance No. 52."

Under the food safety basic law, the Food Safety Commission [10] was established within the Cabinet Office. The roles of the Food Safety Commission include the risk assessment (assessment of the effect of food on health), risk communication, and emergency response. Before an amendment to the specifications and standards for apparatus and containers/packages is enforced, the Ministry of Health, Labor, and Welfare requests the Food Safety Commission for risk assessment.

Global Legislation for Food Packaging Materials. Edited by Rinus Rijk and Rob Veraart

ISBN: 978-3-527-31912-1

The food sanitation law was established under the policy of minimum requirement. Under this law, all food packaging material makers are responsible for their product safety. In the 1960s, package makers, material suppliers, and food producers established industrial safety associations under the support and guidance of the authorities. To complement the food sanitation law, each industrial safety association framed certain voluntary rules, and members of each association were required to comply with these rules. As almost all package makers and material suppliers in Japan joined respective industrial safety associations, these voluntary rules later on became official regulations.

Understanding the food sanitation law and the voluntary rules will help understand how the food safety laws function in Japan.

17.2 The Food Sanitation Law

On December 24, 1947, the food sanitation law (Law No. 233) was promulgated. To complement the food sanitation law, "The Food Sanitation Law Enforcement Regulations" (No. 23, July 1948) and "The Food Sanitation Law Enforcement Ordinance" (No. 229, August 1953) [5, 6] were established.

The food sanitation law actually regulates food safety and part of the regulation sets requirements for apparatus and packages/containers.

The following are the main articles on food packages/containers.

17.2.1 Articles of the Food Sanitation Law

Chapter 1 General Provision
Article 1 [Purpose]

> The purpose of this law is to prevent accident or outbreak of health hazards caused by eating or drinking of foods and to strive for protection of people's health, by establishing necessary regulations and doing management from the viewpoint of public health to ensure food safety.

The last amendment to the Purpose (2003) "to strive for protection of people's health" was added to the main purpose after the occurrence of food-poisoning accidents and BSE.

Article 2 [The duty of nation, prefectures, etc.]
Article 2 describes the duties of the nation, prefectures, and cities that include in principle the availability of education and facility to examine food sanitation.

Article 3 [The duty of persons in food business, etc.]
Article 3 requires business operators in food production to get knowledge on techniques to secure safety of food and food additives. Business operators should

maintain a system to trace materials to allow a quick withdrawal from the market in case of a food sanitation hazard.

Article 4 [Definition]

In this article, various definitions are given; for example,

> "Foods" mean all eating and drinking foods excluding drugs and quasidrugs that are regulated by the Pharmaceutical Affairs Law (No. 145, 1960).
> Terms such as food additives, natural flavoring agent, apparatus, packages/containers, food hygiene, business and registered laboratories that are used in this law are defined.
> Definition of apparatus: tablewares, utensils, and machines or apparatus are the articles that come in direct contact of food or food additives. These do not include machines or equipment used in agriculture and fisheries.
> Definition of packages/containers: articles used for packaging foods or food additives and containing foods or food additives for delivery to the user.

Chapter 2 Foods and Food Additives

This chapter sets requirements for food and food additives, and may concern not only compositional requirements but also hygienic handling of food and contamination of food, for example, with pesticides.

Chapter 3 Apparatus and Packages/Containers

This chapter sets the general requirements for apparatus and packages/containers as detailed in the following sections.

Article 15 [Principals for handling of apparatus and packages/containers for sales]

> Any apparatus and package/container used in business should be clean and sanitary.

Article 16 [Prohibition of sales of harmful or poisonous apparatus and packages/containers]

> Any apparatus and packages/containers which may be injurious to human health by containing toxic or harmful substances or sticking these substances or any which may be injurious to human health by contacting foods or food additives with harmful effect are prohibited for sale or manufacture for sale or import or use in business.

This is a very important article. Even if any product complies with the specifications and standards and if the product is injurious to human health, then the product violates this law.

Article 17 [Prohibition of sale of specific apparatus]

In case a specific apparatus or package/container is found injurious to human health, the Minister of Health, Labor, and Welfare may prohibit the sale or import or manufacture for sales or use of such article in business after seeking the opinion of the Pharmaceutical Affairs and Food Sanitation Council. Specific apparatus or packages/containers that cause injury to human health are covered under the following categories:

1) Apparatus or package/container prohibited by Article 16.
2) Apparatus or package/container that is not in compliance with specifications and standards established by the Minister of Health, Labor, and Welfare under Article 18.

Article 18 [Establishment of specifications and standards of apparatus and packages/containers]

> The Minister of Health, Labor and Welfare may be establishing the specifications and standards of apparatus or packages/containers which are used in business or are on sale or materials of apparatus or packages/containers, after hearing the opinion of "the Pharmaceutical Affairs and Food Sanitation Council," from the viewpoint of public health.

Under this article, the specifications and standards for foods, food additives, and so on were established by the Ministry of Health, Labor, and Welfare in 1959, and the ordinance for specifications and standards for milk and milk products was promulgated by the Ministry of Health, Labor, and Welfare in 1951.

Apparatus and packages/containers and their materials not in compliance with the relevant specifications and standards shall not be used in food contact applications.

Chapters 4–6 concern the labeling and advertisement, official papers on food additives, and plan of inspection and guidance, which is drafted yearly by the Ministry of Health, Labor, and Welfare. These chapters mainly refer to foods and food additives but may also be applicable to apparatus and packages/containers.

Chapter 7 Inspection
Article 27 [Notification of importing foods]

> The person who will import foods or food additives or apparatus or packages/containers to use for business or sale should notify the Minister of Health, Labor and Welfare on each import following the ordinance of the Ministry of Health, Labor and Welfare.

The requirements mentioned are specified under Article 32, "The Food Sanitation Law Enforcement Regulations."

Articles 28–30 set the requirements for inspection of documentation or site. In addition, it is described that the government and each prefecture are responsible for establishing facilities to perform examination of foods, food additives, apparatus, and packages/containers.

Chapter 8 The conditions for registered laboratories are laid down. Such laboratories may perform analysis for food safety.

Chapters 9–11 Plan of inspection and guidance.
A tentative English translation of the food sanitation law appears at the web site: http://www.mhlw.go.jp/english/topics/foodsafety/index.html.

17.3
Specifications and Standards for Food and Food Additives (Notification No. 370)

In 1948, the first specifications and standards were established by the Ministry of Health and Welfare as "Notification No. 54" and the methods of inspection were established as "Notification No. 106."

In 1959, "The Specifications and Standards for Foods and Food Additives" (Notification No. 370) were established superseding Notifications 54 and 106.

17.3.1
Contents

The following is the content of this notification.

Chapter 1 Foods

(A) Standards for components of general foods
(B) Standards for manufacturing, processing, and cooking of general foods
(C) Standards for preservation of general foods
(D) Standards for each food

Chapter 2 Food Additives

(A) General rules
(B) General methods of inspection
(C) Reagents
(D) Specifications for component and standards for preservation of each additive

Chapter 3 Apparatus and Packages/Containers

(A) General specifications for apparatus and packages/containers and their materials
(B) General methods for inspection of apparatus or packages/containers
(C) Reagents
(D) Each specification distinguished by materials for apparatus or packages/containers or their materials
(E) Each specification distinguished by foods, type of apparatus, or packages/containers
(F) Standards for manufacture of apparatus and packages/containers

Chapter 4 Toys

(A) Specifications for toys and their materials
(B) Standards for manufacture of toys

Chapter 5 Sanitizers

17.3.2 Abstract of Restrictions for Packages/Containers

A tentative English translation of "The Specifications and Standards for Foods and Food Additives" (Notification No. 370) appears at the web site: http://www.mhlw.go.jp/english/topics/foodsafety/index.html.

An abstract of restrictions laid down by this notification for packages/containers is given in the next section.

17.3.2.1 General Restrictions (Abstracts)

General restrictions with regard to apparatus and packages/containers are mainly described in A: General specifications for apparatus and packages/containers and their materials.

The main restrictions are as follows:

1) Tin used for plating that is in direct contact of food should not contain lead more than 0.1%.
2) The metal containing more than 0.1% of lead and/or more than 5% of antimony should not be used for manufacture of apparatus and packages/containers that come in direct contact with foods.
3) Solder used for manufacturing apparatus and packages/container should not contain more than 0.2% of lead.
4) Apparatus and packages/containers should not contain synthetic chemical colorants that are not on No. 1 list of Food Sanitation law Enforcement regulations. But the case of no fear of migration into foods by design of apparatus or packages/containers is excluded.

   [Reference]

   Fluorescent materials used to manufacture paper napkins were in trouble in the 1970s.

   The methods of inspection of fluorescent materials are notified as under:

   (1) Notice from the Ministry of Health and Welfare (Food Sanitary Division) 1971 No. 244.
   (2) Notice from the Ministry of Health, Labor, and Welfare (Standards and Evaluation Division, Inspection and Safety Division) 2004 No. 0107001.
5) The resin mainly made of polyvinyl chloride that contains bis(2-ethylhexyl)phthalate as an ingredient should not be used to manufacture apparatus or packages/containers that come in contact with foods including oil or fatty foods. But the case of no fear of migration into foods by design of apparatus or packages/containers is excluded.

17.3.2.2 Specifications for each Material (Abstracts and Summary)

Restrictions and standards for apparatus or packages/containers are mainly described in D. Each specification is distinguished by materials for apparatus or packages/containers or their materials.

Classification of Foods for Migration Testing and Conditions of Migration Testing

In this section and the section B, "General Methods of Inspection for Apparatus or Packages/Containers," classifications of foods for migration testing and conditions of migration testing are described. They are as follows:

1) **Classification of foods**

Foods are classified into four types by each character.
Oily and fatty foods (>20% fat component)
Alcoholic foods (>1% ethyl alcohol)
Foods without oil or fat, fatty foods, and alcoholic foods (>pH 5)
Foods without oil or fat, fatty foods, and alcoholic foods (<pH 5 and pH 5)

2) **Conditions of use**

Conditions of use for apparatus or packages/containers are classified into two classes.
Condition 1: <100 °C and 100 °C
Condition 2: >100 °C

3) **Simulants for each migration testing**

(a) Consumption quantity of $KMnO_4$
Simulant: water

(b) Residue quantity by evaporation and migration of bisphenol A for polycarbonate

Food types	Simulant
Oily and fatty foods	Heptane
Alcoholic foods	20% ethyl alcohol
Foods without oil or fat, fatty foods, and alcoholic foods (>pH 5)	Water
Foods without oil or fat, fatty foods, and alcoholic foods (<pH 5 or pH 5)	4% acetic acid

(c) Migration of As, Cd, Pb, and heavy-metal testing (such as Pb)
Simulant (without metal cans): 4% acetic acid

Simulant (metal cans):	Foods (>pH 5):	water
	Foods (<pH 5 or pH 5):	0.5% citric acid

(d) Migration of formaldehyde and phenol
Simulant: water

(e) Migration of Sb and Ge for polyethyleneterephthalate
Simulant: 4% acetic acid

(f) Migration of methyl metacrylate for polymethyl metacrylate
Simulant: 20% ethyl alcohol

(g) Migration of caprolactam for nylon
Simulant: 20% ethyl alcohol

(h) Migration of Zn for rubber
Simulant (without utensil for feeding milk): 4% acetic acid
Simulant (utensil for feeding milk): water
(i) Migration of epichlorohydrin for metal can
Simulant: pentane
(j) Migration of vinyl chloride for metal can
Simulant: ethanol

5) **Preparation methods for migration testing**
Testing solutions are prepared by the following general method without apparatus and packages/containers made of glass, enamel, ceramic, and utensil for feeding milk made of rubber and metal cans.
(a) General method
Samples are washed by water.
Volumes of simulants are 2 ml per 1 cm^2 of sample area.
The time and temperature of soaking samples are 30 min and 60 °C, respectively, without Case 1 and 2.
Case 1: In case of using water or 4% acetic acid as simulant and use conditions for apparatus or packages/containers at over 100 °C, the time and temperature of soaking samples are 30 min and 95 °C, respectively.
Case 2: In case of using heptane as simulant, the time and temperature of soaking samples are 60 min and 25 °C, respectively.
(b) Specific method

- Apparatus and packages/containers made of glass, enamel, and ceramic
Case 1: the depth of the apparatus and packages/containers is over 2.5 cm or 2.5 cm, but enameled which volume is one over 3 l volumes is excluded.
After washing with water, 4% acetic acid is filled in apparatus and packages/containers.
The time and temperature of soaking are 24 h and room temperature, respectively, in a dark place.
Case 2: the depth (height) of apparatus and packages/containers is under 2.5 cm or impossible to be filled and enameled one which volume is over 3 l volumes.
After washing with water, apparatus and packages/containers are immersed in 4% acetic acid to cover all area. (In case of enameled one over 3 l volume, test piece is used.)
The time and temperature of soaking are 24 h and room temperature, respectively, in a dark place.
- Utensil for feeding milk made of rubber
After washing with water, utensil for feeding milk is immersed in simulant. Volumes of simulants are 20 ml per 1 g of sample weight.
The time and temperature of immersion are 24 h and 40 °C, respectively.
- Metal cans

Samples are washed with water.
If possible, metal cans are filled with simulant heated to 60 °C and covered with glass dish and kept for 30 min at 60 °C.

If filling is impossible, then, metal cans are soaked in simulant, at a ratio of 2 ml per 1 cm^2 of sample area, heated to 60 °C, and kept 30 min at 60 °C.

In case of using water as simulant and conditions of use for apparatus or packages/containers are over 100 °C, the required time is 30 min and temperature 95 °C.

In case of using heptane or pentane as simulant, the required time and temperature of immersing sample are 60 min and 25 °C, respectively.

In case ethanol is used as simulant to measure vinyl chloride, the time and temperature of immersing sample are 24 h and under 5 °C, respectively.

Specifications Distinguished by Materials for Apparatus or Packages/Containers or Their Materials

1) **Apparatus or packages/containers made of glass or enamel or ceramic**
   (a) Restriction items of material test
   None
   (b) Restriction items of migration test
   Case 1: the depth of apparatus and packages/containers is over 2.5 cm or more

Glass

Cadmium: Apparatus for heat cooking	0.05 μg/ml and under
Volume is under 600 ml	0.5 μg/ml and under
Volume is under 3 l and/or over 600 m	0.25 μg/ml and under
Volume is and/or over 3 l	0.25 μg/ml and under
Lead: Apparatus for heat cooking	0.5 μg/ml and under
Volume is under 600 ml	1.5 μg/ml and under
Volume is under 3 l and/or over 600 ml	0.75 μg/ml and under
Volume is and/or over 3 l	0.5 μg/ml and under

Ceramics

Cadmium: Apparatus for heat cooking	0.05 μg/ml and under
Volume is under 1.1 l	0.5 μg/ml and under
Volume is under 3 l and/or over 1.1 l	0.25 μg/ml and under
Volume is and/or over 3 l	0.25 μg/ml and under
Lead: Apparatus for heat cooking	0.5 μg/ml and under
Volume is under 1.1 l	2 μg/ml and under
Volume is under 3 l and/or over 1.1 l	1 μg/ml and under
Volume is and/or over 3 l	0.5 μg/ml and under

Enameled

Cadmium: Volume is under 3 l	0.07 μg/ml and under
Lead: Volume is under 3 l	
Apparatus for heat cooking	0.4 μg/ml and under
Apparatus without heat cooking	0.8 μg/ml and under

Case 2: the depth of apparatus and packages/containers is under 2.5 cm or impossible to be filled.

Glass and ceramics

Cadmium: 0.7 μg/cm^2 and under
Lead: 8 μg/cm^2 and under

Enameled

Cadmium: Apparatus for heat cooking	0.5 μg/cm^2 and under
Apparatus without heat cooking	0.7 μg/cm^2 and under
Lead: Apparatus for heat cooking	1 μg/cm^2 and under
Apparatus without heat cooking	8 μg/cm^2 and under

Case 3: the depth of enameled apparatus and packages/containers is over and/or 2.5 cm and volume is over and/or 3 l
Cadmium: 0.5 μg/cm^2 and under
Lead: 1 μg/cm^2 and under

2) **Apparatus or packages/containers made of synthetic resins**

(2-1) General specifications

All apparatuses or packages/containers made of synthetic resins should comply with general restrictions.

(a) Restriction items of material test

Cadmium: 100 μg/g and under
Lead: 100 μg/g and under

[Reference]
The purpose of this restriction is to prohibit adding the materials made of lead or cadmium. Therefore, even if concentration is under this value, it is prohibited to add the compounds made of lead or cadmium.

1. Notice from the Ministry of Health and Welfare (Food Sanitary Division) 1973 No. 541

(b) Restriction items of migration test

Heavy metals (such as Pb): 1 μg/ml and under
Consumption quantity of $KMnO_4$: 10 μg/ml and under

Synthetic resins mainly made of phenol resins, melamine resins, and urea resins are excluded from consumption quantity of $KMnO_4$.

(2-2) **Specific specifications**

In addition to general specifications, there are specific specifications for specific synthetic resins.

Specific resins are often described "mainly made of #### resins." This means base polymer resins including #### over 50%.

(2-2)-1 Specific specification for apparatus or packages/containers mainly made of phenol resins or melamine resins or urea resins:

(a) Restriction items of material test
None

(b) Restriction items of migration test
Phenol: 5 μg/ml and under
Formaldehyde: fit for specific formaldehyde test
Residue quantity by evaporation: 30 μg/ml and under

(2-2)-2 Specific specifications for apparatus or packages/containers mainly made of formaldehyde resins without phenol resins or melamine resins or urea resins:

(a) Restriction items of material test
None
(b) Restrictions items of migration test
Formaldehyde: fit for specific formaldehyde test
Residue quantity by evaporation: 30 μg/ml and under

(2-2)-3 Specific specifications for apparatus or packages/containers mainly made of polyvinyl chloride resins:

(a) Restriction items of material test
Dibutyl tin compounds: 50 μg/g and under
Cresol phosphate: 1 mg/g and under
Vinyl chloride: 1 μg/g and under

[Reference]

The purpose of this restriction is to prohibit addition of materials made of dibutyl tin or cresol phosphate. Therefore, even if concentration is under this value, it is prohibited to add materials made of dibutyl tin or cresol phosphate.

1. Notice from the Ministry of Health and Welfare (Food Sanitary Division) 1973 No. 541

(b) Restrictions items of migration test
Residue quantity by evaporation: 30 μg/ml and under
In the case of heptane simulants and use condition is under 100 °C or 100 °C: 150 μg/ml and under

(2-2)-4 Specific specifications for apparatus or packages/containers mainly made of polyethylene and polypropylene resins:

(a) Restriction items of material test
None
(b) Restriction items of migration test
Residue quantity by evaporation: 30 μg/ml and under
In the case of heptane simulant and use condition is 100 °C and under, 150 μg/ml and under

(2-2)-5 Specific specifications for apparatus or packages/containers mainly made of polystyrene resins:

(a) Restriction items of material test
Volatiles: 5 mg/g and under
(Total of styrene, toluene, ethyl benzene, isopropyl benzene, and propyl benzene)
In the case of forming styrene and using hot water for cooking
Total: 2 mg/g and under
Styrene: 1 mg/g and under
Ethyl benzene: 1 mg/g and under

(b) Restriction items of migration test
Residue quantity by evaporation: 30 μg/ml and under
In the case of heptane simulant: 240 μg/ml and under

(2-2)-6 Specific specifications for apparatus or packages/containers mainly made of polyvinylidene chloride resins:

(a) Restriction items of material test
Barium: 100 μg/g and under
Vinylidene chloride: 6 μg/g and under
(b) Restriction items of migration test
Residue quantity by evaporation: 30 μg/ml and under

(2-2)-7 Specific specifications for apparatus or packages/containers mainly made of polyethylene terephthalate resins:

(a) Restriction items of material test
None
(b) Restriction items of migration test
Antimony: 0.05 μg/ml and under
Germanium: 0.1 μg/ml and under
Residue quantity by evaporation: 30 μg/ml and under

(2-2)-8 Specific specifications for apparatus or packages/containers mainly made of polymethyl metacrylate resins:

(a) Restriction items of material test
None
(b) Restriction items of migration test
Methyl metacrylate: 15 μg/ml and under
Residue quantity by evaporation: 30 μg/ml and under

(2-2)-9 Specific specifications for apparatus or packages/containers mainly made of nylon resins:

(a) Restriction items of material test
None
(b) Restriction items of migration test
Caprolactam: 15 μg/ml and under
Residue quantity by evaporation: 30 μg/ml and under

(2-2)-10 Specific specifications for apparatus or packages/containers mainly made of polymethylpentene resins:

(a) Restriction items of material test
None
(b) Restriction items of migration test
Residue quantity by evaporation: 30 μg/ml and under
In the case of heptane simulant: 120 μg/ml and under

(2-2)-11 Specific specifications for apparatus or packages/containers mainly made for polycarbonate resins:

(a) Restriction items of material test
Bisphenol A (including phenol and *p-tert*-butyl phenol): 500 μg/g and under
Diphenyl carbonate: 500 μg/g and under
Amines (total of triethylamine and tributylamine): 1 μg/g and under
(b) Restrictions items of migration test
Bisphenol A (including phenol and *p-tert*-butylphenol): 2.5 μg/ml and under
Residue quantity by evaporation: 30 μg/ml and under

(2-2)-12 Specific specifications for apparatus or packages/containers mainly made of polyvinyl alcohol resins:

(a) Restriction items of material test
None
(b) Restrictions items of migration test
Residue quantity by evaporation: 30 μg/ml and under

(2-2)-13 Specific specifications for apparatus or packages/containers mainly made of polylactic acid resins:

(a) Restriction items of material test
None
(b) Restrictions items of migration test
Total lactic acid: 30 μg/ml and under
Residue quantity by evaporation: 30 μg/ml and under

3) **Apparatus or packages/containers made of rubbers:**
(3)-1 Apparatus or packages/containers made of rubber without utensil for feeding milk

(a) Restriction items of material test
Cadmium: 100 μg/g and under
Lead: 100 μg/g and under
Mercaptoimidazoline (only rubbers containing chloride): fit to specific test of mercaptoimidazoline
(b) Restriction items of migration test
Heavy metals (such as Pb): 1 μg/ml and under
Zinc: 15 μg/ml and under
Phenol: 5 μg/ml and under
Formaldehyde: fit for specific formaldehyde test
Residue quantity by evaporation: 60 μg/ml and under

(3)-2 Utensils for feeding milk made of rubber

(a) Restriction items of material test
Cadmium: 10 μg/g and under
Lead: 10 μg/g and under

(b) Restriction items of migration test
Heavy metals (such as Pb): 1 μg/ml and under
Zinc: 1 μg/ml and under
Phenol: 5 μg/ml and under
Formaldehyde: fit for specific formaldehyde test
Residue quantity by evaporation: 40 μg/ml and under

4) **Specifications for metal cans**

(a) Restriction items of material test
None
(b) Restriction items of migration test
Arsenic (as As_2O_3): 0.2 μg/ml and under
Cadmium: 0.1 μg/ml and under
Lead: 0.4 μg/ml and under
Phenol: 5 μg/ml and under (noncoated cans are excluded)
Formaldehyde: fit for specific formaldehyde test (noncoated cans are excluded)
Residue quantity by evaporation: 30 μg/ml and under (non-coated cans are excluded)
In case of coated cans with natural oil coatings including over 3% zinc and the simulant is heptane: 90 μg/ml and under
In case of coated cans with natural oil coatings including over 3% zinc and the simulant is water: if residue quantity by evaporation is >30 μg/ml, quantity of soluble chloroform should be 30 μg/ml and under
Epichlorohydrin: 0.5 μg/ml and under
Vinyl chloride: 0.05 μg/ml and under

17.3.2.3 Specifications for Packages/Containers for Specific Food Type (Abstract)

For specific food types, there are some restrictions for packages/containers.

Specifications for Packages/Containers for Packed Foods that were Sterilized by Pressure Heating (Canned Foods and Bottled Foods are Excluded)

1) Packages/containers should have a barrier against light and permeation of gas to ensure that the quality of foods will not change by degradation of oil or fat.
2) On the test of filling water and sterilizing, there should be no damage, no deformation, no coloring of water, and no change of color.
3) Restriction items of strength test
Compression-resistance testing
Heat-sealing strength testing
Dropping testing.

Specifications for Packages/Containers for Beverages (Juices Used as Ingredient are Excluded)

1) Packages/containers for beverages should be packages/containers made of glass or metal or synthetic resins or paper processed with synthetic resins or aluminum foil processed with synthetic resins or combination of two or more previous materials
2) Restriction of packages/containers made of glass
   (a) In the case of repeat use, packages/containers should be transparent
   (b) Restriction items of strength test (paper covers are excluded)
      Continued pressure-resistance testing (for carbonated soft drinks)
      Vacuum-resistance testing (for hot filling beverages)
      Water filling testing (for others without carbonated soft drinks and hot filling beverages)
3) Restriction of packages/containers made of metal
   (a) Restriction items of strength test
      Pressure-resistance testing (for pressured products)
      Vacuum-resistance testing (for vacuumed products)
   (b) In the case of using material without metal for opening parts
      Restriction items of strength test
      Pinhole testing
      Burst strength testing
      Stickling resistance testing
4) Restriction of packages/containers made of synthetic resins or paper processed with synthetic resins or aluminum foil processed with synthetic resins
   (a) Restriction of synthetic resins for direct contact with foods
      Synthetic resins for direct contact with foods should be those synthetic resins that are regulated by specific specifications. But aluminum foils processed with synthetic resins for sealing are excluded.
   (b) Restriction items of strength test
      Dropping testing
      Pinhole testing
      Sealing testing (for packages/containers made of paper processed with synthetic resins with hot-sealing)
      Compression-resistance testing (for packages/containers made of paper processed with synthetic resins and aluminum foil processed with synthetic resins)
   (c) In the case of sealed products by crown cork
      Restriction items of strength test
      Continued pressure-resistance testing (for carbonated soft drinks)
      Vacuum-resistance testing (for hot-filling beverages)
      Water filling testing (for others without carbonated soft drinks and hot-filling beverages)

5) Restriction of packages/containers made of combination of two or more previous materials.
   Packages/containers should comply with the following strength tests:
   Dropping testing
   Pinhole testing
   Sealing testing (for heat-sealed products)
   Vacuum-resistance testing (for hot-filling beverages)
   Water filling testing (for others without hot-filling beverages)

17.3.2.4 Standards for Manufacturing of Apparatus and Packages/Containers (Abstract)

1) The specified cattle back born should not be used as the raw materials for manufacturing apparatus and/or packages/containers. But the oil and fat delivered from the specified cattle back born which treated with hydrolysis and/or saponification and/or transesterification under high temperature and pressure are excluded.
2) Polylactic acid including D-lactic acid over 6% should not be used in materials for manufacturing apparatus and/or packages/containers with use condition over 40 °C. But the case of use condition "under and/or 100 °C, under and/or 30 min" or "under and/or 66 °C, under and/or 2 h" is excluded.

17.4 The Ordinance of Specifications and Standards for Milk and Milk Products

In 1951, "The Ordinance of Specifications and Standards for Milk and Milk Products" was established by the Ministry of Health and Welfare as "The Ordinance of No. 52."

This ordinance was positioned as the special law for "The Food Sanitation Law Enforcement Ordinance" (No. 229, August 1953) and "The Food Sanitation Law Enforcement Regulations" (No. 23, July 1948).

17.4.1 Articles of the Ordinance of Specifications and Standards for Milk and Milk Products (Abstracts)

Article 1 [Application of this Ordinance]

For milk and milk products and the products mainly made of milk, enforcement of Article 9.1, Article 11.1, Article 13.2, Article 13.3, Article 18.1, and Article 19 of "The Food Sanitation Law" were followed by this ordinance.

Specifications and standards that are not described in this ordinance are followed by the "Specifications and Standards for Foods and Food Additives, etc." (Notification No. 370).

Article 2 [Definition of milk and milk product]

Under this ordinance, milk is defined as raw milk, milk, special milk, raw goat milk, sterilized goat milk, raw sheep milk, milk making adjustment for constituents of milk, low-fat milk, no-fat milk, and processed milk.

Under this ordinance, milk products are cream, butter, butter oil, cheese, condensed milk made of condensed whey oil, ice-cream, concentrated milk, no-fat concentrated milk, condensed milk with no sugar, no-fat condensed milk with no sugar, condensed milk with sugar, no-fat condensed milk with sugar, whole powder milk, no-fat powder milk, powder cream, powder whey, powder whey condensed protein, powder butter milk, powder milk with sugar, formulated powder milk, fermented milk, lactic acid bacteria drinks, and milk drinks.

Article 3 [Specifications and standards]
For milk and milk products, specifications and standards for constituents, manufacturing methods, processing methods, sanitary control methods, apparatus, packages/containers, and materials for manufacture of apparatus and packages/containers were described in attached list.

Articles 4–6 describe filing methods.

Article 7 [Labeling]
Milk and milk products should be labeled in compliance with the food sanitation law, Article 19.

Specifications and standards for labeling are described in Article 7.2.

In this article, milk and milk products that are possible to store and sell at room temperature are defined.

17.4.2 Attached List (Specification and Standards for Milk and Milk Products)

Contents of the attached list are the following:

1) The kind of inhibited decease
2) Specifications and standards of constituent, manufacturing, processing, and storing of milk and milk products
3) Standards of methods of manufacturing, processing, sanitary control on HACCAP
4) Specifications and standards for apparatus, packages/containers, their materials, and methods of manufacturing

17.4.3 Specifications and Standards for Apparatus, Packages/Containers, Their Materials, and Methods of Manufacturing in Attached List (Abstract)

1) Specifications of apparatus for milk and milk products
   (1) Apparatus is easy to sanitize
   (2) Materials that are in direct contact with foods do not cause rust or are protected against rust
   (3) Machine for subdivision, filling, sealing is easy to sanitize and is protected from contamination
2) Specifications for packages/containers for milk and milk products, their materials, and standards of manufacturing methods

In this list, specifications and standards of packages/containers are described for milk, special milk, sterilized goat milk, milk making adjustment for constituent part of milk, low-fat milk, no-fat milk, processed milk, cream, fermented milk, lactic acid bacteria drinks, milk drinks, and formulated powder milk.

So, packages/containers for other milk products should follow the specifications and standards for foods and food additives (Notification No. 370).

17.4.3.1 Specifications and Standards for Apparatus, Packages/Containers, Their Materials and Methods of Manufacturing for Milk, Special Milk, Sterilized Goat Milk, Milk Making Adjustment for Constituent Part of Milk, Low-Fat Milk, No-Fat Milk, Processed Milk, and Cream (Group 1)

Packages/Containers that may be Used for Group 1

1) Glass bottle
2) Packages/containers made of synthetic resins (synthetic resins include only polyethylene, ethylene-1-alkene copolymer, nylon, polypropylene, and polyethylene terephthalate)
3) Package/container made of processed paper with synthetic resins (synthetic resins include only polyethylene, ethylene-1-alkene copolymer, polyethylene terephthalate)
4) Metal can (use only for cream)
5) Package/container made of a combination of synthetic resins and processed paper with synthetic resins
6) Package/container made of a combination of synthetic resins and/or processed paper with synthetic resins and/or metals (used only for cream)

Restriction Items of Glass Bottle

(a) Glass bottle should be bright and uncolored
(b) Inside diameter of bottle nozzle is 26 mm or over 26 mm

Restriction Items of Package/Container Made of Synthetic Resins and Package/Container Made of Processed Paper with Synthetic Resins

1) Restriction of migration
   (1) Preparing methods of migration testing
      (a) Package/container may be filled with simulants

After washing with water, package/container is filled with simulants warmed to 60 °C and kept at this temperature for 30 min. In case of using heptane as simulants, temperature is 25 °C and time is 60 min.

(b) Package/container may not be filled with simulants

After washing with water, samples are immersed in simulant warmed to 60 °C and kept at this temperature for 30 min. In case of using heptane as simulant, temperature is 25 °C.

The volume rate of simulants is 2 ml per 1 cm^2 of samples area.

(2) Migration restrictions
(a) Heavy metals (such as Pb): 1 ppm and under Simulant: 4% acetic acid
(b) Residue quantity by evaporation: 15 μg/ml and under Simulant: 4% acetic acid

In case of packages/containers for only cream and using polyethylene, ethylene-1-alkene copolymer as direct-contact food layer

Residue quantity by evaporation: 75 μg/ml and under Simulant: heptane
(c) Consumption quantity of $KMnO_4$: 5 ppm and under Simulant: water
(d) Antimony (only polyethylene terephthalate) 0.025 ppm and under Simulant: 4% acetic acid
(e) Germanium (only polyethylene terephthalate) 0.05 ppm and under Simulant: 4% acetic acid

2) Restriction items of strength test
Burst-resistance testing
Seal-strength testing

3) Restriction of materials
(1) Part of direct contact to foods should be polyethylene or ethylene-1-alkene copolymer or polyethylene terephthalate
(2) In synthetic resins, with direct contact of foods, the use of additives is prohibited.
In polyethylene or ethylene-1-alkene copolymer, following additives are excluded from this prohibition:
(a) Stearic acid, Ca (only when compliant with Japanese Drug Official Sheet) 2.5 g and under/kg
(b) Glycerin fatty acid ester (only when compliant with “The Specifications and Standards for Foods and Food additives, etc.” Notification No. 370). 0.3 g and under/kg
(c) Dioxide titanium (only when compliant with “The Specifications and Standards for Foods and Foods additives, etc.” Notification No. 370)

4) Restrictions of synthetic resin material
Synthetic resins that are in direct contact with foods should comply with next items.

4-1 Polyethylene and ethylene-1-alkene copolymer
(1) Quantity rate of extracts by hexane: 2.6% and under
(2) Soluble quantity by xylene: 11.3% and under
(3) Arsenic (as As_2O_3): 2 ppm and under
(4) Heavy metal (such as Pb): 20 ppm and under

4-2 Polyethylene terephthalate

Lead: 100 ppm and under
Cadmium: 100 ppm and under

5) Restriction of container/package for milk that may be used to store and sell milk at room temperature should have a barrier to light and permeation of gas.

Restriction Items of Metal Can

1) Restrictions of migration
   (1) Preparation methods of migration testing
   The methods of migration are the same as mentioned in Item 4.3.1.3.1.
   (2) Migration restrictions
   (a) Arsenic (such as As_2O_3): 0.1 ppm and under Simulant: 4% acetic acid
   (b) Heavy metals (such as Pb): 1 ppm and under Simulant: 4% acetic acid

The items below are applicable when a synthetic resin is in direct contact with foods.

   (c) Residue quantity by evaporation: 15 μg/ml and under Simulant: 4% acetic acid
   (d) Consumption quantity of $KMnO_4$: 5 ppm and under Simulant: water
   (e) Phenol fit for specific phenol test Simulant: water
   (f) Formaldehyde fit for specific phenol test Simulant: water

1) Restrictions for synthetic resin material
   Synthetic resins that come in direct contact with food should comply with the following requirements:
   (1) Lead: 100 ppm and under
   Cadmium: 100 ppm and under
   Composition of polyvinyl chloride resins, which are in direct contact with foods, should comply with following items:
   (2) Dibutyl tin: 50 ppm and under
   (3) Cresol phosphate: 1000 ppm and under
   (4) Vinyl chloride: 1 ppm and under

17.4.3.2 Specifications and Standards for Apparatus, Packages/Containers, Their Materials and Methods of Manufacturing for Fermented Milk, Lactic Acid Bacteria Drink, Milk Drink (Group 2)

Packages/Containers that may be Used for Group 2 Include

1) Glass bottle
2) Packages/containers made of synthetic resins
3) Packages/containers made of processed paper with synthetic resins
4) Packages/containers made of processed aluminum foil with synthetic resins
5) Metal can
6) Packages/containers made of combination with synthetic resins and processed paper with synthetic resins and processed aluminum foil with synthetic resins and metals

Restrictions for Glass Bottle

(a) Glass bottle should be bright

Restrictions for Package/Container Made of Synthetic Resins and Package/Container Made of Processed Paper with Synthetic Resins and PackaContainer Made of Processed Aluminum Foil with Synthetic Resins

1) Restrictions of migration
   (1) Preparation methods of migration testing.
   The methods of migration are the same as mentioned in Item 4.3.1.3.1.
   (2) Migration restrictions
      (a) Heavy metals (such as Pb): 1 ppm and under Simulant: 4% acetic acid
      (b) Residue quantity by evaporation: 15 μg/ml and under Simulant: 4% acetic acid
      (c) Consumption quantity of $KMnO_4$: 5 ppm and under S imulant: water

If a package/container is made mainly of polyethylene terephthalate, the package/container shall also comply with the following limits:
      (d) Antimony: 0.025 μg/ml and under Simulant: 4% acetic acid
      (e) Germanium: 0.05 μg/ml and under Simulant: 4% acetic acid

2) Requirement for strength test
   Burst-resistance testing or stickling strength testing
3) Restrictions of materials
   (1) Part in direct contact with foods should be polyethylene or ethylene-1-alkene copolymer or polystyrene or synthetic resins mainly made of polypropylene or synthetic resins mainly made of polyethylene terephthalate.
   But the case of aluminum foil processed with synthetic resins for sealing is excluded from this requirement.
4) Restrictions for synthetic resin materials
   Synthetic resins that are in direct contact with foods should comply with the following items:
   (1) Polyethylene or ethylene-1-alkene copolymer or synthetic resins mainly made of polypropylene
      (a) Quantity rate of extracts by hexane: 2.6% and under
   (Synthetic resins mainly made of polypropylene 5.5% and under)
      (b) Soluble quantity by xylene: 11.3% and under
   (Synthetic resins mainly made of polypropylene 30% and under)
      (c) Arsenic (such as As_2O_3): 2 ppm and under
      (d) Heavy metal (such as Pb): 20 ppm and under
   (2) Polystyrene
      (a) Volatiles: 1500 ppm and under
   (Total of styrene, toluene, ethyl benzene, isopropyl benzene, and *n*-propyl benzene)
      (b) Arsenic (as As_2O_3): 2 ppm and under
      (c) Heavy metal (as Pb): 20 ppm and under
   (3) Synthetic resins mainly made of polyethylene terephthalate
      (a) Lead: 100 ppm and under
   Cadmium: 100 ppm and under

5) Restriction of container/package for milk that may be used to store and sell milk at room temperature should have a barrier to light and permeation of gas.

Restrictions for Metal Cans

Same as those of Section 4.3.1.4.

Restrictions for Aluminum foil Processed with Synthetic Resins for Sealing

1) Restrictions of migration
   (1) Methods of migration testing
   Samples are immersed using apparatus that heats the simulant to 60 °C and keeps that temperature for 30 min. The volume rate of simulant is 2 ml per 1 cm^2 of samples area.
   (2) Restrictions of migration
   (a) Heavy metals (such as Pb): 1 ppm and under Simulant: 4% acetic acid
   (b) Residue quantity by evaporation: 15 μg/ml and under Simulant: 4% acetic acid
   (c) Consumption quantity of $KMnO_4$: 5 ppm and under Simulant: water
   (d) Phenol fit for specific phenol test Simulant: water
   (e) Formaldehyde fit for specific phenol test Simulant: water
2) Requirement for strength test
   Burst-resistance testing
3) Restrictions for synthetic resin material
   Synthetic resins that are in direct contact with food should comply with following restrictions:
   (1) Arsenic (such as As_2O_3): 2 ppm and under
   (2) Lead: 100 ppm and under
   Cadmium: 100 ppm and under
   Composition of polyvinyl chloride resins that are in direct contact with food should comply with the following specifications:
   (3) Dibutyl tin: 50 ppm and under
   (4) Cresol phosphate: 1000 ppm and under
   (5) Vinyl chloride: 1 ppm and under

17.4.3.3 Specifications and Standards for Apparatus, Package/Container, Their Materials and Methods of Manufacturing for Formulated Powder Milk (Group 3)

Packages/Containers that may be Used for Group 3

1) Metal can (include the use of synthetic resins for sealing of opening parts).
2) Laminated packages/containers made of synthetic resins (laminated with aluminum foils, laminated with aluminum and paper, and/or cellophane).
3) Packages/containers made of a combination with metal can and laminated packages/containers made of synthetic resins.

Restrictions of Packages/Containers

1) Restrictions of migration (for only synthetic resins without coatings)
   (1) Preparing methods of migration testing
      (a) Package/container that can be filled with simulant

After washing with water, package/container is filled with simulant warmed to 60 °C and kept at that temperature for 30 min.

      (b) Package/container that cannot be filled with simulant

After washing with water, samples are immersed in simulant warmed to 60 °C and kept at that temperature for 30 min. The volume rate of simulant is 2 ml per 1 cm^2 of sample area.

   (2) Restrictions of migration
      (a) Heavy metals (such as Pb): 1 ppm and under Simulant: 4% acetic acid
      (b) Residue quantity by evaporation: 15 μg/ml and under Simulant: 4% acetic acid
      (c) Consumption quantity of $KMnO_4$: 5 ppm and under Simulant: water
      (d) Antimony (only polyethylene terephthalate): 0.025 ppm and under Simulant: 4% acetic acid
      (e) Germanium (only polyethylene terephthalate): 0.05 ppm and under Simulant: 4% acetic acid
2) Requirement for strength test
   Burst-resistance testing
3) Restrictions of materials
   (1) Part in direct contact with food should be polyethylene or ethylene-1-alkene copolymer or polyethylene terephthalate.
   Polyethylene or ethylene-1-alkene copolymer that is in direct contact with food, the use of additives is prohibited.
4) Restrictions for synthetic resin material
   Synthetic resins that are in direct contact with food should be complied with the following requirements:
   4-1Polyethylene and ethylene-1-alkene copolymer
      1. Quantity rate of extracts by hexane: 2.6% and under
      2. Soluble quantity by xylene: 11.3% and under
      3. Arsenic (such as As_2O_3): 2 ppm and under
      4. Heavy metal (as Pb): 20 ppm and under
   4-2Polyethylene terephthalate
      (1) Lead: 100 ppm and under
      Cadmium: 100 ppm and under

17.5 The Food Safety Basic Law and Relationship with the Food Sanitation Law

In 2003, the food safety basic law was enforced. This law was promulgated in the backdrop of the BSE outbreak that was a turning point and was placed as food safety assessment regulation, recently.

The food safety basic law and the Food Safety Commission are described exactly in English on the web site of the Commission, English page. Web address: http://www.fsc.go.jp/english/index.html/.

Relationship with the food sanitation law is simply illustrated as follows:

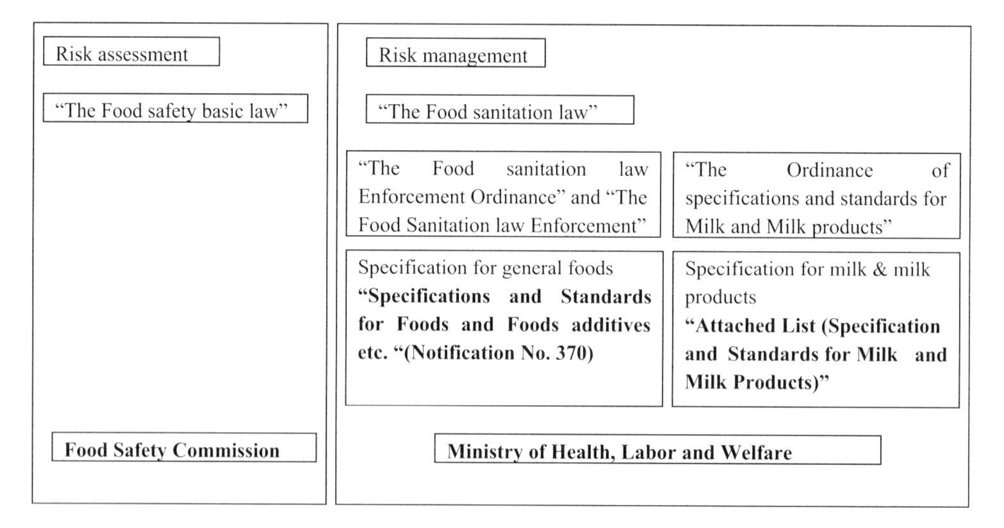

When the Ministry of Health, Labor, and Welfare amends the "Specifications and Standards for Foods and Food Additives, etc.," the Ministry requests the Food Safety Commission to conduct a risk assessment.

In the past, requests were made on the following three issues concerning the apparatus and packages/containers:

- Risk assessment of the chemical, recycled polyethylene terephthalate (two issues)
- Risk assessment of polylactic acid resins
- Risk assessment of polyethylene terephthalate for addition to the synthetic resin list that may by used for packages/containers of milk (group 1) as direct food contact material.

17.6 Industrial Voluntary Rules

Since the 1960s, the use of plastics in the materials of food apparatus and packages/containers was increasing. In order to regulate the use of plastics, the food sanitation law was amended, but it was not enough to cover the whole gamut of issues because the law was established under the policy of minimum requirement. Package makers, material suppliers, and food contact material producers established their industrial safety associations under the support and guidance of the authorities to complement the food sanitation law. Each industrial safety association framed voluntary rules and all members of these associations adhered to these voluntary rules. In Japan, almost all packaging makers and material suppliers joined industrial safety associations, and later on these voluntary rules were authorized as national law. In the food sanitation law, there are few descriptions of base synthetic resins, especially starting materials, and additives. To cover these points, many industry voluntary rules have their own set of positive lists or negative lists of materials and additives.

The main industry voluntary rules of materials are listed in the table below.

Name of association	Material	Established in year	Type of list
Japan Hygienic PVC Association	Polyvinyl chloride	1967, June	Positive list
Hygienic Association of Poly Vinylidene Chloride	Polyvinylidene chloride	1977, June	Positive list
Japan Hygienic Olefin & Styrene Plastics Association	PE, PP, PS, PET, PBT, PC, and so on (28 synthetic resins)	1973, September	Positive list
Japan Printing Ink Makers Association	Inks	1952, April	Negative list
Japan Petrolatum Wax Industry Association	Petrolatum wax	1974, April	Positive list
Japan Adhesive Industry Association	Adhesives	1974, June	Negative list
The Japan Rubber Manufactures Association	Rubbers	1950, July	Positive list

The positive lists of Japan Hygienic PVC Association and Japan Hygienic Olefin & Styrene Plastics Association [11, 12] are very popular in Japan. These positive lists define base polymer and restrictions of additives, and there are restrictions of starting materials and restrictions of characterization, migration, and so on, on the line of US CFR21 regulations.

These basic rules provide a list of materials that are permitted for use:

1) Materials listed on CFR21 (USA) and FCNs are permitted.
2) Materials listed only on additives list of the EU directives are permitted for additives.
3) Direct foods additives are permitted.
4) Materials listed on regulations of the UK, Germany, Italy, Holland, Belgium, and France are permitted (the monomers' list of the EU directives is not included).

A new material is assessed by each committee of the industrial safety association. But assessment methods are very similar to the FDA or the EU (migration test and migrant's safety level).

A food contact material supplied by the members of these industrial safety associations should be in compliance with industry voluntary rules and registered with the industry safety associations. These associations publish confirmation sheets of compliance describing the registration number and use conditions. For articles to be used, the confirmation sheets of compliance can be issued to members. This system is very unique and useful for assurance of food safety.

17.7 Sheets of Confirmation with Compliance

There are many analytical laboratories that are registered with authorities in Japan. These laboratories issue inspection sheets that describe the results of migration tests in accordance with regulations set for testing (e.g., the inspection sheet of Notification No. 370 on testing). These sheets are often used for confirmation of compliance with regulations, and for importing a material this sheet is often attached to the papers required. Together with inspection sheets by analytical laboratories, the confirmation sheets of compliance with voluntary rules set by industry safety associations are often used for ensuring food safety.

17.8 Conclusions

In Japan, safety of food apparatus and package/container is assured by regulations and industry voluntary rules. At present, however, globalization is expanding, and many foods are imported into Japan. Now, it has become necessary to harmonize with global regulations and to improve systems and regulations in Japan.

Recently, a scientific study on health and welfare with respect to food apparatus and package/container has been started by the NIHS under Dr. Kawamura [13] in collaboration with many industry safety associations and supported by the Ministry of Health, Labor, and Welfare. In this study, issues concerning paper, metal can, and so on have been looked into and now synthetic resins have been taken up. Results of this study will reflect in the improvement of regulations.

References

1 The Constitution of Japan, May 5, 1947.

2 The Food Sanitation Law, Law No. 233 December 24, 1947; last amendment Law No. 49 June 5, 2009. An English translation is available on the web site: http://www.mhlw.go.jp/english/topics/foodsafety/index.html.

3 The Food Safety Basic Law, Law No. 48 May 23, 2003. An English translation is available on the web site: http://www.fsc.go.jp/english/index.html/.

4 The Pharmaceutical Affairs Law, Law No. 145 August 10, 1960; last amendment Law No. 84 June 21, 2006.

5 The Food Sanitation Law Enforcement Ordinance, Ordinance No. 229 August 31, 1953; last amendment Ordinance No. 309 June 7, 2000.

6 The Food Sanitation Law Enforcement Regulations, Regulation No. 23 July 13, 1948; last amendment Regulation No. 119 June 4, 2009.

7 The Specifications and Standards for Foods, Food Additives, etc. Notification No. 370 December 28, 1959; last amendment Notification No. 325 June 4, 2009.

8 The Ordinance of Specifications and Standards for Milk and Milk Products, Ordinance No. 52 December 27, 1951; last amendment Ordinance No. 132, 2007.

9 The Ministry of Health, Labor, and Welfare, Japan, Homepage http://www.mhlw.go.jp/.

10 Food Safety Commission, Japan, Homepage http://www.fsc.go.jp/.

11 Japan Hygienic PVC Association, Homepage http://www.jhpa.jp/.

12 Japan Hygienic Olefin & Styrene Plastics Association http://www.jhospa.gr.jp/.

13 Specifications, standards and their testing methods for apparatus and packages/containers, by Dr. Yoko Kawamura (2006).

18
China Food Contact Chemical Legislation Summary

Caroline Li and Sam Bian

18.1
Introduction

Chinese regulations on food containers and packaging materials have gone through tremendous changes over the past few years.

In this chapter, the current status and the future trend of the regulation will be discussed.

18.2
Current Status

The definition of the food container and packaging materials according to Chinese regulation is "paper, bamboo, wood, metal, enamel, porcelain, plastic, rubber, natural fiber, chemical fiber and glass products for packaging and containing food as well as paint contacting food, including machines, pipes, conveyor belts, containers, appliances and tableware, and so on."

The main law governing the regulations on food containers and packaging materials in China is the China Food Hygiene (Sanitary) Law (CFHL).

The CFHL was approved at the 16th conference of the 8th session of the Standing Committee of the National People's Congress. It came into force on October 30, 1995.

The framework of CFHL is as follows [1]:

First Chapter: General rule
Second Chapter: Food hygiene
Third Chapter: Food additive hygiene
Fourth Chapter: Hygiene requirement for food container, food packaging material, tools and equipment used for food production
Fifth Chapter: Food hygiene standards and guidelines
Sixth Chapter: Food hygiene management
Seventh Chapter: Food hygiene supervision
Eighth Chapter: Legal liability
Ninth Chapter: Supplementary articles

Global Legislation for Food Packaging Materials. Edited by Rinus Rijk and Rob Veraart

ISBN: 978-3-527-31912-1

Requirements for food container and food packaging material are regulated in Chapter 4, "Regulate food container, food packaging material, tools and equipment used for food production." Any packaging article consisting of new material or substances not regulated by GB (national compulsory standard) requires approval from the authority.

18.3
China Food Safety Regulatory Entity

According to the CFHL, Chapter 5, the Ministry of Health (MOH) is responsible for establishing, maintaining, and revising related hygiene standards for food, food additives, and food packing materials. It is also responsible for establishing and approving related food hygiene rules and regulations.

The State Administration of Food and Drug (SFDA) and local provincial health bureau are in charge of monitoring and supervision of the actual implementation of food safety regulations.

According to No. 23 decision of the State Council in 2003 [2], the General Administration of Quality Supervision, Inspection, and Quarantine of the People's Republic of China (AQSIQ) is responsible for granting required permission for production of food and food packaging products. So, in a nutshell, the various regulatory authorities in China are involved at

- National level:
  - MOH (responsible for food hygiene, chemical contamination, food-borne disease control, the permission and inspection for new food, new food contact reason/articles and food contact additive's notification approval, preparation/revision of food hygiene standards, etc.)
  - SFDA (coordination)
  - AQSIQ (for production permission), mainly for the QS permission approval, and imported food contact material compliance verification.
- Provincial level:
  - MOH local departments (supervision and inspection). No FDA involvement.
- Exception:
  - In Shanghai, the FDA has the main responsibility to ensure that China MOH food contact material requirements and AQSIQ's food contact material requirements are implemented.

18.4
Regulations and Rules under the China Food Hygiene (Sanitary) Law

The Food Hygiene Law regulates food container, food packaging material, and tools and equipment used for food production. It lays down rules for plastics, paper, rubber, coating used in food container, food packaging material, and tools and equipment. For example, rules for hygienic administration of plastic articles and substances in contact with food.

Rules on food container and packing material are listed as follows:

- Rule on hygienic administration of plastic articles and substances in contact with food
- Rule on hygienic administration of paper for food packaging use
- Rule on hygienic administration of ceramic foodware
- Rule on hygienic administration of rubber articles in contact with food
- Rule on hygienic administration of aluminum food container
- Rule on hygienic administration of enamel foodware and food container
- Rule on hygienic administration of inner coating of food can
- Rule on hygienic administration of cycloxyl phenolic resin used for food can inner coating
- Rule on hygienic administration of chlorinated PVC food container

The above nine rules on the packing materials indicate the basic/general requirements for these finished articles and the related substance.

18.5
Hygiene Standards on Food Container and Packing Material

In China, food packaging materials must comply with general provisions contained in the Food Hygiene Law, food hygiene rules and regulations, and any relevant hygiene compulsory control standard (usually with titles of GB).

- GB (compulsory)
  - Article
  - Material and substance
  - Resin
  - Paper
  - Rubber
  - Indirect additives
- GB/T (recommended)

There are in total 40 hygiene standards for various food packaging materials; some are listed in Table 18.1.

Commonly used plastics are all covered under those compulsory standards, including PVC, PE, PP, PS, PET, ABS, and so on. If a new resin or new resin finished plastic articles needs to be used in food packing material, the MOH approval is must.

Here, we provide some examples of hygiene control standards (GB). For plastic, there are GB 9691-88 for hygienic standard of polyethylene(PE) resin for food packaging application and GB 9687-88 for polyethylene finished products.

Such hygienic standard requires that food contact packaging is free of harmful substances and provides limits on impurity, heavy metal levels, and so on.

For example,

GB 9691-1988 Hygiene Standards for PE resin used in food packaging

- Physical property: white pellet, must not have the unusually different smell, and smell of foreign matter.

Table 18.1 Compulsory hygiene standards for food packaging materials from the Ministry of Health, PRC.

No.	Name of standards	Standard No.
1	Hygienic standards for polyvinyl chloride resin used in food container and packaging material	GB 4803-1994
2	Hygiene standards for porcelain enamel food container	GB 4804-1984
3	Hygienic standard for epoxy phenolic resin coating for the internal lacquer of food cans	GB 4805-1994
4	Hygienic standard for foodstuff rubber products	GB 4806.1-1994
5	Hygienic standard for rubber nipple	GB 4806.2-1994
6	Sanitary specifications for perchlorovinyl interior coating for food container	GB 7105-1986
7	Hygiene standards for urushiol coating inside food containers	GB 9680-1988
8	Hygiene standards for PVC finished products used in food packaging	GB 9681-1988
9	Hygienic standard for internal coating of food cans	GB 9682-1988
10	Hygienic standard composite laminated food packing bag	GB 9683-1988
11	Hygienic standard for stainless steel food containers and wears	GB 9684-1988
12	Hygiene standards for auxiliary used in food containers and packaging materials	GB 9685-2003
13	Hygienic standards for polyamide epoxy resin used as internal coating of food container	GB 9686-1988
14	Hygienic standard for polyethylene products used as food containers and tablewares	GB 9687-1988
15	Hygienic standards for polypropylene products used as food containers and tablewares	GB 9688-1988
16	Hygienic standard for polystyrene products used as food containers and tablewares	GB 9689-1988
17	Hygienic standard for melamine products used as food containers and tablewares	GB 9690-1988
18	Hygienic standard for polyethylene resin used as food packaging material	GB 9691-1988
19	Hygienic standard for polystyrene resin used as food packaging material	GB 9692-1988
20	Hygienic standard for polypropylene resin used as food packaging material	GB 9693-1988
21	Hygiene standards for aluminum food containers	GB 11333-1989
22	Hygienic standard for anticoherent silicone paint for food container	GB 11676-1989
23	Hygienic standard for water soluble epoxy internal coating of beverage cans	GB 11677-1989
24	Hygienic standard for polytetrafluorethylene used as inner coating of food containers	GB 11678-1989
25	Hygienic standard of raw paper used for food packaging	GB 11680-1989
26	Hygienic standard for polyethylene terephthalate products used as food containers and packing materials	GB 13113-1991
27	Hygienic standard for polyethylene terephthalate resin used as food containers and packing materials	GB 13114-1991
28	Hygienic standard of unsaturated polyester resin and glass fiber reinforced plastics used as food containers and packing materials	GB 13115-1991

Table 18.1 (*Continued*)

No.	Name of standards	Standard No.
29	Hygienic standard for polycarbonate resin used as food containers and packaging materials	GB 13116-1991
30	Hygiene standards for pottery food utensil	GB 13121-1991
31	Hygiene standards for diatomite	GB 14936-1994
32	Hygienic standard for polycarbonate products used as food containers and packaging materials	GB 14942-1994
33	Hygienic standard for bottle sheet and granular materials of polyvinyl chloride for food packaging	GB 14944-1994
34	Hygienic standard for collagen casing	GB 14967-1994
35	Hygienic standard of vinylidene chloride-vinyl chloride copolymer resins food containers and packaging	GB 15204-1994
36	Hygiene standards for PA6I used in food packaging materials	GB 16331-1996
37	Hygiene standards for finished products of nylon used in food packaging materials	GB 16332-1996
38	Hygienic standard for rubber-modified acrylonitrile-butadiene-styrene products used as food containers and packaging materials	GB 17326-1998
39	Hygienic standard for acrylonitrile-styrene products used as food containers and packaging materials	GB 17327-1998
40	Hygienic standard for foodstuff plant fiber of container	GB 19305-2003

- Dry lost weight $\leq$0.15 (%); ignition residual $\leq$0.20 (%), hexane extraction $\leq$0.15 (%).

GB 9687-1988 Hygiene Standards for PE finished products (articles) used in food packaging

- Odor: must not have the unusually different smell.
- Requirements on the related migration property as listed in Table 18.2.

18.6 Hygiene Standards for Food Contact Substance (Indirect Additive)

In China, additives for food packaging materials are defined as "the substances added to improve or assist to improve the quality and property of food containers and packaging materials to meet the expected purpose during the process of production; the assistants added in the process of producing food containers and packaging materials to help the smooth production but not to improve the quality and property of final products are also included [3]."

The standard specifically focused on food contact substance (indirect additive) is the "Hygienic Standards for Adjuvant and Processing Aids in Food Containers and Packaging Materials" (GB 9685-2003). The GB 9685-2003 is practically a food contact additive positive list in China. However, normally the phrase of positive list for food contact substance is not used. The requirements of adjuvant and processing aids in food containers and packaging materials should follow GB 9685 standard.

Table 18.2 The related migration properties.

Testing items	Value
Evaporation residual (mg/l) (4% acetic acid, 60 °C, 2 h)	≤30
Evaporation residual (mg/l) (65% ethanol, 60 °C, 2 h)	≤30
Evaporation residual (mg/l) (hexane, 60 °C, 2 h)	≤60
$KMnO_4$ consumed concentration (mg/l) (water, 60 °C, 2 h)	≤10
Heavy metal (Pb), mg/l (4% acetic acid, 60 °C, 2 h)	≤1
Discolor tests: ethanol	Negative
Discolor tests: buffet oil or transparent oil	Negative
Leaching	Negative

The standard sets forth a list of raw materials permitted for use as adjuvant and processing aids in food containers and packing materials. It is applicable for adjuvant and processing aids in food containers, packing materials, food machinery, and industrial tools for production of plastic, rubber, paint, paper, bonds, printing ink, and so on. The regulated materials are categorized by intended use, such as plasticizers, stabilizers, antioxidants, solvents, colorants, and so on. The standard also limits the scope of the use of materials and the maximum amounts that may be used in China. At present, there are over 60 chemicals on this list. Some are illustrated in Table 18.3.

All categories of substances covered under GB 9685-2003 are listed as follows:

- Plasticizer
- Stabilizer
- Antioxidant
- Lubricant
- Foaming agents
- Solvents
- Antiaging agents
- Adhesives
- Antioxidants
- Impact modifiers
- Fillers
- Coupling agents
- Colorants
- Vulcanization agent
- Vulcanizing accelerators
- Antifogging agents
- Humectant accelerators
- Photoinitiators
- Oil-proof agents

General requirements for substances that can be used for food packaging should follow GB 9685 and related regulations and rules on packing material.

Table 18.3 Some of the categories, range of use of adjuvant and processing aids in food containers and packing materials, and maximum dosages as listed in GB 9685-2003 [3].

Category	Product	Range of use	Maximum dosage/(%)	Remarks
Plasticizers	Dioctyl adipate (DOA)	Plastic, rubber	35	During the combined usage, converted to maximum dosage according to proportion
	Dibutyl phthalate (DBP)	Plastic, rubber	10	
	Di-(2-ethylhexyl)phthalate (DOP)	Plastic, glass paper, paint, bonds, or rubber	40	
	Diisononyl phthalate (DIOP)	Plastic, rubber, or plastic bottle mat	40 for plastic, rubber 50 for plastic bottle mat	Not to be used for products to be in long-term contact with oil
	Dioctyl sebacate (DOS)	Plastic	5	
	Butyl phthalyl butyl glycolate (BPBG)	Paint, bonds	40	
	Diphenylisooctyl phosphate (DPOP)	Food packing materials	40	
	Butyl stearate (BS)	Plastic, rubber	5	
	Epoxidized soybean oil	Food packing materials	According to demands of normal production	Also used as a stabilizer
	Glycerol monostearate	Plastic	According to demands of normal production	Also used as a lubricant
Stabilizers	Calcium stearate	Plastic	5.0	
	Magnesium stearate	Plastic	1.0	
	Calcium diethylene glycol or s, s′-calcium diethylene glycol	Plastic	1.0	
	Zinc stearate	Plastic	3	
Others	. . .			

18.7 Update in China Food Contact Chemical Regulation and Forecast on Future

18.7.1 Update of China Food Contact Chemical Legislation

The present limited list of allowed substance for food packaging use (GB 9685-2003) clearly is not in sync with a fast evolving Chinese market. In 2005, due to mounting pressure from industry, the MOH informed the industry of its desire to look into expanding the current positive list. The MOH agreed upon a one-time only "fast track" expansion of the current positive list, that is, to allow industry to nominate substances to the positive list as long as they meet certain criteria. This was one-time effort only as all future new substances need to go through a formal notification procedure. The AICM (Association of International Chemical Manufacturers) has formed a task force composed of various chemical industry members to technically support the MOH team. The AICM Food Contact Task Force is technically supported by industry associations such as the CEFIC, SPI, and ETAD.

The AICM has submitted a list of substances the industry wants to add to the China positive list. The information on the list includes chemical names, CAS#, application, regulatory status (e.g., cleared by the FDA, EFSA, BfR, JHOSPA, etc. for food contact applications). The initial list was composed of over 3000 substances.

The MOH was concerned that it does not have sufficient resources to handle the evaluation for such a large amount of substances. Some more new criteria were set: only substances that are sold in China would be considered for addition to the new positive list; the substance has to be on the China SEPA chemical inventory (IECSC); the substance has undergone risk assessment by other country authorities; only substances with clearance for food contact applications under at least two regulatory bodies are allowed. This includes FDA, EU (including EFSA, German, and French approval), and JHOSPA clearances. Under these criteria, the list has been shortened to about 959 substances.

On September 09, 2008, the GB9685 - 2008, Hygiene standard for use of additives in food containers and packaging materials, replacing GB9685-2003, has been published. The implementation date is June 1, 2009. The standard is issued by the Ministry of Health of the People's Republic of China, Standardizadition Administration of the People's Republic of China. It includes 959 additive items. For the additives, CAS number, scope of use, maximum level, specific migration limits or maximum concentration, and other restrictions if applicable.

18.7.2 Forecast on the New Food Contact Regulation in China

So what will the new food contact regulation in China be like? The following sections provide a forward-looking view from the authors based on the present best understanding.

18.7.2.1 Expanded Food Contact Additive Positive List

With GB 9685 - 2008, no we additional substances are allowed to be added at the present.

Under the new GB 9685-2008, for each additive, the following information is listed [4].

Name

The substance's CAS

Chemical name of the additive

CAS RN

Scope

添加剂名称 N,N'''-1,2-乙二基二[N-[3-[[4,6-二[丁基（1,2,2,6,6-五甲基-4-哌啶基）氨基]-1,3,5-三嗪-2-基]氨基]丙基]-N,N''-二丁基-N,N''-二（1,2,2,6,6-五甲基-4-哌啶基）-1,3,5-三嗪-2,4,6-三胺

CAS号 106990-43-6

使用范围	最大使用量	特殊迁移量/最大残留量	备注
塑料	PE: 0.1%; PP: 0.1%; PS: 0.4%; AS: 0.4%; ABS: 0.4%		

Plastics

Max. con.

The substance's approved scope

SML

Note

It is worth pointing out that the new GB 9685-2008 specifies the type of plastics for a specific substance. This means the use of a substance is limited to the approved type of plastic (PE, PP, PS, AS, ABS, etc.). For the substance to be used in other plastics, an application to the MOH for additional application needs to be submitted.

According to the present best understanding, if a substance is indicated with Max concentration and SML (specific migration limit), both limits need to be complied with (meaning the worst case between Max concentration and SML will be observed).

GB 9685-2008 was implemented since June 1st 2009. In June 2009, a notice regarding implementation of food safety law was issued by the Ministry of Industry & Information Technology. The Industry was reguested to start "clean up process", to check whether the food contact additives used are listed in the national hygiene standards, and self declare the additive not listed, to Ministry of Health (MOH) within one year from the date of the issue of the notice. The self-declaration can be used, at the same time, as nomination, for new additives. On December 4, 2009, the MOH issued a circular, providing some details regarding the information required for the self-declaration of new additives. This includes the information of regulatory status in other regulatory schemes, application, usage lavel, etc.

18.7.2.2 The New Food Contact Substance Notification Procedure

A draft food contact substance/resin notification procedure has been prepared by the MOH. The draft is under review and waiting for approval from the MOH. For details please see Appendix A (personal communication with Dr. Fan Yongxian from China MOH).

A concept similar to TOR is introduced. In the draft proposal, if migration is <10 ppb, procedure steps (1)–(6) in the proposal are needed, statement that no or low safety concern due to minimal migration is needed, and no toxicology tests are required but assurance that no CMR substance included as impurity may be required.

If a material already has other country/region clearance for food contact, it is easier to get approval in China; the principal is "not to repeat toxicology testing and assessment already done." Nevertheless, execution details are not available yet.

Proprietary information: All substances on GB 9685 are public information and only these substances shall be used for food contact application. If a new substance is notified and approval is given, the approval will be published by the MOH on the Internet, with a specification. Other companies that do not go for the notification procedure can also sell a substance on the market if the substance meets the specifications published on the MOH web site. This is until the next round of GB 9685 renewal. Once the GB 9685 is renewed, all newly approved substances will be made public. GB 9685 is expected to be renewed every 3 years, or so, based on current understanding.

Unlike the US FDA's FCN program, no commitment is given from the MOH on time frame for completing a review and approving a notification.

It is not clear how much MOH will charge for the notification, but the fee is not expected to be very high. Companies can hire consultants to undertake notification procedure on their behalf and pay charges.

18.7.2.3 Future Steps

Similar to GB 9685, the MOH plans to consolidate other nine GBs on various types of food contact material. Almost all Chinese standards pertaining to plastic food contact resin/food contact articles were to be reviewed in 2007–2009; for example, PE(GB 9691-1988 and GB/T5009.58-1996), PS (GB 9692-1988 and GB/T5009.59-2003), PP (GB 9693-1988 and GB/T5009.71-1996), PC(GB 13116-1991 and GB 14942-1994) were expected to be updated in 2010.

Together with food contact additive standards' updating, Chinese authorities also plan to establish related food contact SML determination methods.

18.8 Other Food Packaging Material-Related Regulation in China: Requirements for Local Manufacturing and Import of Food Contact Materials

18.8.1 Requirements for Local Manufacturing of Food Contact Materials

The AQSIQ published several rules regarding QS (quality safety) certificate requirements for all food containers, packaging finished articles, food processing tools/equipment, and so on in 2006.

QS certificate requirements apply to the finished food contact articles, not for food contact additives or food contact resins. The certificate is intended to control the product quality and food safety risk level of food contact articles manufactured in China.

According to these rules, if a food contact material belongs to the categories listed in the QS catalog [5, 6], local manufacturers who do not have the QS certificates from AQSIQ are not allowed to produce and sell the food containers/packaging products. To get this QS certificate, manufacturers need to apply to the local AQSIQ agency. An audit of production process and sites by the AQSIQ or its assigned auditors, sample testing, statement that the packaging product is in compliance with food contact regulation, and so on, may be requested for the purpose of issuance of the QS certificate. The QS certificate shall be renewed every 3 years. These rules also require that if the raw materials for manufacturing food containers/packaging are outside the range of the GBs, safety assessment is needed to demonstrate that the use of the materials is safe.

So far, the AQSIQ has published two required QS material catalogs. Public notice No. 2006-133, September 8, 2006 [5] is the catalog for the plastic food packing container. It requires that the various plastic food-packing types comply with relevant GB standards as listed in Table 18.4.

The AQSIQ public notice No. [2007]279 of July 2007 [6] is for the paper food contact article; the catalog includes paper tea bags, paper food packs, paper cups, and so on.

As a result of these recently established QS requirements, suppliers of raw materials for food contact use may have to provide their customers with the compliance status of the raw materials. Obviously, before the GB 9685 was updated it was difficult to issue compliance statement based on the limited list of allowed substances in then GB 9685. Such requirements from AQSIQ may fall in a gray area, and it is arguable whether the product is in compliance with Chinese food contact regulation. However, after GB 9685 is implemented, it is expected that such AQSIQ requirements are complied with more strictly.

18.8.2 Requirements for Importing Food Contact Materials

The AQSIQ has issued rule No. 135 [7] "Inspection and Supervision Regulations for Packaging Containers and Materials for Imported and Exported Foods." This regulation came into force on August 1, 2006.

Packaging containers and materials for imported foods refer to the interior packaging, sales packaging, transportation packaging, and packaging materials of imported foods, which are in contact with or are expected to be in contact with food. The AQSIQ uses a filing system for importers of food and its packages, and does have the authorization to inspect both the food and the packaging products for imported foods. The scope of the inspection and supervision includes the inspection, quarantine, and supervision of the production, processing, storage, sales, and other operations related to exported food packages.

Table 18.4 The QS catalog for the plastic food-packing container [5].

Category	Product unit	Product types	Standard No.	Note
Packaging products	1. Noncomposite membrane bag	1. Polyethylene self-bonding preservation packaging	GB 10457-1989	
		2. Retail package (only food packaging plastic bag)	GB/T 18893-2002	
		3. Polyethylene blow molding film used in liquid packing	QB 1231-1991	
		4. Polyvinylidene chloride (PVDC) sheet casing film used in food packaging	GB/T 17030-1997	
		5. Biaxially oriented polypropylene (BOPP) pearl light film	BB/T 0002-1994	*
		6. High-density polyethylene (HDPE) blow molding film	GB/T 12025-1989	
		7. PE blow molding film used in packaging	GB/T 4456-1996	
		8. BOPET film used in packaging	GB/T 16958-1997	
		9. Uneasily stretched HDPE film	QB/T 1128-1991	
		10. PP blow molding film	QB/T 1956-1994	
		11. BOPP film for heat-seal	GB/T 12026-2000	*
		12. CPE, CPP film	QB 1125-2000	*
		13. Clip Chain valve bag	BB/T 0014-1999	*
		14. Aluminum film used in packaging	BB/T 0030-2004	

2. Composite membrane bag	15. Composite membrane film or bag resistant to cooking and frying	GB/T 10004-1998
	16. BOPP/LDPE composite membrane film or bag	GB/T 10005-1998
	17. BOPA/LDPE composite membrane film or bag	QB/T 1871-1993
	18. Composite membrane film or bag used in preserved Szechuan pickle packing	QB 2197-1996
	19. Plastic composite membrane film or bag used in liquid food packing	GB 19741-2005
	20. Paper base compound materials used in liquid food asepsis packing	GB 18192-2000
	21. Composite membrane film or bag used in liquid food asepsis packing	GB 18454-2001
	22. Paper base compound materials used in liquid food preservation packing (Elopak)	GB 18706-2002
	23.Multilayer composite membrane film or bag	GB/T 5009.60-2003
3. Film	24. PVC hard parcel or film used in food packaging	GB/T 15267-1994
	25. BOPS film	GB/T 16719-1996
	26. PP Extrusion film	QB/T 2471-2000

(Continued)

Table 18.4 (*Continued*)

Category	Product unit	Product types	Standard No.	Note
	4. Woven bag	27. Plastic woven bag	GB/T 8946-1998	
		28. Composite plastic woven bag	GB/T 8947-1998	
Container products	5. Container	29. PE blow molding tub	GB/T 13508-1992	
		30. PET carbonic acid drink bottle	QB/T 1868-2004	
		31. PET nonbubbles drink bottle	QB 2357-1998	
		32. PC drinking water tin	QB 2460-1999	
		33. PET bottle used in hot packing	QB/T 2665-2004	
		34. Flexible plastic folding packing container	BB/T 0013-1999	
		35. Packing container plastic security tip	GB/T 17876-1999	
		36. Plastic feeder, plastic drink cup (pot), plastic bottle model	GB 14942-1994/GB 13113-1991/GB 17327-1998	*
Tool products	6. Food tool	37. Melamine plastic tableware	QB 1999-1994	
		38. Plastic cutting board	QB/T 1870-1993	
		39. Disposable plastic tableware	GB 9688-1988/GB 9689-1988	

Note: Products marked with "*" are suitable for use in food container.

For the purpose of the inspection and supervision of imported food packaging, it needs to be pointed out that the importer is expected to provide information including the following [7]:

- Description of ingredients and additives of the imported food container and packaging materials.
- Inspection and quarantine certificates for the imported food container and packaging materials issued by the foreign authority. The import company and the export company must submit the statement of compliance with the standards and requirements: "This food packaging and material complies with the Food Hygiene Law of the PRC, the Import and Export Commodity Inspection Law of the PRC, and specific implementation regulations. The mentioned food packaging and materials contain no poisonous or harmful substances and are qualified by our self-inspection. We hereby certify that the above-mentioned statements are true and genuine, and we will bear all responsibility in case of false statement." The statement should be signed by the legal representative of the import/export entity concerned.

Overall, for imported food packing articles, the company needs to make sure that

1) all used food contact additives are in compliance with GB 9685-2003 or the draft GB 9685-2008 version;
2) all used resins are in compliance with related resin GB standards of China;
3) the finished articles are in compliance with available GB standards for food contact articles;
4) if any new food contact additive, or any new food contact resin, is found in a finished article, making this kind of statement is not a smart choice. To avoid this situation, a company needs to apply for approval with the MOH.

For example, following are the related requirements for oil packing materials:

- All additives used in oil packaging articles, resins, materials must comply with the GB 9685 standards.
- Food packaging resins (polymers) used in oil packing articles should meet the related GB resin (polymer) standards.
- The oil manufacture must use the food packaging articles from those suppliers who have the required QS certificate for oil plastic packaging.
- The oil manufacture should display the related QS certificate number on the oil packaging.

Appendix A

Draft Food Contact Notification Requirements for New Food Contact Substance or New Food Contact Resins, New Food Contact Articles Prepared by MOH (Personal Communication with Dr. Fan Yongxian from China MOH)

1) Application Form of New FCS.
2) The notified substance/material's general chemical information. If the FCS was prepared from plants or animals and if the above information does not exist, the

main component(s) shall be indicated. If the product is a special product (mixture from reaction), just provide the physical and chemical data of the main components. The required information should include: chemical name, synonyms, chemical structure, molecular formula, molecular weight, physical, and chemical features, CAS RN.

3) Typical and maximum dosages of the FCS in the applications and the applicable condition of use in food contact material. Provide the food contact conditions: applicable food type, temperature, and substrate.
4) Manufacturing process for the notified FCS.
5) Specification of the FCS and analysis method.
6) The migration information, Annex 1
7) The evidence to show that the FCS is in compliance with new chemical regulations (IECSC inventory).
8) Toxicological safety assessment, depending on the migration level, the corresponding concern level 1–5 is defined. For corresponding concern level, following toxicological study has to be performed and a final toxicological assessment report has to be submitted, Annex 2.
9) If available, submit documents on the food exposure analysis done by other countries, including related approval evidence, other country regulations, and other regional standards.
10) Verification report: the quality test report of three lots of product samples, toxic assessment report, hygiene assessment report (issued by province level labs that are accepted by the MOH).
11) 50–100 g sample of the FCS must be submitted.
12) Submit other documents that can assist the approval.

All above-mentioned information/documents should be provided in Chinese and if the information/documents are issued by other countries, a Chinese abstract shall be provided.

Annex 1: Migration information

1) If migration of FCS is in low-concern category (maximum migration less than 0.01 mg/kg), submit national or international expert statements, or other information to explain.
2) If the 100% migration quantity or generally accepted diffusion (e.g., with barrier layer) is expected to be less than 0.05 mg/kg, submit the migration calculation report or declaration.
3) If the related components' maximum migration is over 0.05 mg/kg (including 0.05 mg/kg), the migration testing report should be provided including related detailed information of the migration report.
4) For applying for extension of use of a food contact substance/material already approved previously in China, submit the relevant information or reports, for example, migration report/information, related to the new intended use.

Annex 2: Toxic safety assessment information

1) If each component's migration level is less than 0.01 mg/kg, just submit publicly available toxic data/information.

2) If the component's migration level is between 0.01 and 0.05 mg/kg, three mutagenicity studies (Ames test, mammal cell test (CHO), and chromosomal aberration) must be submitted.
3) If the component's migration level is between 0.05 and 5 mg/kg, three mutagenicity studies (Ames test, mammal cell test (CHO), and chromosomal aberration), and a 90-day repeated oral toxic study must be submitted.
4) If the component's migration level is between 5 and 60 mg/kg, actuate oral toxic study, three mutagenicity studies (Ames test, mammal cell test (CHO), and chromosomal aberration), a 90-day repeated oral toxic study, and a chronic toxicity study must be submitted.

Note: A report from foreign GLP labs or from Chinese institutes, recognized by the MOH is acceptable for new FCS application. In case of non-GLP report, supporting expert verification information is required.

References

1 China Food Hygiene law (CFHL), the Standing Committee of National People's Congress, 1995.10.30.
2 On further strengthening the work of the Food Safety, Notice 2004-23, the State Council, 2004.9.1.
3 GB 9685-2003, *Hygienic Standard for Adjuvant and Processing Aids in Food Containers and Packing Materials,* issued by the Ministry of Health, People's Republic of China, on 2003.9.24. Effective 2004.5.1
4 GB 9685-2008, *Hygienic Standards for Uses of Additives in Food Containers and Packaging Materials,* issued by the Ministry of Health of the People's Republic of China, published on September 09, 2008.
5 Implementation of market permit system for the plastic food packaging and containers, tools, and other products, the AQSIQ Notice No. 2006-133, 2006.9.8.
6 Implementation of market permit system for the paper food packaging and containers, the AQSIQ Notice No. 2007-279, 2007.7.13.
7 "Inspection and supervision regulations for packaging containers and materials for imported and exported foods," the AQSIQ, 2006.8.1.

19
Principal Issues in Global Food Contact: Indian Perspective

Sameer Mehendale

19.1
The Indian Subcontinent: A Study in Contrast

The Indian subcontinent has a population of 1.4 billion, and is continuing to grow at a very high rate, adding every year to its base a population almost equivalent to Australia. The largest percentage of literates and illiterates in the world is found here. India has the maximum number of young-generation people.

On the climatic front, India is found to have one of the world's wettest and driest weather. In this region, climatic conditions vary from high levels of temperature and humidity to low levels of temperature and humidity.

India is one of the largest producers of milk, tea, sugar, wheat, detergents, biscuits, and two-wheelers. And is one of the largest software providers to the world.

India is moving toward a developed economy, with growing contribution of services to GDP. At the same time, it is becoming a global manufacturing outsourcing hub.

19.2
Food Contact Legislation

The food contact legislation was defined for the first time in India by the "Prevention of Food Adulteration Act" of 1954 and Rules of 1955 (Reference: Akalank's BARE ACT).

19.3
General Guidelines

Maximum emphasis and stress is given to packaging labeling.

Global Legislation for Food Packaging Materials. Edited by Rinus Rijk and Rob Veraart

ISBN: 978-3-527-31912-1

19.4
Condition for Sale and License

19.4.1
Conditions for Sale

(1) Every utensil or container used for manufacturing, preparing or containing any food or ingredient of food intended for sale shall be kept at all times in good order and repair and in a clean and sanitary condition. No such utensils or container shall be used for any other purpose.
(2) No person shall use for manufacturing, preparing or storing any food or ingredient of food intended for sale, any utensil or container that is imperfectly enameled or imperfectly tinned or that is made up of such material or is in such a state as to be likely to injure such food or render it noxious.
(3) Every utensil or container containing any food or ingredient of food intended for sale shall at all time be either provided with tight-fitting cover or kept closed or covered by properly fitting lid or by close fitting cover or gauze from dust, dirt and flies and other insects.
(4) No utensil or container used for the manufacture or preparation of or containing any food or ingredient of food intended for sale shall be kept in any place in which such utensil or container is likely by reason of impure air or dust or any offensive, noxious or deleterious gas or substances or any noxious or injurious emanations, exhalation, or effluvium, to be contaminated and thereby render the food noxious.
(5) A utensil or container made of the following materials or metals, when used in preparation, packaging, and storing of food shall be deemed to render it unfit for human consumption.
   (a) Containers that are rusty;
   (b) Enameled containers that have become chipped and rusty;
   (c) Copper or brass containers that are not properly tinned;
   (d) Containers made of aluminum not conforming in chemicals composition to IS: 20 specification for cast aluminum and aluminum alloy for utensils or IS: 21 specification for wrought aluminum and aluminum alloy for utensils.
   (e) Containers made of plastic materials not conforming to the following Indian Standards specifications, used as appliances or receptacles for packing or storing whether partly or wholly, food articles, namely: * IS: Indian Standard
      (i) *IS: 10146 Specification for polyethylene in contact with food stuff;
      (ii) *IS: 10142 Specification for styrene polymers in contact with food stuff;
      (iii) *IS: 10151 Specification for polyvinyl chloride (PVC) in contact with food stuff;
      (iv) *IS: 10910 Specification for polypropylene in contact with food stuff;
      (v) *IS: 11434 Specification for ionomer resins in contact with food stuff;

(vi) *IS: 11704 Specification for ethylene acrylic acid (EAA) copolymer;
(vii) *IS: 12252 Specification for polyalkylene terephathalates (PET);
(viii) *IS: 12247 Specifications for Nylon 6 Polymer;
(ix) *IS: 13601Specifications for ethylene vinyl acetate (EVA);
(x) *IS: 13576 Specifications for ethylene methacrylic acid (EMAA).

(f) Tin and plastic containers once used shall not be reused for packaging of edible oils and fats.

Provided that utensils or containers made of copper though not properly tinned may be used for the preparation of sugar confectionery or essential oils and mere use of such utensils or containers shall not be deemed to render sugar, confectionery or essential oils unfit for human consumption.

(6) No person shall sell compounded asafoetida exceeding 1 kg in weight except in a sealed container with a label.
(7) No person shall sell Hingra without a label on its container upon which is print a declaration in the form specified in Rule 42.
(8) No person shall sell titanium dioxide (food grade) except under Indian Standard Institution certification mark.
(9) No person shall sell salseed far for any other purpose except for BAKERY AND CONFECTIONERY and it shall be refined and shall bear the label declaration as laid down in Rule 42 (T).
(10) Edible common salt or Iodized salt or iron fortified common salt containing anticaking agent shall be sold only in package, which shall bear the label as specified in sub-rule (V) of Rule 42.
(10A) Iron fortified common salt shall be sold only in high-density polyethylene bag (HDPE) (14 mesh, density 100 kg/m^3, unlaminated) package, which shall bear the label as specified in sub-rule (VV) of Rule 42.
(11) No person shall sell lactic acid for use in food except under Indian Standards Institution marks.
(12) The katha prepared by Bhatti method shall be conspicuously marked as "Bhatti Katha."
(13) All edible oils, except coconut oil, imported in crude, raw or unrefined forms shall be subjected to the process of refining before sale for human consumption. Such oils shall bear a label declaration as laid down in Rule 42 (W)
(14) Dried glucose syrup containing sulfur dioxide exceeding 40 ppm shall be sold only in a package, which shall bear the label as specified in sub-rule (X) of Rule 42.
(15) No person shall store or expose for sale or permit the sale of any insecticide in the same premises where articles of food are stored, manufactured, or exposed for sale.

Provided that nothing in this sub-rule shall apply to the approved household insecticides, which have been registered as such under the Insecticide Act, 1968 (46 of 1968)

Explanation: For the purpose of this sub-rule, the word "Insecticide" has the same meaning as assigned to it in the Insecticide Act 1968 (46 of 1968).

(16) Condensed milk sweetened, condensed skimmed milk sweetened, milk powder, skimmed milk powder (partly skimmed milk powder and partly skimmed sweetened condensed milk) shall not be sold except under Indian Standards Institution certification mark.
(17) No person shall sell mineral oil (food grade) for use in confectionary except under Indian Standards Institution certification mark.
(18) No person shall sell confectionary weighing more than 500 gm except in packed condition and confectionery sold in pieces shall be kept in glass or other suitable containers.

Explanation: For the purposes of sub-rules (17) and (18) "confectionery" shall mean sugar boiled confectionary, lozenges and chewing gums and bubble gums.

(19) No person shall manufacture, sell, store, or exhibit for sale an infant milk food, infant formula (milk cereal-based weaning food and processed cereal-based weaning food) except under bureau of Indian standards Institution certification mark.
(20) No person shall sell protein rich atta and protein rich maida except in packed condition mentioning the names of ingredients on the label.
(21) The blended edible vegetable oils shall not be sold in loose form. It shall be sold in sealed packages weighing not more than 5 kg. It shall also not be sold under the common or generic name of the oil used in the blended but shall be sold as "Blended Edible Vegetable Oil." The sealed package shall be sold or offered for sale only under AGMARK certification mark bearing the label declarations as provided under Rule 42 and Rule 44 besides other labeling requirements under these rules.

19.5 Packing and Labeling of Foods

19.5.1 Package of Food to Carry a Label

Every package of food shall carry a label and unless otherwise provided in these rules, there shall be specified on every label:

(a) The name, trade name, or description of food contained in the package, provided that the name, trade name, or the description of food given on the package of food shall not include the name of any food or ingredient prefixed or suffixed to it, if such food ingredient is not the main ingredient of the final food product.
(b) The names of ingredients used in the product in descending order of their composition by weight or volume as the case may be.

Provided that in case of artificial flavoring substances, the label may not declare the chemical names of flavors, but in the case of natural flavoring substances or natural-identical flavoring substances, the common name of flavors shall be mentioned on the label.

Provided also that whenever gelatin is used as an ingredient, a declaration to this effect shall be made on the label by inserting the word "Gelatin-Animal Origin."

Provided also that when any article of food contains whole or part of any animal including birds, fresh water or marine animals or eggs or product of any animal origin, but not including milk or milk products, as an ingredient:

(a) A declaration to this effect shall be made by a symbol and color code so stipulated for this purpose to indicate that the product is a nonvegetarian food. The symbol shall consist of a brown color filled circle having a diameter not less than the minimum size specified in the table given below, inside the square with brown outline having side double the diameter of the circle, as indicated in clause (16) of sub-rule (ZZZ) of Rule 42.

	Area of principal display panel	Minimum size of diameter in mm
1.	Up to 100 cm square	3
2.	Above 100 cm square up to 500 cm square	4
3.	Above 500 cm square up to 2500 cm square	6
4.	Above 2500 cm square	8

(b) The symbol shall be prominently displayed:
   (1) On the package having contrast background on principal display panel.
   (2) Just close in proximity to the name or brand name of the product.
   (3) On the labels, containers, pamphlets, leaflets, advertisements in any media.

Provided that where any article of food contains egg only as nonvegetarian ingredient, the manufacturer, or packer or seller may give declaration to this effect in addition to the said symbol.

Provided also that in case of any bottle containing liquid milk or liquid beverage having milk as an ingredient, soft drink, carbonated water, or ready-to-serve fruit beverages, the declarations with regard to addition of fruit pulp and fruit juice as well as the "date of manufacture" and "best before date" shall invariably appear on the body of the bottle.

Provided also that in case of returnable bottle, which are recycled for refilling, where the label declarations are given on the crown, the declaration referred to in the above proviso, with regard to addition to the fruit pulp and fruit juice shall be enforced as per the schedule given below. The bottles on which the year of manufacture is not embossed, the date of replacing such bottle shall be the first day of April 2008.

Acidity regulator, acids, anticaking agent, antifoaming agent, antioxidant, bulking agent, color, color retention agent, emulsifier, emulsifying salt, firming agent, flour treatment agent, flavor enhancer, foaming agent, gelling agent, glazing agent; humectants, preservative, propellant, raising agent, stabilizer, sweetener, thickener.

Provided also that for declaration of flavors on the label the class of flavors, namely, natural flavors and natural flavoring substances or natural identical flavoring substances or artificial

Flavoring substances, as the case may be, shall be declared on the label.

Provided further that when statement regarding addition of colors or flavors is displayed on the label in accordance with Rule 24 and Rule 64BB, respectively, addition of such colors or flavors need not be mentioned in the list of ingredients

Provided also that in case both color and flavor are used in the product, one of the following combined statements in capital letters shall be displayed, just beneath the list of ingredients on the label attached to any package of food so colored and flavored namely:

1) CONTAINS PERMITTED NATURAL COLOUR(S) AND ADDED FLAVOUR(S)

OR

2) CONTAINS PERMITTED SYNTHETIC FOOD COLOUR(S) AND ADDED FLAVOUR(S)

OR

3) CONTAINS PERMITTED NATURAL AND SYNTHETIC FOOD COLOUR(S) AND ADDED FLAVOUR(S)

OR

4) CONTAINS PERMITTED NATURAL*/AND* SYNTHETIC* COLOUR(S) AND ADDED FLAVOUR(S)

(For the period up to and inclusive of September 1, 2001)

Provided further that whenever any article of food contains whole or part of any animal including birds and fresh water or marine animals or eggs as an ingredient, a declaration to this effect shall be made by a symbol and color code so stipulated for this purpose to indicate that the product is nonvegetarian food. The symbol shall consist of a circle with a single chord passing through its center from top left hand side to the right diagonally as indicated below.

Packaging label requirements are given for very few products, as in the ones mentioned below.

19.5.1.1 Oils

The package, label or the advertisement of edible oils and fats shall not use the expressions "super refined," "Extra Refined," "Macro-Refined," "Double-refined," "Ultra-Refined," "Anti-cholesterol," "Cholesterol Fighter," "Soothing to Heart," "Cholesterol Friendly," Saturated Fat Free," or such other expressions that are an exaggeration of the quality of the product.

19.5.1.2 Milk and Infant Food Substitutes

A statement "MOTHERS MILK IS BEST FOR YOUR BABY" in capital letters is a prerequisite on the label.

19.5.1.3 Packaged Drinking Water

Every package of mineral water shall carry the following declaration in capital letters: NATURAL MINERAL WATER.

It is important to note these rules and regulations need to be adhered to strictly. In summary, there needs to be adequate emphasis put on differentiating between vegetarian and nonvegetarian very clearly. Stress needs to be put on natural and synthetic colors and flavors. It is also a mandatory requirement that the languages should be in Hindi and English only. Another important requirement is that there should be no crowding of text matters on the label.

19.6 Indian Standards for Direct Food Contact

Based on the BIS (Bureau of Indian Standards) standards 9833-1981, this standard lists permitted pigments and colorants for use with plastics and may be regarded as safe for use in contact with food stuffs, pharmaceuticals and drinking water.

19.7 Methods of Analysis and Determination of Specific and Overall Migration Limits

Based on the BIS standards 9845:1998, this standard prescribes the method of analysis for determination of overall migration of constituents of single or multi-layered heat-sealable films, single homogeneous nonsealable films, finished containers and closures for sealing as lids, in the finished form, performed, or converted form.

BIS Standard 10171:1986 – gives the guidelines on suitability of plastics for food.
BIS Standard 10146:1982 – gives the following plastic resins: polyethylenes, polypropylenes, ionomers, acid copolymers, nylon, polystyrene, polyester, EVA, and so on.

19.8 Acceptability Criteria

Due to the shortage of trained manpower with the FDA (Food and Drug Application), there is limited awareness with these government bodies. As a result, limited audits are conducted due to lack of packaging experts and knowledge on the required labeling attributes. There is a lot of focus on food ingredients and labeling compared to packaging labeling, which is significantly low. There are no stringent rules and regulations for packaging for this food contact legislation to be implemented. The Indian corporate is very well aware of this. All the packaging suppliers are asked to certify their products for food contact. At the same time, testing labs are available for conducting migration tests, and PVC is allowed for use in direct food contact.

19.9 Future

The whole system is still in a nascent stage; however, at the same time, the awareness is growing. Here in India the media is very strong, hence, compliance with the food contact legislation can be ensured. There is lot of focus on HACCP and BRC compliance coming into focus. Batch traceability is another aspect of food contact legislation gaining significant importance.

20
Southeast Asia Food Contact

Legislation Update

Caroline Li and Sumalee Tangpitayakul

Over the past decades, some Southeast Asian countries have established various regulations for food containers and packaging materials, although most of the focus of regulatory bodies is still on food. Regulations for food containers and packaging materials in Southeast Asian countries usually provide general guidelines that prohibit any package or container that yields toxic, injurious, or hazardous substance, and put forward limitations on heavy metals and other hazardous substances.

In this chapter, regulations for food containers and packaging materials in Singapore, Malaysia, and Thailand will be discussed.

20.1
Singapore

The Singapore regulations on food containers and packaging materials are the Food Regulations 1988 that came into operation on October 1, 1988, specified under Chapter 283, Section 56 [1] of the Sale of Food Act. The regulation is currently governed under the Agri-food and Veterinary Authority of Singapore.

The definition of container under the regulation "includes any form of packaging of food for sale as a single item, whether by way of wholly or partly enclosing the food or by way of attaching the food to some other article and in particular includes a wrapper or confining band [1]."

The Singapore food regulations have the following requirements regarding containers for food under Part III, 37 [1].

The following PVC package/containers are prohibited:

1) If package or container yields, or is likely to yield more than 0.05 ppm vinyl chloride monomer; or

Global Legislation for Food Packaging Materials. Edited by Rinus Rijk and Rob Veraart

ISBN: 978-3-527-31912-1

2) If package or container yields, or is likely to yield, compounds known to be carcinogenic, mutagenic, teratogenic, or any other poisonous or injurious substance.
3) If package or container may release lead, antimony, arsenic, cadmium, or any other toxic substance to food.

There are specific requirements on the level of lead in ceramic foodware. The allowed level of lead is dependent on types/shapes of the container. For example, it only allows if

1) the maximum amount of lead in any one of six units examined is not more than 3.0 mcg of lead/ml of leaching solution in the case of a flatware with an internal depth of not more than 25 mm;
2) the maximum amount of lead in any one of six units examined is not more than 2.0 mcg of lead/ml of leaching solution in the case of a small hollowware with a capacity of less than 1.1 l but excluding cups and mugs;
3) the maximum amount of lead in any one of six units examined is not more than 1.0 mcg of lead/ml of leaching solution in the case of a large hollowware with a capacity of 1.1 l or more but excluding pitchers;
4) the maximum amount of lead in any one of six units examined is not more than 0.5 mcg of lead/ml of leaching solution in the case of cups and mugs; and
5) the maximum amount of lead in any one of six units examined is not more than 0.5 mcg of lead/ml of leaching solution in the case of pitchers [1].

The regulation has also a requirement that lead piping shall not be used for beer, cider, or other beverages or liquid food.

20.2 Malaysia

Malaysian regulations on food containers and packaging materials are the Food Regulations 1985 [2], specified under the Food Act 1983. The regulation is governed by the Ministry of Health.

Sections 27–36 (A) of the regulation provides specific requirements on food package/containers. The details are as follows:

> 27. Use of Harmful Packages Prohibited
>
> It is prohibited to use food container/packaging which yields or could yield to its contents, any toxic, injurious or tainting substance, or which contributes to the deterioration of the food."
>
> 28. Safety of Packages for Food
>
> It is prohibited to use food container/packaging either capable of imparting lead, antimony, arsenic, cadmium or any other toxic substance to any food prepared, packed, stored, delivered or exposed in it, or is not resistant to acid

> unless the package, appliance, container or vessel satisfies the test described in the Thirteenth Schedule.

Under 13th schedule there are tests for food container/packaging for storage of food and for cooking. For food container/packaging for storage, the food container/packaging is filled with leaching solution (4% acetic acid in water v/v) for 24 h under room temperature. The leaching solution should contain less than 0.2 ppm of antimony, arsenic, cadmium, individually, and <2 ppm of lead for the container/packaging to be used for storage.

For food container/packaging for cooking, the container with leaching solution will be heated to 120 °C and boiled for 2 h. Then, the container with leaching solution will be kept at room temperature for 22 h. The leaching solution should contain less than 0.7 ppm of antimony, arsenic, cadmium, individually, and <7 ppm of lead for the food container/packaging to be used for cooking.

Under Section 29: Use Of Polyvinyl Chloride Package Containing Excess Vinyl Chloride Monomer Prohibited, "It is prohibited to use food container/packaging made of polyvinyl chloride that contains more than 1 mg/kg of vinyl chloride monomer." Furthermore, the regulation prohibits sale of food in package if the food itself contains more than 0.05 mg/kg of vinyl chloride monomer.

Malaysian regulation has strict restrictions regarding recycling of packages/containers. It is prohibited to use recycled package for certain foods such as sugar, flour, and edible oil. Package for product of swine origin shall not be used for food of nonswine origin.

If a nonbottle package has been used for food, it is prohibited to bring it in contact with food again, unless an extra layer is brought between the recycled plastic and the food.

Any bottle that has previously been used for alcoholic beverage or shandy shall not be used for any food, other than alcoholic beverage and shandy.

However, certain kind of recycling for similar products is allowed. For example, polycarbonate containers of not less than 20 l in size that have previously been used for natural mineral water may be used for the same purpose. Glass bottles that have been used for alcoholic beverage or shandy can be used for the same purpose. The same applies to boxes or crates for vegetable/fruit.

Recycling of a packaging material previously used for another food product is prohibited for milk, soft drink, alcoholic beverage or shandy, vegetable, fish or fruit, and polished rice.

> 34. Presumption As To The Use Of Any Packages
>
> For the purposes of regulations 32 and 33, where a package, appliance, container, or vessel containing food bears any mark or label belonging to another food it shall be presumed that such package, appliance, container or vessel has been used for that particular food as shown by such mark or label.

The regulation makes it very clear that toys, coins, and so on are not allowed to be placed on food. However, the following are allowed: article for measuring the recommended quantity of food to be consumed, provided that such article is sterile, label, and sachet of reduced iron powder for the purpose of absorbing oxygen.

The regulation has the following requirements regarding "reduced iron powder":

1) The reduced iron powder . . . shall be enclosed in a sachet in such a manner that the oxygen absorber will not contaminate, taint, or migrate into the food.
2) The sachet itself and its label shall compose of material that will not contaminate, taint, or migrate into the food.A list of chemicals allowed to be in the sachet of reduced iron powder is provided in the regulation. It includes calcium chloride, calcium hydroxide, iron oxide, and so on.
3) The sachet of reduced iron powder shall be labeled with the words "OXYGEN ABSORBER" or any words having similar effect. The caution statements "DO NOT EAT CONTENTS" and "CONTAINS IRON POWDER" should be included on the label.

20.3 Thailand

The Thai regulations on food containers and packaging materials include several notifications issued under the Food Act of B.E. 2000 (1979) [3], by the Ministry of Public Health. In general, the food container must conform with the following quality or standard: must be clean; must not give out substance to contaminate the food and therefore likely to be harmful to health; must not contain "pathogenic micro-organisms"; and must give out no color to contaminate the food. It also specifies that the container must have never been used before, unless it is glass, ceramic, or plastic. It should have never been previously used for fertilizer, poisonous substance, or substance likely to be harmful to health. However, there are specifications of packaging depending on type of material used.

There are three notifications under this Food Act, which are as follows:

1) **Notification No. 92 B.E. 2528 (1985):** Prescription for quality or standard for food containers, use of food containers, and prohibition of use of things as food containers. This notification specifies the migration limits of lead and cadmium that leach from ceramic and enameled metal containers. The limits are specific to container/vessel shapes, for example, small deep vessels, large deep vessels, and so on. There is specific definition of various container/vessel shapes in the regulation. For example, 2.5 mg/l of lead and 0.25 mg/l of cadmium are allowed for infant food containers. On the other hand, 7 and 0.7 mg/l of lead and cadmium, respectively, are allowed for shallow vessels. The definition of the shallow vessel is "vessels of a depth not more than 25 mm when measured vertically from the deepest point internally to the horizontal level of the topmost part of the rim." This notification is expected to be soon revised in order to be in line with international standards.

2) **Notification No. 295 B.E. 2548 (2005) [3]:** Qualities or standards for container made of plastic. This notification regulates 12 types of plastic food packaging. General requirements of this regulation are very similar to that of Notification No. 92 as described in Clause 3. For example, this regulation requires that the container made from plastic must be clean; must not give out substance to contaminate the food and therefore likely to be harmful to health; must not contain "pathogenic microorganisms"; and must not give out any color to contaminate the food.

According to Clause 4 of Notification No. 295, the plastic packaging materials must conform with the specifications described in this notification, which include specifications both for material and for migration test. Notification No. 295 puts forward specifications for containers made of plastic. Specifications under Notification No. 295 are divided into two categories, the first part sets limits for heavy metals such as lead and cadmium in the plastic itself, and other hazardous chemicals that can migrate into food depending on the type of plastic used such as vinyl chloride monomer from polyvinyl chloride, bisphenol A from ploycarbonate, and so on (see Table 20.1).

The second part sets limits for substances that migrate into food simulants. This includes phenol, formaldehyde, and so on (see Table 20.2). The limits are again specific to a particular type of plastic, for example, polyvinyl chloride, polyethylene polypropylene, polystyrene, and so on. Clause 5 of Notification No. 295 states that "The analysis of qualities or standards of dispersion of plastic containers shall be carried out by the methods prescribed by Food and Drug Administration." Based on the present best understanding, the migration tests are done using four food simulants: water for food with $pH > 5$, 4% acetic acid for food with $pH < 5$, *n*-heptane for fatty food, and 20% ethanol for alcoholic food.

Clause 6 of Notification No. 295 specifies that plastic containers used for containing milk or milk products shall be made of polyethylene, ethylene, 1-alkene copolymerized resin, polypropylene, polystyrene, or polyethylene terephthalate. Additional limits for substances extracted by *n*-hexane and xylene for polyethylene, ethylene, 1-alkene copolymerized resin, and polypropylene are also described in Appendix 1.

Clause 7 of Notification No. 295 prohibits the use of colored plastic containers to contain food, except in the following cases [3]:

(a) laminate plastics, only the layer not coming into direct contact with food;
(b) plastics used for containing fruits with peel;
(c) containers made of reused plastic, for which approval has been obtained for containing fruits with peel.

The regulation prohibits recycling of certain food containers.

> Use of a container which has previously been used to pack or wrap a fertilizer, poisonous substance or substance likely to be harmful to health as a food container is prohibited.

Table 20.1 Qualities and standards for plastics.

Detail / Type of plastic [a]	Maximum level (Milligram per 1 Kilogram)											Plastic used for containing milk or milk product which type of plastic on the contact side are:			
	Polyvinylchloride	Poly ethylene Polypropylene	Polystyrene	Polyvinylydene chloride	Polyethylene terephtalate	Polycarbonate	Nylon (pa)	Polyvinyl alcohol	Polymethy methacrylate	Polymethyl pentene	Melamine	Polyethylene or Ethylene 1-alkene copolymerized resin	Poly propylene	Polystyrene	Polyethylene terephthalate
(1) Lead	100	100	100	100	100	100	100	100	100	100	100	-	-	-	100
(2) Heavy metal (calculated as lead)	-	-	-	-	-	-	-	-	-	-	-	20	20	20	-
(3) Barium	-	-	-	100	-	-	-	-	-	-	-	-	-	-	-
(4) Dibutyltin compound	50	-	-	-	-	-	-	-	-	-	-	-	-	-	-
(5) Cresyl phosphate	1 000	-	-	-	-	-	-	-	-	-	-	-	-	-	-
(6) Vinyl chloride monomer	1	-	-	-	-	-	-	-	-	-	-	-	-	-	-
(7) Volatile substance; toluene, ethylbenzene, isopropylbenzene, normal propylbenzene and styrene	-	-	5 000 – 2 000[b]	-	-	-	-	-	-	-	-	-	-	1 500	-
(8) Vinylidene choride	-	-	-	6	-	-	-	-	-	-	-	-	-	-	-
(9) Asenic	-	-	-	-	-	-	-	-	-	-	-	2	2	2	-
(10) Extracted substance by normal hexane	-	-	-	-	-	-	-	-	-	-	-	26 000	55 000	-	-
11) Substance dissolv in xylene	-	-	-	-	-	-	-	-	-	-	-	113 000	300 000	-	-
(12) Bisphenol a (included phenol and p-t-butylphenol)	-	-	-	-	-	500	-	-	-	-	-	-	-	-	-
13) Diphenolcarbonate	-	-	-	-	-	500	-	-	-	-	-	-	-	-	-
(14) Amine (tri-ethalene and tri-butylamene)	-	-	-	-	-	1	-	-	-	-	-	-	-	-	-
(15) Cadmium	100	100	100	100	100	100	100	100	100	100	100	-	-	-	100

Extracted from the "Notification of the Ministry of Public Health (No. 295) B.E. 2548 (2005)." *Remark*: do not analyze.

a) Other types of plastics that do not determine qualities or standards shall have qualities or standards according to Food and Drug Administration.

b) In case of use at a temperature higher than 100 °C, but the quantity of styrene shall not exceed 1000 mg/1 kg and that of ethyl benzene shall not exceed 1000 mg/1 kg.

Table 20.2 Qualities and standards of dissemination.

Type of plastic[a] / Detail	Maximum level (Milligram per 1 Cube decimeter of reagent)											Plastic used for containing milk or milk product which type of plastic on the contact side are:			
	Polyvinylchloride	Poly ethylene Polypropylene	Polystyrene	Polyvinylydene chloride	Polyethylene terephtalate	Polycarbonate	Nylon (pa)	Polyvinyl alcohol	Polymethy methacrylate	Polymethyl pentene	Melamine[c]	Polyethylene or Ethylene 1-alkene copolymerized resin	Poly propylene	Polystyrene	Polyethylene terephthalate
(1) Phenol	-	-	-	-	-	-	-	-	-	-	Not detect	-	-	-	-
(2) Formaldehyde	-	-	-	-	-	-	-	-	-	-	Not detect	-	-	-	-
(3) Antimony	-	-	-	-	0.05	-	-	-	-	-	-	-	-	-	0.025
(4) Germanium	-	-	-	-	0.1	-	-	-	-	-	-	-	-	-	0.05
(5) Heavy metal (calculated as lead)	1	1	1	1	1	1	1	1	1	1	1	1	1	1	1
(6) Potassium permanganate used for reaction	10	10	10	10	10	10	10	10	10	10	10	5	5	5	5
(7) Residue substances which is evaporate in water (in case of foods with acidity exceeding 5)	30	30	30	30	30	30	30	30	30	30	-	-	-	-	-
(8) Residue substances which is evaporate in 4% concentrated acetic acid (in case of foods with acidity less than 5)	30	30	30	30	30	30	30	30	30	30	30	15	15	15	15
(9) Residue substances which is evaporate in 20% concentrated alcohol (in case of alcoholic foods)	30	30	30	30	30	30	30	30	30	30	-	-	-	-	-
(10) Residue substance from volatile matters in normal heptane (in case of lipid oil and food contains lipid)	150	150 30[b]	240	30	30	30	30	30	30	120	-	75[d]	-	-	-
(11) Bisphenol a (phenol and p-t-butyl phenol) extracted by water (in case of food with acidity exceed 5)	-	-	-	-	-	2.5	-	-	-	-	-	-	-	-	-
(12) Bisphenol a (phenol and p-t-butyl phenol) extracted by 4% concentrated acetic acid (in case of food with acidity less than 5)	-	-	-	-	-	2.5	-	-	-	-	-	-	-	-	-

(Continued)

Table 20.2 (*Continued*)

Type of plastic* / Detail	Maximum level (Milligram per 1 Cube decimeter of reagent)											Plastic used for containing milk or milk product which type of plastic on the contact side are:			
	Polyvinylchloride	Poly ethylene Polypropylene	Polystyrene	Polyvinylydene chloride	Polyethylene terephtalate	Polycarbonate	Nylon (pa)	Polyvinyl alcohol	Polymethy methacrylate	Polymethyl pentene	Melamine	Polyethylene or Ethylene 1-alkene copolymerized resin	Poly propylene	Polystyrene	Polyethylene terephthalate
(13) Bisphenol a (phenol and p-t-butyl phenol) extracted by 20% concentrated ethanol (in case of alcoholic food)	-	-	-	-	-	2.5	-	-	-	-	-	-	-	-	-
(14) Bisphenol a (phenol and p-t-butyl phenol) extracted by normal heptane (in case of lipid oil and food contains lipid)	-	-	-	-	-	2.5	-	-	-	-	-	-	-	-	-
(15) Caprolactame	-	-	-	-	-	-	15	-	-	-	-	-	-	-	-
(16) Meta crylate	-	-	-	-	-	-	-	-	30	-	-	-	-	-	-

Extracted from the "Notification of the Ministry of Public Health (No. 295) B.E. 2548 (2005)." *Remark*: do not analyze.

a) Other types of plastics that do not determine qualities or standards shall have qualities or standards according to the Food and Drug Administration.

b) In case of use at temperature higher than 100 °C, analyze at a temperature of 95 °C for 30 min.

c) For milk and creamy milk products.

d) In case of use at a temperature higher than 100 °C.

Additional requirement in the regulation includes that the "Use of a container which is made for packing other thing which is not a food or which bears a design or any statement that may cause a misconception with respect to the material parts of the food contained therein as a food container is prohibited."

3) **Notification No. 117 B.E. 2532 (1989):** Specific to feeding bottles, for storing milk, or other liquid for consumption by infants and children, which consist of bottle, lid rubber teat, and rubber teat cover. The bottle, rubber teat, rubber cover shall be clean and shall have no color contaminating the food. In case the bottle is made of plastic, the plastic shall be of polycarbonate, which can withstand boiling heat. The regulation further specifies that lead and cadmium in the plastic should be <20 ppm, and migration of heavy metals, potassium permanganate, by water or 4% acetic acid depending on the pH of milk, should be within limits specified. For bottles made of other types of plastic, approval must be obtained from the Thai FDA. The rubber teat shall withstand boiling heat and comply with limits of lead and cadmium (10 ppm each); the quantity allowed for nitrosamine is 0.01 mg/kg for bottles made of rubber. Similar to plastics, there is a limit (in mg/1 dm^3 of the dissemination solution) on heavy metals, phenol, formaldehyde, and residue substances that evaporate in water under dissemination.

It is worth mentioning that there are a set of industrial product standards, developed by the TISI (Thai Industrial Standard Institute) under the Ministry of Industry [4]. TISI standards provide guidelines on quality and other properties of products and related processes. TISI developed both mandatory and voluntary Thai Industrial Standards (TIS) to suit the need and the growth of industry, trade, and economy of the country. Standards are developed according to the government policy of consumer protection, industrial promotion to be competitive in world market, environmental protection, and natural resources' preservation.

Product certification according to TIS: Product certification schemes of TISI consists of two types with different certification marks: voluntary certification mark and mandatory certification mark.

Voluntary TIS

TISI Mandatory

TISI standards cover a vast range of products and are not limited to food contact packaging materials. There are a few TISI standards that aim at food packaging. For example, TIS 564-2546 (2003) for ceramic ware, such as porcelain, in contact with food; TIS 17-2532 (1989) for unplasticized polyvinyl chloride pipes for drinking water services (compulsory standard); and so on. Not all the TISI standards are mandatory; for plastic packaging, only TIS 1136-2536 (1993) for cling film is compulsory. Other list of "compulsory" TISI standards can be found on the TISI web site.

According to the Thai Industrial Product Standard Act B.E. 2511 (1968), "and person who manufactures industrial products which are required by the Royal Decree to conform with the standard must produce an evidence to a competent official for inspection and receive a license from the Council. The application for a license, the inspection and the issue of a license shall be in accordance with the rules and procedure prescribed in the Ministerial Regulation." The same seems to apply to import and sale.

20.4
Conclusions

There is no provision that specifies substances that may or may not be used in food packaging (i.e., a "positive" or "negative" list) for Singapore and Malaysia. Thailand is in the process of establishing a positive list for polymers, additives, catalysts, and so on. Generally speaking, these countries require that the mentioned standards in each country's regulations are complied with and the levels of vinyl chloride monomer, heavy metals, and other substances considered to be "hazardous" do not exceed the limits. The packaging/container is generally considered safe (e.g., clean) under these countries' regulations provided that product standards, if applicable and compulsory, are followed.

References

1 (2005) Sale of Food Act (Chapter 283), in *Food Regulations*, Revised edition, SNP Corporation Ltd, Legal Publishing.
2 Ministry of Health, Malaysia, official web site http://fsis.moh.gov.my/fqc/ReferenceBooks/ActRule.asp?FAC_ID=21.
3 Thai Food and Drug Administration web site http://www.qmaker.com/fda/new/web_cms/subcol.php?SubCol_ID=77&Col_ID=14.
4 TISI web site http://www.tisi.go.th/standard/catalog.html.

21
Legislation on Food Contact Materials in the Republic of Korea

Hae Jung Yoon and Young Ja Lee

21.1
Introduction

Problems related to food safety have never been eliminated although they have been there since time immemorial. When occurrences of sanitary hazards that are immanent in food chain are noticed, concerns may extend to include food packaging along with processes of cultivation, harvest, manufacture, circulation, and consumption. It is supported with the Codex Alimentarius Commission (CAC) definition of food hygiene as "all conditions and measures necessary to ensure the safety and suitability of food at all stages" [1], and therefore it should be to secure food safety that necessary minimum food standards related to risk prevention are set in anticipation.

While the Ministry of Health, Welfare, and Family Affairs (MIHWAF) has the authority to legislate amendments to the Food Sanitation Act (FSA) [3, 4] and its implementing Presidential Decree and the Ministerial Ordinance to the Food Sanitation Act, the Korea Food and Drug Administration (KFDA) is the principal government agency in the Korean food system responsible for ensuring safe and wholesome food.

21.2
Food Sanitation Act

The Food Sanitation Act is the legal basis for the food safety-related work. Passed by the National Assembly, the fundamental objective of the FSA is to contribute to the improvement of national health by preventing hygienic dangers and harm caused by food and by improving the quality of food; this Act is supported by various educational and administrative norms including provision for establishing food standards as the minimum requirements to assure food quality and prevent food-related risk. Articles 7 and 9 of the FSA lay down standards for foods and their related consequential entities, for example, food additives, specifying criteria for manufacturing, processing, usage, or storage. These standards not only specify raw materials or components considered beneficial and necessary but also specify unsuitable substances.

Global Legislation for Food Packaging Materials. Edited by Rinus Rijk and Rob Veraart

ISBN: 978-3-527-31912-1

These standards and specifications stipulate that foods and food additives should keep their determined quality all along the food chain, till reaching the consumer. Manufacturers and distributors must observe these provisions through each step of manufacturing and marketing processes. The Commissioner of the KFDA notifies publicly the amendments to standards and specifications for foods and food additives as ordinances and their collected forms, called the Food Code or Food Additive Code [3, 5, 6], that are provided periodically as hard copies for the convenience of food manufacturers, inspectors, and other stakeholders in food business.

The quality of food is certified only by the Commissioner who recognizes the importance of food hygiene and food safety. When a violation is found, it becomes the object of punitive provisions with ban on sale, recall, and destruction. The FSA also stipulates provisions for handling and management of food and food additives, as well as equipment and facilities. Article 4 of the FSA lays down six basic criteria for prohibition of sale of foods and food additives.

1) Those that are rotten or immature and are injurious to human health.
2) Those that contain or bear toxic or injurious substances (except those substances that are recognized with no apprehension of injury to human health).
3) Those that are either contaminated with or suspected to be contaminated with pathogenic microorganisms that are injurious to human health.
4) Those that are due to uncleanness and admixing or addition of extraneous substance are injurious to human health.
5) Those that are provided by unauthorized or unqualified manufacturer or distributor who does not have license.
6) Those the import(s) of which is prohibited or those for which declaration(s) of import is not complete.

In regard to food packaging, there is a peculiar terminology defined in Article 2 of the FSA. The term "equipment" means apparatus, such as tableware, cookware, and machines, implements, and other things used for collecting, manufacturing, processing, preparing, storing, transporting, displaying, and delivering or taking food or food additives, which come in direct contact with food or food additives, excluding such machines, implements, and other things as used for harvesting in agriculture and fishery. The term "container and package" means articles that are used for holding or packaging food or food additives and are offered with them at the time of their delivery. In addition, Article 8 of the FSA prohibits the sale or manufacture for sale of food containers, packaging materials, and equipment that contain or bear any harmful or toxic substance that may be injurious to human health.

The Presidential Decree and the Ministerial Ordinance lay down more detailed guidelines on how the FSA is to be implemented. Especially, the ordinance includes the provisions for conducting food-related business in Korea, including the relevant penalties for compliance failure.

The national food safety strategy influences both the legislation and the organizational structure of enforcement that usually meet the resources of country's socioeconomic and political environment [2]. Besides MIHWAF, there are other government food control bodies also that are responsible for establishing regulations and standards for food-related work including food import. The Ministry for Food, Agriculture, Forestry, and Fisheries (MIFAFF) is responsible for more than 20 Acts, ordinances, and relevant documents on agricultural products including livestock and dairy products. Readers interested in details may consult the MIFAFF web site [7], which is available only in Korean language. Because of peculiar nature of packaging substances that pose potential risk, regulations on the packaging-related products overlap one another. While the Ministry of Environment restricts waste and recycling mainly by "Promotion of Saving and Recycling of Resources ACT" and "Toxic Chemical Control Acts," the Ministry of Knowledge Economy is responsible not only for establishing trade policy related to export and imports of goods in general but also for implementing "Quality Management and Safety Control of Industrial Products Acts."

21.3 Food Code

The Food Code based on the FSA stipulates the necessary minimum of food safety concerns that should be secured. Except for meat, poultry, and dairy products that are regulated by the MIFAFF, the KFDA, which is responsible for setting and implementing standards and specifications for food in general, conducts the Food Code that stipulates standards and specifications for manufacturing, processing, usage, cooking, storage of food, and equipment, container, and packaging for food products.

The Food Code was enacted in 1966 and has continued to evolve ever since. It consists of nine chapters that specify the standards of governing food products with maximum residue levels of pesticides, antibiotics, and radioactive ray standards, as well as testing methods. Annex to the Food Code contains labeling application on food, food additives, and equipment, and containers and packaging for food products. By no means, it is discussed here with regard to some selected provision for food packaging.

21.3.1 General Provision

The first chapter of the Food Code is "General Provision" that provides 27 basic application statements with respect to general matters such as scope of the code, food categorization, and selection of analytical methods.

The code classifies food products into three hierarchical categories:

- **Food Class**: the precedence of classification of 22 food groups according to general food descriptions and examples are "confectionery," "sugar products," and so on.
- **Food Species**: the subordinate rank of food class, and examples are "processed grain," " rice cake," "dried confectionery," and so on.
- **Food Type**: The basic categories that require mandatory labeling such as "candy," "Kangjung (Korean cracker, oil-and-honey pastry)," and so on.

Food species without food type sometimes can be considered "food type" so that individual standards and specifications for 126 food types are delineated into 22 food classes.

The appropriate test method for determination of compatibility with standard and specification are given in the next chapter; however, the 24th statement of this chapter stipulates that Pass/Fail determination should be principally performed and judged according to the test method specified in the Food Code. A method can also be used if it is judged more precise than the method specified in the Food Code. However, when the test result is suspected, it shall be performed and judged only through the specified method in the Food Code.

21.3.2 Common Standard and Specification for Food in General

Chapter 3 of the Food Code includes requirements for raw materials, manufacture, and process applying to general food. It gives an eye to two packaging-related statues: (1) Container(s) and/or package(s) produced by the declared manufacturer who intends to operate packaging business shall only be put to use; however, it may be waived if product manufacturer directly manufactures container and/or package in order to pack his own product. (2) Container(s) and package(s) that are collected for reuse shall be confirmed whether any impurity is removed before reuse. This is applicable only to glass bottles.

21.3.3 Equipment, Containers, and Packaging for Food Products

Substances that consist of food contact materials are not considered as food additives. However, food contact material prior to its use or sale for food packaging should require premarket approval under the provisions of Article 9 of the FSA, and Chapter 7 of the Food Code delineates the relevant statutes. Two different types of regulations for containers and packaging materials have been stated: general standard and material-specific specifications.

The general standards that apply to all containers and packaging materials primarily address various food contact applications.

1) The shape or structure of equipment, container, or package shall protect its content from contamination, physically or chemically.
2) The specification of residue after evaporation may be waived if food ingredient such as starch and glycerin is applied on the food contact surface of a container or package.
3) Solder shall not be used in the manufacture or repair of equipment, container, and packaging on the food contact surface.
4) Only electrodes made of iron, aluminum, platinum, titanium, or stainless steel shall be used when electric current flows directly into food.

5) Food contact surface of equipment, container, and packaging, which is made of copper or copper alloy, shall be properly treated with tin coating or polishing so as not to be hygienically harmful. However, it may be exempted for a lustrous contact surface, with a peculiar characteristic of noncorrosiveness.
6) In case of using synthetic colorants in the manufacture of equipment, container, and packaging, colorants permitted as food additives shall be used. However, this requirement is abandoned when the colorant is added to melted glaze, glass, or enamel or when it can be demonstrated that there is no migration to food.
7) Printing inks shall be completely dried when manufacturing container and package. A plastic package that changes shape when contents are loaded shall not contain toluene more than 2 mg/m^2. Above all, any food contact surface shall not be printed.
8) Di-(2-ethylhexyl)phthalate (DEHP or DOP) shall not be used in the manufacture of equipment, container, and package. However, it is abandoned when there is no apprehension that di-(2-ethylhexyl)phthalate is migrated into food.
9) Di-(2-ethylhexyl)adiphate (DEHA or DOA) shall not be used in the manufacture of a wrap or cling sheet.
10) Di-*n*-butyl-phthalate (DBP) and benzyl-*n*-butyl-phthalate (BBP) shall not be used in the manufacture of a baby milk bottle.

There have been established eight material-specific standards:

- plastics
- cellophane, regenerated cellulose
- rubber
- paper and paper board
- metal (including metal cans)
- wood
- glass, ceramics, enamel, and pottery
- starch

Specifications for these materials do not generally identify substances that may be used to manufacture materials, but rather identify end tests that must be conducted on materials to ensure they meet the specifications established by the KFDA. These tests include heavy metal limits, total residue after evaporation under specified migration conditions, and residue limits from specified materials. Forty-one kinds of plastics have been approved for food packaging use and their specifications have been established (Table 21.1).

Specifications include either residue limits of monomers, additives and byproducts or limits of migration into food simulant from specific plastic resins that are well known for their hazardous characteristics, such as bisphenol A as the suspected endocrine disrupter. Plastic resin-coated metal can should comply with specifications provided in Table 21.2.

All types of fluorescene brightener are prohibited for use in paper and paperboard, as well as materials that contain not less than 70% of starch that would be in direct

Table 21.1 Approved plastic materials for food packaging use and their specifications (December 12, 2007).

Plastics	Specification	
	Residue limit from material (mg/kg)	Limit of migration into food simulant (mg/l)
1. Polyvinyl chloride (PVC)	Lead (100), cadmium (100), vinyl chloride monomer (1), dibutyltin compound (50), ester of cresol phosphate (1000)	Heavy metals (1)[a], potassium permanganate consumed (10)[b], residues after evaporation (30, but 150 in heptane)
2. Polyethylene and polypropylene (PE/PP)	Lead (100), cadmium (100)	Heavy metals (1)[a], potassium permanganate consumed (10)[b], residues after evaporation (30, but 150 in heptane)
3. Polystyrene (PS)	Lead (100), cadmium (100), volatile substances (5000, in the case of forming PS: 2000)	Heavy metals (1)[a], potassium permanganate consumed (10)[b], residues after evaporation (30, but 240 in heptane)
4. Polychlorovinylidene (PVDC)	Lead (100), cadmium (100), vinylidene chloride (6), barium (100)	Heavy metals (1)[a], potassium permanganate consumed (10)[b], residues after evaporation (30),
5. Polyethylene terephthalate (PET)	Lead (100), cadmium (100)	Heavy metals (1)[a], potassium permanganate consumed (10)[b], residues after evaporation (30), antimony (0.05), germanium (0.1), terephthalic acid (7.5), *iso*-phthalic acid (5.0)
6. Phenolformaldehyde (PF)	Lead (100), cadmium (100)	Heavy metals (1)[a], residues after evaporation (30), phenol (5.0), formaldehyde (4.0)
7. Melamineformaldehyde (MF)	Lead (100), cadmium (100)	Heavy metals (1)[a], residues after evaporation (30), phenol (5.0), formaldehyde (4.0), melamine (30)
8. Ureaformaldehyde (UF)	Lead (100), cadmium (100)	Heavy metals (1)[a], residues after evaporation (30), formaldehyde (4.0), phenol (5.0)
9. Polyacetal, polyoxymethylene (POM), polyformaldehyde	Lead (100), cadmium (100)	Heavy metals (1)[a], residues after evaporation (30), formaldehyde (4.0)

10. Acryl resin	Lead (100), cadmium (100)	Heavy metals (1)[a], potassium permanganate consumed (10)[b], residues after evaporation (30), methylmethacrylate (15, applied only when contained more than 50% of methylmethacrylate)
11. Polyamide/Nylon (PA/Nylon)	Lead (100), cadmium (100)	Heavy metals (1)[a], potassium permanganate consumed (10)[b], residues after evaporation (30), caprolactam (15)
12. Polymethylpentene (PMP)	Lead (100), cadmium (100)	Heavy metals (1)[a], potassium permanganate consumed (10)[b], residues after evaporation (30, but 120 in heptane)
13. Polycarbonate (PC)	Lead (100), cadmium (100), bisphenol A (500), D-phenylcarbonate (500), amines (1.0)	Heavy metals (1)[a], potassium permanganate consumed (10)[b], residues after evaporation (30), bisphenol A (includes phenol and *p*-tertiary butylphenol: 2.5)[c]
14. Polyvinylalchol (PVA)	Lead (100), cadmium (100)	Heavy metals (1)[a], potassium permanganate consumed (10)[b], residues after evaporation (30)
15. Polyurethane (PU)	Lead (100), cadmium (100)	Heavy metals (1)[a], potassium permanganate consumed (10)[b], residues after evaporation (30), isocyanate (0.1)
16. Polybutene-1 (PB-1)	Lead (100), cadmium (100)	Heavy metals (1)[a], potassium permanganate consumed (10)[b], residues after evaporation (30, but 120 in heptane)
17. Butadiene (BDR)	Lead (100), cadmium (100)	Heavy metals (1)[a], potassium permanganate consumed (10)[b], residues after evaporation (30, but 240 in heptane)
18. Acrylonitrile-butadiene styrene (ABS)/acrylonitrile styrene (AS)	Lead (100), cadmium (100), volatile substances (5000)	Heavy metals (1)[a], potassium permanganate consumed (10)[b], residues after evaporation (30, but 240 in heptane), acrylonitrile (0.02)
19. Polymethacrylstyrene (MS)	Lead (100), cadmium (100), volatile substances (5000)	Heavy metals (1)[a], potassium permanganate consumed (10)[b], residues after evaporation (30, but 240 in heptane), methylmethacrylate (15)
20. Polybutylene terephthalate (PBT)	Lead (100), cadmium (100)	Heavy metals (1)[a], potassium permanganate consumed (10)[b], residues after evaporation (30)
21. Polyarylsulfone (PASF)	Lead (100), cadmium (100)	Heavy metals (1)[a], potassium permanganate consumed (10)[b], residues after evaporation (30)
22. Polyarylate (PAR)	Lead (100), cadmium (100)	Heavy metals (1)[a], potassium permanganate consumed (10)[b], residues after evaporation (30)

(*continued*)

Table 21.1 (*Continued*)

Plastics	Specification	
	Residue limit from material (mg/kg)	Limit of migration into food simulant (mg/l)
23. Hydroxybutyl polyester (HBP)	Lead (100), cadmium (100)	Heavy metals (1)[a], potassium permanganate consumed (10)[b], residues after evaporation (30)
24. Polyacrylonitrile (PAN)	Lead (100), cadmium (100)	Heavy metals (1)[a], potassium permanganate consumed (10)[b], residues after evaporation (30), acrylonitrile (0.02)
25. Fluoro resins (FR)	Lead (100), cadmium (100)	Heavy metals (1)[a], potassium permanganate consumed (10)[b], residues after evaporation (30)
26. Polyphenylene ether (PPE)	Lead (100), cadmium (100)	Heavy metals (1)[a], potassium permanganate consumed (10)[b], residues after evaporation (30)
27. Ionomer	Lead (100), cadmium (100)	Heavy metals (1)[a], potassium permanganate consumed (10)[b], residues after evaporation (30)
28. Ethylenevinylacetate (EVA)	Lead (100), cadmium (100)	Heavy metals (1)[a], potassium permanganate consumed (10)[b], residues after evaporation (30)
29. Methylmethacrylate-acrylonitrile-butadiene-styrene (MABS)	Lead (100), cadmium (100), volatile substances (5000)	Heavy metals (1)[a], potassium permanganate consumed (10)[b], residues after evaporation (30), methylmethacrylate (15), acrylonitrile (0.02)
30. Polyethylenenaphthalate (PEN)	Lead (100), cadmium (100)	Heavy metals (1)[a], potassium permanganate consumed (10)[b], residues after evaporation (30)
31. Silicone		Heavy metals (1)[a], potassium permanganate consumed (10)[b], residues after evaporation (30)
32. Epoxy resin	Lead (100), cadmium (100), amines (1.0)	Heavy metals (1)[a], potassium permanganate consumed (10)[b], residues after evaporation (30), bisphenol A (includes phenol and *p*-tertiary butylphenol: 2.5)[c], diglycidyl ether of bisphenol A (includes diglycidyl ester of bisphenol A dichloride and diglycidyl ether of bisphenol A dihydrate: 1.0), diglycidyl ether of bisphenol F (includes diglycidyl ether of bisphenol F dichloride and diglycidyl ether of bisphenol F dehydrate: 0.5)

33. Polyetherimide	Lead (100), cadmium (100), bisphenol A (includes phenol and *p*-tertiary butylphenol: 500)	Heavy metals (1)[a], potassium permanganate consumed (10)[b], residues after evaporation (30), bisphenol A (includes phenol and *p*-tertiary butylphenol: 2.5)[c]
34. Polyphenylenesulfide (PPS)	Lead (100), cadmium (100)	Heavy metals (1)[a], potassium permanganate consumed (10)[b], residues after evaporation (30),
35. Polyethersulfone (PES)	Lead (100), cadmium (100)	Heavy metals (1)[a], potassium permanganate consumed (10)[b], residues after evaporation (30)
36. Poly(cyclohexane-1,4-dimethylene terephthalate (PCT)	Lead (100), cadmium (100)	Heavy metals (1)[a], potassium permanganate consumed (10)[b], residues after evaporation (30), antimony (0.05)
37. Ethylenevinylalcohol (EVOH)	Lead (100), cadmium (100)	Heavy metals (1)[a], potassium permanganate consumed (10)[b], residues after evaporation (30)
38. Polyimide (PI)	Lead (100), cadmium (100)	Heavy metals (1)[a], potassium permanganate consumed (10)[b], residues after evaporation (30)
39. Polyetheretherketone (PEEK)	Lead (100), cadmium (100)	Heavy metals (1)[a], potassium permanganate consumed (10)[b], residues after evaporation (30, but 150 from heptane)
40. Polylactide, polylactic acid (PLA)	Lead (100), cadmium (100)	Heavy metals (1)[a], potassium permanganate consumed (10)[b], residues after evaporation (30)
41. Polybutylene succinate-*co*-adipate: (PBSA)	Lead (100), cadmium (100)	Heavy metals (1)[a], potassium permanganate consumed (10)[b], residues after evaporation (30, exempt when contained starch)

a) Measured lead content by colorimetric analysis.
b) Volume (ml) of 0.01 N potassium permanganate reacted with 100 ml of testing solution.
c) Bisphenol A should be not more than 0.6 mg/l.

Table 21.2 Migration limits from metal can (mg/l).

Lead	Not more than 0.4
Cadmium	Not more than 0.1
Arsenic	Not more than 0.2
Nickel	Not more than 0.1
Chromium	Not more than 0.1
Phenol	Not more than 5.0
Formaldehyde	Not more than 4.0
Vinyl chloride monomer	Not more than 0.05
Epichlorohydrin	Not more than 0.5
Bisphenol A	Not more than 0.6
Diglycidyl ether of bisphenol A[a)]	Not more than 1.0
Diglycidyl ester of bisphenol F[b)]	Not more than 1.0
Residues after evaporation	Not more than 30[c)]

a) Includes diglycidyl ester of bisphenol A dichloride and diglycidyl ether of bisphenol A dihydrate.
b) Includes diglycidyl ester of bisphenol F dichloride and diglycidyl ether of bisphenol F dihydrate).
c) Not more than 90 mg/l from *n*-heptane for cans coated with natural oil coating containing 3% or more zinc oxide, otherwise not more than 150 mg/l from *n*-heptane).

contact with food. Most recently, residue limits of certain fungicides have been established for wooden chopsticks including disposable ones.

Samples taken for testing should be in a ready-for-use state, and surfaces intended to come in contact with food in actual use should be tested. For migration test, the surface to volume ratio would be used according to the shapes of samples. In case of a sample holding a specified volume, for example, a bottle, the bottle should be tested with the specified volume of simulant. However, a sample that cannot hold a liquid, 1 cm^2 of contact surface area would be immersed in 2 ml of the specified simulant, while a sample with flat shape-like film or sheet would be applied in a special cell that can easily hold the surface in contact with the simulant. Unless the condition is specified, for example, intended for use at over 100 °C, the sample would be immersed in the specified simulant for 30 min at 60 °C. Table 21.3 presents the representative testing conditions for actual use of material.

21.4 Data and Information Required for Submission of Food Contact Material Prior to Use

In case of packaging materials not covered by the Food Code, manufacturers must obtain premarket approval from the KFDA. A submission to the KFDA must include the name and contact information of the manufacturer of the material, the product name, identification by the chemical name of the product, a description of its manufacturing process and its intended use, material and migration specification, and test methods that are used to ensure compliance with these specifications.

Table 21.3 Simulants for migration test according to various food types.

Food types	Food simulants	Test conditions unless specified
Fatty food	*n*-Heptane	25 °C for 1 h
Alcoholic beverage	20% Ethanol	60 °C for 30 min
Others		
Below pH 5	4% Acetic acid	60 °C for 30 min
Over pH 5	Distilled water	60 °C for 30 min

After reviewing the submitted documents, which should be complete within 14 working days, a provisional approval shall be notified to the manufacturer or supplier of a food contact material. The provisional approval shall be maintained unless new data demonstrate that the intended use of the material is not safe. Subsequently, the provisional approval may be proceeded to obtain proclamation to be appended to the Food Code. Contact information for the division responsible is as follows:

Food Packaging Division,
Food Safety Evaluation Department, KFDA
#194, Tongil-no, Eunpyung-gu, Seoul 122-704, Korea

21.5 Labeling Standard for Food and Food Additives

Article 10 of the FSA specifies that all food, food additives, and packages are required to be labeled with necessary information in Korean, and Labeling Standard for Food (LSF) provides the relevant stipulations. Specifically, Article 6 of LSF requires the care label for the sake of consumer safety on plastic wrap that should be used only under the food temperature at 100 °C and must not directly contact fatty food surface. In addition, additives used in plastic wrap, such as plasticizers, stabilizers, and antioxidants should be declared.

Labels should bear the information such as the name and address of the manufacturer where products may be returned or exchanged when they are found defected. However, this requirement may be waived if product manufacturer himself manufactures container and/or package to pack his own product.

Containers or packages that can be recycled must carry a "separation and discharge" sign. In accordance with "Promotion of Saving and Recycling of Resources Act," containers or packages that are made using paper, metal, glass, and plastic materials must be marked with a "separation and discharge" sign to facilitate the recycling of wastes. The sign should indicate the type of material the package is composed of. For example, "PET," "HDPE," "PVC," or "Other" should be indicated for containers or packaging made of plastic materials. For metals, either iron or aluminum should be indicated.

21.6 Requirements for Importing Food Contact Materials

Besides the Korea Customs Service (KCS), responsible for ensuring that all necessary documentation is in place before the product is released to the local market, there are several government agencies involved in the import clearance process. KFDA inspection results are transmitted thus shortening KCS clearance time by electronically connecting the import food network system of the KFDA to the system of the KCS.

The importer or the importer's representative should submit the "import declaration for food and food additives" to regional KFDAs that conduct inspection of a given food and additive, so that document inspection, organoleptic inspection, laboratory inspection, and random sampling examination are conducted accordingly. If a product complies with the Food Code, the KFDA issues a certificate for import so that an importer can import the product that is recognized by the KFDA Commissioner as safe, subject only to a document inspection unless selected for the random sampling examination. If a product does not comply with the Food Code, the KFDA will notify the applicant and the regional customs office about the nature of the violation. The importer decides whether to destroy the product or return the shipment to the country of origin. Recent amendment adds that if a product is filed as incompatible, then it becomes subject to a mandatory laboratory test and it should be proved that there has been no violation on five consecutive import occasions. However, minor violation, for example, incorrect labeling, may be corrected and the importer may reapply for inspection. In general, laboratory inspection takes place within 5 days.

21.7 The Future Direction for Packaging Material Regulations

In parallel with implementation of KFDA's mission to continuously strengthen the present regulations, the KFDA endeavors to develop scientifically firm risk assessment systems for harmful substances in food packaging materials. When a new technology, including processing techniques, emerges, a new health issue always evolves along with it, and a need to develop various risk management options would lead to establishing effective amendments to standards and specifications. At present, the KFDA has a broad interest; for example, substances that pose potential risks such as contaminants in recycled materials and migration factors in relation to various food types. Because the positive list on food contact substances has been adopted in advanced countries, the KFDA put its deliberation on harmonizing it with the present food packaging material regulatory system. It is foreseen that building a positive list of food contact substances would demand voluntary cooperation from relevant interest groups, including research groups that may provide both biological properties of substance and consumer exposure information.

References

1 Codex Alimentarius Commission (2006) *Procedural Manual*, 16th edn. World Health Organization and Food and Agriculture Organization of the United Nations.
2 FAO/WHO (2000) *Assuring Food Safety and Quality: Guidelines for Strengthening National Food Control Systems*, FAO/WHO.
3 KFDA web site at www.kfda.go.kr in Korean.
4 MIHWAF web site at www.mw.go.kr in Korean.
5 Food and Drug Administration, Food Code, 2007, Moonyung-sa, Seoul, Korea.
6 Korea Food and Drug Administration, Additive Code 2007, Moonyung-sa, Seoul, Korea.
7 MIFAFF web site at www.mifaff.go.kr in Korean.

22
Australia and New Zealand

Robert J. Steele

22.1
Introduction

Australia and New Zealand are stable, culturally diverse, and democratic societies with skilled workforces and strong competitive economies. The population of both countries is less than 25 million, and large distances to their export markets affect the breadth and depth of their food contact legislations.

Post-World War II, both countries developed strong agricultural industries with markets initially in the United Kingdom and Europe. Export growth is now more focused on Asia with their burgeoning populations and middle classes.

The food law in both countries is now harmonized through a historic agreement signed between both countries on December 6, 2002 [1]. This agreement outlined the following objectives:

- Providing safe food controls for the purpose of protecting public health and safety;
- Reducing the regulatory burden on the food sector;
- Facilitating the harmonization of Australia and New Zealand's domestic and export food standards and their harmonization with international food standards;
- Providing cost-effective compliance and enforcement arrangements for industry, government, and industry;
- Recognizing that responsibility for food safety encompasses all levels of government and a variety of portfolios; and supporting the joint Australian and New Zealand efforts to harmonize food standards.

An excellent review of the recent changes to the Australian and New Zealand food laws can be found in Ref. 2. This journal describes the evolution of the Australian and New Zealand food standards: although there is discussion of contaminants and natural toxicants [3], a notable omission is any discussion of food contact legislation in Australia and New Zealand.

Global Legislation for Food Packaging Materials. Edited by Rinus Rijk and Rob Veraart

ISBN: 978-3-527-31912-1

Table 22.1 Overview of the Australian food industry.

	2001–2002	2002–2003	2003–2004	2004–2005	2005–2006
Fisheries production ($b)	27.7	32.5	31.9	33.9	30.2
Value added, food processing ($b)	16.4	16.6	17.3	17.5	na
Share of GDP (%)	1.9	1.8	1.8	1.8	na
Food and liquor retailing turnover ($b)	76.8	81.9	88.7	91.8	97.4[a]
Share of total retailing (%)	46.2	46.2	46.2	46.7	47.4
Value of food exports ($b)	22.6	22.4	24.0	24.1	23.3
Value of food imports ($b)	6.0	6.0	6.7	7.0	8.2

a) Indudes an imported value for horticulute production in 2006–06.

22.2 Australia

The food industry plays an essential role in Australia's economic and social wealth. About 20% of manufacturing sales and services income come from the food industry where 206,000 Australians are engaged. Most food sold in Australia is grown and supplied by Australian farmers who export almost two-thirds of their agricultural produce.

In 2007–2008, Australia exported $US17 billion worth of food compared to food imports in the same year of $US 7 billion, despite the effects of drought [4]. Meat and grains have consistently been the two largest export categories, with meat accounting for 30% of the value of food exports in 2006–2007 and grains nearly 15%. Wine and dairy exports have also significantly grown in recent years, with wine accounting for nearly 13% of exports in 2006–2007 and dairy nearly 10% (Table 22.1).

Australia's major markets for exports are Japan and the United States, making up 20 and 13%, respectively. Since 1990–1991, there has also been an increased share of exports going to Indonesia (from 2% in 1990–1991 to 7% in 2006–2007), the Republic of Korea (4–8%), New Zealand, and the United Kingdom (both 2–5%). Australia has benefited from its capacity to supply high-quality food products to Pacific Rim countries and to more distant markets such as Saudi Arabia and the United Arab Emirates.

22.3 Australia's Regulatory Framework

In 1901, the six states of Australia formed a federation and formed the Commonwealth of Australia. The Constitution made provision for a national level of government referred to as the Commonwealth, with legislative power exercised through a federal Parliament comprised of a Senate and a House of Representatives.

The Australian Constitution provides for the powers of the Commonwealth to be exercised at three levels:

- power is conferred on the Parliament;
- executive power, to assent to and administer laws, and to carry out the business of government, is conferred on the Governor-General, Ministers of State, departments, other government agencies, and the defense forces;
- judicial power is vested in the High Court of Australia and other courts established by the Parliament.

The former six colonies became six states. Each retained its own Parliament and initially exercised legislative powers to control food production and distribution. Only much later did a truly coherent nationwide set of food standards develop under the Commonwealth Government.

The Commonwealth Parliament can only make laws in relation to a range of subjects specified in the Constitution. Major areas include taxation, defense, external affairs, trade, and immigration. The Commonwealth now has the capacity to regulate food businesses despite the apparent limitations in the Constitution. It does so in close cooperation with the states and territories and New Zealand. It was not always so. Until 1975 each Australian state and territory and New Zealand had responsibility for the regulation of the food supply.

Initially, the states and territories based their food laws on the 1860 UK Act for the Prevention of Adulteration of Food and Drink. As a result, there developed a wide range of discrepancies and differences. It was not until 1954 that the Commonwealth Government made its first attempt to harmonize food law. The Food Standards Committee was established through the National Medical and Research Council, with aim of developing a national set of food standards that would be adopted by each state and territory. By 1974, a comprehensive set of prescriptive standards had been published, but there was far from unanimous adoption.

This lack of uniformity became an unnecessary barrier to trade between the states and with the growth of franchised food outlets and large international food companies trading in several if not all states and territories, the urgent need for reform was realized. In 1975, the Model Food Act was developed; however, again it was not uniformly adopted by the states. In 1997, the Commonwealth Government initiated a major review of food legislation with the view to reduce regulatory load and to make compliance with the regulations easier while maintaining the high standard of the Australian food supply.

The development of an effective national food safety regulatory system in Australia was driven by a number of imperatives [5] including

- to reduce inconsistencies and inefficiencies with state and territory legislation;
- to reduce the cost of food regulation on the food industry;
- to respond to an increase in food borne illness;
- to counter the perception that existing requirements are ineffective in reducing the growing burden of food-borne illness; and
- to have legislation consistent with world trade obligations.

By 2000, states and territories formally agreed to a national food safety regulatory system.

The system consisted of

- nationally consistent food acts;
- mandatory standards for food safety practices, food premises, and equipment;
- a "model" standard for "food safety programs"; and
- supporting infrastructure projects to assist with its implementation.

The primary objective was to make food businesses responsible for the safety of the food they handle and sell by adopting a preventive approach to managing food safety.

The states and territories implemented this system and the food regulation system is based on a partnership between governments, consumers, and industry to develop joint food standards.

22.4
Enforcement

Regulation without enforcement is a waste of resources that poses unacceptable risks to consumers, while imposing unfair financial imposts on responsible food corporations.

While the setting of standards has been harmonized between nations, states, and territories, the enforcement of these regulations has not been uniform and in some cases simply was not done. In 1994, the Australian Office of Regulation Review conducted a survey of the government agencies involved in enforcement of food regulations [6]. The survey found that while over 600 agencies had responsibility for enforcing domestic food regulation basically under the states and territories because of the constitutional arrangements. Each agency had a slightly different approach to enforcement – essentially the allocation of responsibility at the various levels of governments. Local government, which bears much of the responsibility for enforcing the food regulations, as such is not recognized specifically in the Constitution and was established under legislation of the individual states.

Under changes to the Food Act from January 1, 2008 [7] local governments' responsibilities are

- to be appointed as "enforcement agencies," that is, councils are required to carry out regular inspections and enforcement for the retail and food service sectors;
- to investigate food complaints;
- to respond to urgent food safety recalls;
- to report on key food regulation activities to the relevant State Food Authority;
- to inspect retail food businesses regularly except for low-risk businesses.

Relevant enforcement agencies are now detailed on the Food Standards Australia web site. Food recalls, when a noncompliant food product is found to be on the market, is coordinated by Food Standards Australia New Zealand (FSANZ) but not in New Zealand. Recalls occur in consultation with the senior food officers or their deputies in the states and territories, and a sponsor that is usually the product's supplier, for example, the manufacturer or the importer.

22.5 The State Food Authority

Publishes summary reports on its web site. The aim is to reduce food-borne illness by controlling the most important food handling practices, to improve consistency among councils, and to increase public awareness about food regulation. A recent change in the Authority's operation has been the initiative to "name and shame" noncompliance to food safety regulations [8].

22.6 New Zealand

Technically, New Zealand is a constitutional monarchy; however, for all intents and purposes, the Queen acts as a titular head of State as the government is formed from a democratically elected House of Representatives. The government advises the head of State, the Queen. Although the Queen is the source of all legal authority in New Zealand, she acts through her representative, the Governor-General, appointed on the advice of the Prime Minister. The Queen and her representative act on the advice of the government in all but the most unusual circumstances. Although it has no codified constitution, the Constitution Act 1986 is the principal formal statement of New Zealand's constitutional structure.

Agriculture is the main export industry in New Zealand with dairy products accounted for over 20% ($US 5 billion) of total merchandise exports, and the largest company of the country, Fonterra, a dairy cooperative and New Zealand's largest corporation, controls almost one-third of the international dairy trade. Other agricultural items are meat 13.2%, wood 6.3%, fruit 3.5%, and fishing 3.3%.

The importance of agriculture to the New Zealand economy can be shown by the composition of its exports (Table 22.2).

Table 22.2 Exports of food commodities.

	2000	2005
($NZ millions)	26 111	30 618
Milk powder, butter, and cheese	3895	4924
Meat and edible offal	3379	4577
Fruit	972	1212
Fish, crustaceans, and molluscs	1230	1134
Casein and caseinates	806	651

Table 22.3 Summary statistics of New Zealand grape wine industry[a].

	1990	2000	% change
Number of wineries	131	358	173
Total vine area (hectares)	5800	12 194	110
Producing area (hectares)	4880	9752	100
Average yield (tones per hectare)	14.4	8.9	62
Wine production (million liters)	54.4	60.1	10
Wine exports (million liters)	4	19.2	380
Wine exports ($US million)	11.4	104.9	507.6
Domestic sales of NZ wine (million liters)	39.2	40	2
Imported wine (million liters)	4.5	28.6	535
Imported wine ($ million)	27.8	127.3	358

a) Data from MAF, http://www.maf.govt.nz/mafnet/rural-nz/profitability-and-economics/producer-boards/review-of-wine-legislation/winedisc-0.2.htm (last accessed 28.01.2010).

New Zealand also has an expanding wine industry with substantial increase in vines planted in the past decade (Table 22.3).

22.7 New Zealand's Regulatory Framework

In New Zealand, food is regulated under the Food Act 1981 and delegated legislation under that Act [9]. The Food Act 1981

- defines relevant terms, such as, food and sale;
- outlines prohibitions on sale (including unfit food);
- prohibits misleading labeling and advertising;
- provides powers of enforcement and offences;
- contains provisions to make regulations and food standards.

The Food Safety Regulations 2002, New Zealand Food Standards, and the Dietary Supplements Regulations 1985 were promulgated under the Food Act 1981. Under Section 11C of the Food Act 1981, the Minister of Food Safety has the power to issue food standards that set minimum requirements for the quality and safety of food for sale. Section 11E of the Food Act 1981 sets out the preconditions that must be considered before the Minister can issue a food standard.

The regulations are designed to

- protect public health;
- avoid pointless restrictions on trade;
- harmonize with international food standards and agreements, in particular, the Australia New Zealand Joint Food Standards Agreement.

The New Zealand (Australia New Zealand Food Standards Code) Food Standards 2002 is the legal instrument that incorporates the Australia New Zealand Food Standards Code into the New Zealand law.

22.8 Relationship with Codex

The Codex Alimentarius Commission (Codex) is the international food standards setting body recognized by the World Trade Agreements on Sanitary and Phytosanitary and Technical Barriers to Trade as being the reference point for food standards applied in international trade with the objectives of protecting the health of consumers and ensuring fair practices in the food trade. These requirements are written into the agreement between Australia and New Zealand for developing their Food Standards Code.

Through their input to the work of Codex, Australia and New Zealand support the use of science in standards development as fundamental to ensuring the safety of food supply and recognizes the importance of using science to validate food standards.

Evidence-based risk assessments are used to develop regulatory measures to lower the risk of food-borne hazards. These regulations are made after a suitable risk analysis [10] has been conducted. The development of food standards for Australia and New Zealand is based on the procedures recommended by the international food standard setting body.

While food standards should also be evidence based, it should be noted that for food contact standards the availability of robust scientific evidence that any of the 11 000 or so chemicals used to make the wide range food contact materials is hard to find because

- it is difficult to demonstrate the absence of risk;
- animal studies, while ethically more acceptable than human studies, are often not applicable to human consumption;
- epidemiological studies are expensive and can be very difficult to interpret;
- the composition of food contact materials is changing over time.

22.9 Food Contact Legislation

The Australia New Zealand Food Standards Code does not specify details of materials permitted to be added to or used to produce food packaging materials. The extent of food contact regulation in the FSANZ Code is essentially a single Standard 1.4.3 [12]. Maximum levels for packaging contaminants such as tin in canned foods at 250 mg/kg) and vinyl chloride at 0.01 mg/kg are detailed in Standard 1.4.2 (Australia only). Other chemicals in packaging materials that could be of concern for public health and safety when present above a certain level can be detailed and controlled as required.

A general provision that *adulterated food shall not be sold or presented for sale* also covers circumstances that may not be covered under the above provisions. Standard 1.4.3 states:

> Articles and materials may be placed in contact with food, provided such articles or materials, if taken into the mouth, are not
>
> (a) capable of being swallowed or of obstructing any alimentary or respiratory passage; and
> (b) otherwise likely to cause bodily harm, distress or discomfort.

Articles and materials are fairly widely defined as "any materials in contact with food, including packaging material, which may enclose materials such as moisture absorbers, mould inhibitors, oxygen absorbers, promotional materials, writing or other graphics."

Standard 1.4.3 aims to both protect consumers and allow manufacturers of articles and materials in contact with food freedom to develop new food contact materials without onerous approval processes or restrictions on what substances can be used. The definition of "food additive" in the FSANZ Code is limited to those substances that are intentionally added to a food and that perform a technological function in the final food. A substance present in a food contact material that is intended to migrate into food would require an explicit permission should it meet the definition of a food additive. This would include substances used in active and intelligent packaging such as oxygen or ethylene scavengers, antimicrobial substances, or articles to indicate the extent of ripening in fruit.

Industry can find further guidance from the Editorial note to Standard 1.4.3 that refers to the Australian Standard for plastic materials for food contact use, AS2070-1999 [13]. This standard provides industry a guidance according to which plastic materials may be used for food contact. The standard refers to compliance with either US or EC regulations that are in a state of continuous flux. Approved plastics in the United States are now included in a list of notifications to which the US FDA does not object. The EU, based on evaluations of the European Food Safety Authority (EFSA), has a "positive list" through the Plastics Directive, which is accommodated in AS2070. Reference to US regulations and EC Directives gives some scope for manufactures of plastics for food contact use. However, manufacturing of food contact materials in the Asian subcontinent limits the food industry to those packaging manufacturers that comply with US or EC regulations. If such manufacturers do not export to Europe or the United States, then obtaining the relevant evidence of compliance may be difficult and expensive.

22.10 Recycled Material

Section 4.2.1 of AS2070 states "Post-consumer recycled material shall not be used in direct contact with food." This statement is in conflict with the primary goal of AS2070, namely, to harmonize with the international standards. Recycled materials do not require premarket approval for use in the United States, although the US FDA offers advice, in the form of a guidance, for recycled plastic suitable for food contact applications [14]. The EU Regulation (EC) No. 282/2008 [18] requires authorization of the recycling process while only food-grade materials are allowed in the recycling process.

In 1999, the Australian and New Zealand Environment and Conservation Council Ministers, local government, and packaging companies agreed to develop a National Packaging Covenant [15]. The Covenant is a voluntary initiative, by government and industry, to reduce the environmental effects of packaging and facilitate the reuse and recycling of used packaging materials. In May 2009, the Covenant had 732 signatories. The reuse and recycling of packaging materials, therefore, encompasses the use of recycled plastics in contact with food that is specifically prohibited in the current AS2070-1999. It may be worthwhile to review this standard with a view to remove the blanket prohibition on postconsumer recycled plastics for food contact use.

22.11 Food Recall Examples Under Section 1.4.3 of the FSANZ Code

1) Kraft Foods Limited conducted a voluntary food recall of a ready-to-eat meal as a precautionary health measure. This recall was made because some of the products were found to contain *small pieces of rigid, blue plastic* that originated from the manufacturing process.
2) National Foods announced an immediate, voluntary recall of its range Yoplait Go-Gurt and Yoplait Smackers tube products as a result of a *packaging defect*. The packaging defect means there is a risk of *small pieces of clear plastic film* separating from the outside of the tube that may pose a choking hazard, especially to small children.
3) Patak's Foods Ltd (UK) through its Australian distributor (General Mills Australia Pty Ltd) recalled all 540 g jars of Rogan Josh Simmer. The sauces potentially contained glass fragments.
4) Heller Tasty Ltd recalled its lamb and mint sausages after *small pieces of soft blue plastic* were found in them.
5) Ceres Enterprises Ltd, Venerdi Ltd, Organic Bakeworks Ltd, and Paraoa Bakehouse recalled a number of foods that may contain *broken glass particles*.

22.12 Conclusions

There is an implied low risk of food-borne illness from food contact materials as evidenced by the extent of the FSANZ food contact regulations and enforcement. Apart from packaging material fragments finding their way into finished goods, no food recalls from noncompliant food contact materials or contamination from those packaging materials have been initiated since the inception of FSANZ. It is unclear if this results from a high level of compliance by the food industry with Section 1.4.3 or from the difficulty of determining if any given packaging material actually compliant. A 2004 survey of the xenoestrogen bisphenol A (BPA) (16) from canned foods showed that the levels of BPA identified in canned foods were unlikely to be of concern to adult health [16]. However, recent results [17] from a study of students consuming cold beverages from polycarbonate bottles showed that the geometric mean of urinary BPA increased by 69%. These results may warrant further attention to the migration of chemicals from food contact materials.

References

1 COAG Senior Officials Working Group (2000) An enhanced food regulatory system for Australia. The report of the COAG Senior Officials Working Group on Food Regulation responding to The Food Regulation Review (Blair) Report and related food legislation, Canberra. See also Council of Australian Governments Communique 6 December 2002, URL: http://www.coag.gov.au/intergov_agreements/docs/food_regulation_agreement_2002.rtf.

2 Winger, R. (2003) Australia New Zealand Food Standards Code, *Food Control*, **14** (6), 355 (Note: all articles in this Volume 14 (6) are dedicated to the development of the FSANZ Food Standards).

3 Abbott, P., Baines, J., Fox, P., Graf, L., Kelly, L., Stanley, G., and Tomaska, L. (2003) Review of the regulations for contaminants and natural toxicants. *Food Control*, **14** (6), 383–389.

4 Bragatheswaran, G. and Lawrence, L. (2009) *Overview of the Australian food industry*, 2006–2007 in Australian Food Statistics 2007. ABARE Australian Government Department of Agriculture, Fisheries and Forestry URL: http://www.daff.gov.au/__data/assets/pdf_file/0003/680745/foodstats2007.pdf.

5 Healy, M. Brooke-Taylor, S., and Liehne, P. (2003) Reform of food regulation in Australia and New Zealand. *Food Control*, **14** (6), 357–365.

6 Industry Commission, Office of Regulation Review (Nov. 1995) *Enforcing Australia's* Food Laws. Information paper, Canberra.

7 See for example NSW Food Authority (May 2009) URL: http://www.foodauthority.nsw.gov.au/localgovernment/food-regulation-partnership/.

8 NSW Food Authority (May 2009) Name & Shame list grows across NSW. Media Release URL: http://www.foodauthority.nsw.gov.au/aboutus/media-releases/mr-26-may-09-name-and-shame-list-grows-across-nsw/.

9 New Zealand Food Safety Authority (May 2009) URL: http://www.nzfsa.govt.nz/labelling-composition/publications/regulation-of-food-in-nz/index.htm.

10 WHO/FAO (1995) *Report of the Joint FAO/WHO Expert Consultation on Risk Analysis*, World Health Organization, Geneva.

11 Food Standards Australia New Zealand. URL: http://www.foodstandards.gov.au/.

12 FSANZ, URL: http://www.foodstandards.gov.au/thecode/foodstandardscode/standard143articlesa4245.cfm.

13 Australian Standard AS2070-1999. (1999) *Plastic Materials for Food Contact Use*, Standards Australia 1, the Crescent Homebush, NSW 2140.

14 US FDA CFSAN (August (2006)) *Use of Recycled plastics in Food Packaging: Chemistry Considerations* Guidance for Industry.

15 The National Packaging Covenant (2009) URL: http://www.packagingcovenant.org.au/.

16 Thompson, B.M. and Grounds, P.R. (2005) Bisphenol A in canned foods in New Zealand: an exposure assessment. *Food Additives and Contaminants*, **22** (1), 65–72.

17 Carwill, J.L., Luu, H.T., Bassett, L.S. Driscoll, D.A., Yuan, C., Chang, J.Y., Ye, X., Calafat, A.M., and Michels, K.B. (May 2009) Use of polycarbonate bottles and urinary bisphenol A concentration. *Environmental Health Perspectives*, URL: http://dx.doi.org/. doi: .

18 Commission Regulation (EC) 282/2008 of 27 March 2008 on recycled plastic materials and articles intended to come into contact with food and amending Regulation (EC) No. 2023/2006, OJ L 86, 28.3.2008, p. 9.

Index

a
acceptability criteria 343
acceptable daily intake (ADI) 230
active/intelligent materials and articles 15
– specific measure 15
Administration of Quality Supervision, Inspection, and Quarantine (AQSIQ) 320, 329
– public notice No. [2007]279 329
Administrative Acceptability of the Petition (AAP) 42, 64
adsorbant resins, resolution 58
Adulteration Act 243
ALARA principle, *see* as low as reasonably achievable (ALARA) principle
allergic reactions 41
aluminum 137
– alloys 95, 97
– conditions of use 137
– decree no. 76 of 18/4/2007 137
– foil 215, 312
– – restrictions 312
American Iron and Steel Institute (AISI) 134
analytical methods, validation 212
analytical test procedures 265
Andean Community 271
animal health and welfare (AHAW) 28
antimicrobials substance 14, 39, 74, 223, 376
antioxidants 7, 34–36, 74, 89, 213, 223, 233, 247f., 258, 324, 341, 365
artificial sausage casings 79f.
as low as reasonably achievable (ALARA) principle 144
Association of International Chemical Manufacturers (AICM) 326
– food contact task force 326
atomic absorption spectrometer (AAS) 217
atomic emission spectrometer (AES) 217
Australia 369–371
– enforcement 372
– food industry, role 370
– food standards
– – code 374
– – evolution 369
– major markets 370
– national food safety regulatory system, development 371
– office of regulation review 372
– regulatory framework 370

b
bacterial mutagenicity assay 237
baking papers, sensory testing 215
barrier layer concept 128
benzo(*a*) pyrene 216, 260
– content 216
beverages 305
– packages/containers, specifications 305f.
biobased polymers 7
biodegradable polymers 7
bisphenol A (BPA) 377
bisphenol A diglycidyl ether (BADGE) 13, 162
bisphenol F diglycidyl ether (BFDGE) 13
black iron 94f.
Bureau of Indian Standards (BIS), standards 9833-1981 343
business operators 152–155
– guidance 153

c
cadmium, migration 218
Canada Agricultural Products Act 245
Canadian Center for Toxicology (CCT) 229

Global Legislation for Food Packaging Materials. Edited by Rinus Rijk and Rob Veraart

ISBN: 978-3-527-31912-1

Canadian Food Inspection Agency (CFIA) 244, 247, 252
– enforcement activities 252
– regulated product 247
carbon black 215, 260
carcinogenic substances 148, 225
cast iron, metallic/organic coating 97
cellulose
– films 265f.
– materials 158, 260, 264f.
CEN methods 18, 38
– protocols 38
Center for Food Safety and Applied Nutrition (CFSAN) 224
CEPE 162, 166, 172
– lists 172
– web site 172
ceramic 88, 94, 119f., 135
– articles 156
– French regulation 94
– metals, maximum limits 156
Chemical Abstract Service Registry Number 236
China 326, 328
– expanded food contact additive positive list 327
– food contact chemical legislation summary, current status 319
– food contact chemical regulation, forecast on future 326
– food contact materials, importing requirements 328
– food packaging material-related regulation 328
– food safety regulatory entity 320
– local manufacturing, requirements 328
– new food contact regulation, forecast 326
– new food contact substance notification procedure 327
– SEPA chemical inventory 326
China food hygiene law (CFHL) 319f.
– framework 319
– regulations and rules 320
CMRs, use 167
coatings 57, 100
– French regulation 100f.
– materials 267
– resolution 57f.
– – chronological development 57
– – content 57
– – documents inventory 57
Code of Practice 162–165, 168–171, 173
– contents 163
– main points 164
– – declaration of conformity 168
– – dual use additives 167
– – good manufacturing practice 165
– – multilayer coatings 168
– – risk assessment 171
– – scope 164
– – substance lists 166
– – substances, restrictions/testing 168
Codex Alimentarius Commission (CAC) 355, 375
– food hygiene, definition 355
colorants 74, 102, 223
– definition 74
– French regulation 102
coloring matters 101
– French regulation 101–104
Commission Directive 2002/72/EC 146
Commission Directive 2004/19/EC 145
Commission Directive 93/10/EEC 147, 165
commodities 67f., 73
– basic requirements 68f.
– definition 67
– ordinance 71
Commonwealth Government 371
Commonwealth Parliament 371
Community legislation 1–6, 8, 12, 19, 24
– food legislation 17
– horizontal legislation 2–6
– – framework regulation 2–5
– – good manufacturing practice 5
– material 11
– – active/intelligent materials and articles 13
– – BADGE/BFDGE/NOGE in coated materials, plastics, and adhesives 13
– – ceramic articles 12
– – regenerated cellulose film 12
– – rubber teats/soothers 12
– overview 3
– recycled plastics 10
– rules 2
– specific measures 6–11
– – general requirements in plastics directive 6–8
– – plastics 6–10
Community Reference Laboratory (CRL) 10, 19
compliance, declaration 16, 168
constituents policy 225
Constitution Act 1986 373
Consumer Safety Officer (CSO) 238
consumption factor (CF) 184
contact conditions, selection 199
– food simulants 199–203

– time and temperature conditions 203–205
conventional migration test conditions 37
conveyor belts 77
– Balata 77
– Gutta–Percha 77
copper, objects 99
cork stoppers, resolution 58
– chronological development 58
– documents inventory 58
– resolution, content 58
Council of Europe (CoE) 19f., 49f., 52f., 63, 143, 166f.
– activity 50–52
– future 63f.
– guidelines 52–54
– – adoption procedure 52
– partial agreement, aim 50
– political organization 49
– resolutions 24, 52–54
– – adoption procedure 52
– – resolution AP 154, 173
– – resolution ResAP 148
– technical documents 52–54
– – adoption procedure 52
cross-linking agents 232
cumulative estimated dietary intake (CEDI) 184, 237
cyclohexane, UV absorption 216

d

D reduction factor (DRF) 202
dairy products, regulations 245
declaration of compliance 169
declaration of interest (DoI) 30
de minimus principle 175
Denmark, legislation in 151
– in-house documentation 152
– public control 152
direct food contact 263, 343
– adhesives for 263
– Indian standards 343
Directorate General for Health and Consumers 138
dispersion coatings, types 74
dissemination 351
– qualities 351–352
– standards 351–352
drinking water, testing 214
Drugs Act 243

e

elastomeric materials 262
– food contact articles 76
– packages, technical regulation 262
enamel 119f.
epoxy plastics, definition 112
epoxy regulation (EC) no. 1895/2005 161f.
estimated daily intake (EDI) 237
Europe 27
– petitioning requirements 27
– safety assessment 27
European Commission (EC) 47, 50
– food and veterinary office 18
European directives 86, 286
– integration 86–89
European economic area (EEA) 20, 153
– agreement 153
European Food Safety Authority (EFSA) 5, 8, 16, 27–30, 38, 42f., 110, 128, 176, 180, 193, 376
– advisory forum 27
– CEF panel 28
– element 28
– FCM, evaluation, checklist 43
– implementation 28
– preparation 28
– remit 28–30
– – scientific bodies 29
– – supporting units 30
– role in safety evaluation of food contact materials 28–30
– safety assessment 30f.
– – assumption 30
– – data requirement 30–42
– – general information on substance 32
– – principle 31
– – supplied data within submission 31–42
European Free Trade Association (EFTA) 153
European parliament and council regulation (EC) no. 1935/2004 28
European professional organizations 46, 50, 53
European Standardization Committee 37
European Union (EU) 17, 20, 180, 182, 188
– activity 52
– *vs.* CoE 52
– conventional assumption 180
– directives 145
– – 2007/19/EC, compliance declaration requirements 153
– food contact materials, control 17
– – business operators, role 17
– – European Commission, role 18, 149
– – member states, role 17
– – official control analysis 18
– – sampling methods 18
– food law 1, 127
– legislation 152

-- types 1
-- national decree 152
- legislative documents 163
- member states 10
- plastics directive 144
- project 189
- regulations 71, 151
-- 76/893/CEE and 89/109/ CEE 126
-- regulation (EC) no. 1935/2004/CE 127, 144, 154
-- regulation (EC) no. 282/2008 131, 376
- risk assessment 182
- wide research project 24
- regulatory system 190
evidence-based risk assessments 375
exposure assessment 51, 178–190
- data exposure 179f.
- factors to consider 181
- food contact articles, estimating exposure to migrants 181
-- deterministic 182
-- probabilistic (stochastic) modeling 183
-- simplistic 182
-- US FDA approach 184
- packaging of foodstuffs 185
-- concentration data 187–189
-- containing the migrant 189
-- dietary surveys, overview 185
exposure-based approach 193

f

Food Additive Petitions (FAP) 238
fat reduction factor (FRF) 10, 149, 202
Federal Environment Agency 75
Federal Food, Drug and Cosmetic Act (FDCA) 223
- exemptions 226
-- basic polymer/resin doctrine 232f.
-- functional barrier 228
-- GRAS 229
-- housewares 230–232
-- mixture doctrine 233
-- no migration 226–228
-- prior sanction 228
-- threshold of regulation 229f.
federal food law 141
Fick's laws of diffusion 188
finished product (FP) 35, 79, 110f., 118, 120–122, 130f., 133, 197, 247f., 261, 268
- colorants purity 217
- color release 216
- compliance, declaration 217–219
- contact conditions 199–205
- migration experiments 205–211
- organoleptic testing 214, 215
-- microwave application 214
-- oven conditions 215
- polycyclic aromatic hydrocarbons (PAHs), absence 215f.
- residual content 212
-- analytical determination 212
-- worst-case calculations 212
- supporting documentation 217–219
-- ceramics 218
-- materials with BADGE 219
-- plastic materials 218f.
Finland 151
- food contact materials, legislation 151f.
Finnish Customs Authority 152
Finnish Food Safety Authority 152
Fish Inspection Act 245
flavors, additives (food) contact materials exposure task (FACET) 190
- FACET industry group (FIG) 193
-- packages 193
-- structure 190
- project 194
-- trade associations 192
fluorescent materials 296
- brightener, types 359
food 9, 52, 131, 141, 161, 201, 297, 340, 365
- alcohol-containing 201
- classification 297
- commodities, exports 373
- consumption databases 193
- consumption surveys 189
- contamination 198
-- migration 198
-- source 198
- conventional classification 52, 131
- direct contact, code of practice for coatings 161
- extraction properties 9
- food package to carry label 340
- labeling standard 365
- Legislative Act 141
- packing and labeling 340
- products, packages 214
- purity 223
- regulations 272, 372
- safety 244, 291
-- basic law *vs.* food sanitation law 313f.
-- Commission, roles 291
-- food safety basic law 291
-- food sanitation law 291
-- regulation of Canada 244
- stuffs industry 167

– type 297, 304
– – packages/containers specifications 304
– – simulant 297
Food Act 346, 348, 372, 374
– notifications 348
food additive 224, 295, 365, 376
– code, *see* food code
– definition 224, 376
– petition process 224, 234, 238
– regulations 224, 229
– safety 184
– standards 295–306, 365
Food and Drug Administration 168, 224, 233, 349
– application 343
– CFSAN supports 230
– good manufacturing practice (GMP) 225
– GRAS determination 230
– regulatory burden 233
– rules and regulations 224, 229
– – food additive, definition 224f.
– system 168
– USA CFR-21 regulations 286
– web site 224
Food and Drug Law Institute 231
Food and Drug Regulations 245–247, 250
Food and Drugs Act 243–246, 252
food balance sheets (FBS) 186
food code 356f., 364, 366
– common standard and specification for food 358
– food products 358
– – equipment, containers, and packaging 358
– – hierarchical categories 357
– general provision 357
– types 358
food contact approaches 71, 167, 207
– applications 227, 234, 236, 238, 265, 267
– – porous ceramic materials 267
– development 167
– legislation 110, 289, 375
– – global Israeli approach 288f.
– – guiding principles 290
food contact materials (FCMs) 1, 7, 13, 35, 40, 42, 44, 67, 83, 85, 87, 128f., 142, 153, 175, 197, 214, 217, 223, 315, 323, 329, 334, 364
– basic principles 84
– Community legislation, on products 4
– compliance 197
– – demonstration 197–199
– – testing 197
– data and information required for submission 364
– EFSA working group 31
– EU directives 87
– EU regulation 87
– evaluation process 42
– – re-evaluation 44–46
– hygiene standards 323
– French regulation 83
– functions 14
– legislation 151
– Legislative Act 141
– manufacturers 142
– material classes, legal assessment 71–81
– maximum dosages 334
– migrants, estimating risks 175
– migration 200
– – determination 210
– packaging 321
– plastics, harmonization of legislation 8
– production 153
– – chain 128
– – public control 153
– regulations, implementation issues 289f.
– role 1
– rules 131
– sensory testing 214
– SI-5113 provisions 286f.
– use 334
– WG submits 44
food contact materials and articles 125f., 144–148
– active/intelligent materials 147
– ceramic, glass, enamel/analogue materials 147
– inks 148
– metals and alloys 147
– paper and board materials 147
– paraffin, waxes, and colorants 148
– plastic materials 144
– – migration limits 145
– – recycled plastic materials 146
– regenerated cellulose, use 146
– silicone materials 148
– Italian legislation 126
– principles 125
– synthetic scheme 126
food contact notification (FCN) system 184, 224f., 233f., 236
– Drug Administration Modernization Act 233
– food additive regulations 234
– food contact substances 233
– – food additive regulation system 233
– prenotification consultations 235

– program 234
– requirements for 235–237
– – chemical identity 236
– – comprehensive summary 236
– – environmental assessment (EA) 237
– – estimation of dietary intake 237
– – intended conditions of use 236
– – intended technical effect 236
– – toxicity information 237
– submitters 236
food container 319–321, 325
– definition 319
– hygiene standards 321
– requirements 320
– rules 321
food-grade PCR-PET packages 261
food hygiene 270, 355
– Codex Alimentarius Commission (CAC) definition 357
food laws 4, 17, 369, 371
– hygiene law 320f.
– labeling 4
– 1860 UK Act 371
food packaging 103, 126, 245, 251, 264, 333
– applications 251
– articles 333
– food contact coatings 170
– inks, French regulation 103f.
– production 189
– regulations 126
– TTC 171
food packaging law 223, 243
– Canada 243
– United States 223
food packaging materials 225, 246f., 320
– compulsory hygiene standards 322
– definition 323
– plastic materials 360–364
– requirements 320
Food Safety Legislation 255
– Central America 275
– – Belize 277
– – Costa Rica 276
– – Cuba 277
– – El Salvador 276
– – Guatemala 276f.
– – Honduras 277
– South America 255
– – Andean Community 271–274
– – Chile 271
– – MERCOSUR 255–269
– – Mexico 274f.
– – Venezuela 269f.
Food Safety Regulations 2002 374
Food Sanitation Act (FSA) 189, 355–357
food sanitation law 292–295
– enforcement ordinance 292, 306
– enforcement regulations 292, 294
food simulants 138, 199
– alcohol-containing food 201
– alternative fatty food simulants 201
– directive 85/572/EEC 200
– dry products 201
– future developments 203
– migration, food type 199
– reduction factors 201, 202
– – DRF 201f.
– – FRF 202
– – total reduction factor (TRF) 202
– substitute fatty food simulants 201
Food Standards Australia New Zealand (FSANZ) code 372, 375–377
– food contact regulations 377
– food recall examples 377
– – under section 1.4.3 377
Food Standards Committee 371
food surveillance surveys 188
framework directive 1, 127
– adoption 1
framework regulation 2, 4–7, 67, 109
– amendments 2
– principles 6f.
– regulation (EC) no. 1935/2004 2, 127, 138, 157, 162f., 165, 171, 173, 175, 194
– requirements 5
freeware software packages 211
French Food Safety Agency 85
French regulation 93, 101
functional barrier concept 9, 148
– compliance, verification with migration 9

g

gas chromatography with mass detection (GC-MS) 198
Gaussian distributions 187
generally recognized as safe (GRAS) 225
Germany, national legislation in 67
– basic requirements on commodities 68
– BfR recommendations 69–71
– commodities definition in LFGB 67
– commodities ordinance 70
– recommendations 19
glass 62, 94, 119, 134, 267
– ceramics 119
– French regulation 94

– guidelines 62
– – chronological development 62
– – documents inventory 62
– – resolution, content 62
– migration tests 135
– packages 267
– type 134, 267
global food contact 337
– principal issues 337
– – food contact legislation 337
– – general guidelines 337
– – Indian subcontinent 337
– – sale and license condition 338–340
good manufacturing practice (GMP) 4–6, 132, 144, 152
– principles 5
– production, stages 5
– regulation (EC) no. 2023/2006 161, 165
– rules 4
Grupo Mercado Comun (GMC) 255
– resolutions 268

h
hazard assessment 176
– Cramer class I/II/III 177
hazard-based approach 175
health Canada guidance 250
health products and food branch (HPFB) 244, 247
– food directorate 244, 247
– jurisdiction 247
high-density polyethylene bag (HDPE) 339
high-performance liquid chromatography with MS detection (HPLC-MS/MS) 199
hollow glassware, coating materials 80
hot-filter paper 78
household budget surveys (HBS) 186
household metallic articles 96
Hygiene Standards 321
– for Adjuvant and Processing Aids in Food Containers and Packaging Materials 323

i
imported food packaging 329
– inspection/supervision 329
indirect food contact, adhesives 263
inductive coupled plasma (ICP) 217
industrial voluntary rules 314–316
inertness, principle 7, 14, 69
Insecticide Act 1968 339, 340
International Life Science Institute (ILSI) 171, 181
– Europe, decision tree 178
ion exchange 58
– resolution 58
– – chronological development 59
– – content 59
– – documents inventory 58
isocyanate-based adhesives 270
Israel 287f.
– approved test laboratories 287, 288
– food contact regulation 286
– new standard 286
– packaging materials, imports 288
– voluntary standards 285
Italy 125
– corpus of laws 125
– national legislation 125, 129, 138
– – decrees on general principles 126, 127
– – decrees on specific materials 127
– – list/text 138
– – ministerial decree 21 march 1973, amendments 127
– regulation 135

j
Japan Hygienic PVC Association 315

k
Korea customs service (KCS) 366
Korea Food and Drug Administration (KFDA) 355, 364, 366
– commissioner 356
Kosher regulations 289

l
labeling standard for food (LSF) 365
legal analytical methods 133
legal systems 19
– authorized substances system 19
– no specific legislation system 19
– premarket approval system 19
– recommendations system 19
– structure 244
– – agencies 244
– – laws 244
legislation process 283
limit of detection (LOD) 187

m
Malaysian regulations 346f.
mammal cell test 335
material
– safety data sheets 248
– specific rules 70
– specific standards 359
mathematical modeling 211, 249
Meat Inspection Act 245

MERCOSUR
– legislation 259f., 268
– resolutions 269
metal can 304, 310, 312
– restrictions 310, 312
– specifications 304
metallic materials 266
– legislation 266
– packaging 94
–– aluminum 95
–– metal-coated steel *see* tinned iron
–– noncoated steel, *see* black iron
–– stainless steel 94
–– tin 95
metals and alloys 61
– guidelines 61
–– chronological development 61
–– content 61
–– documents inventory 61
microcrystalline parafins 216
microcrystalline waxes 77
MIFAFF web site 357
migration 206
– cells, types 206
– experiments 205
–– aqueous/fatty food simulants 208
–– contact methods 205–207
–– mathematic modeling 211
–– MPPO 210
–– overall migration 208
–– specific migration 210
– model, development 189
– preparation methods 298
– related parameters 198
– testing 10, 37, 298
–– reduction factors applicable 10
minimum inhibitory concentration (MIC) 39
Minister of Food Safety 374
Ministry for Food, Agriculture, Forestry, and Fisheries (MIFAFF) 357
Ministry of Agriculture and Forestry 152f.
Ministry of Health (MOH) 127, 320, 326
Ministry of Health, Welfare, and Family Affairs (MIHWAF) 355f.
Ministry of Industry and Trade 288
Ministry of Knowledge Economy 357
Model Food Act 371
modified polyphenylene oxide (MPPO) 210
– overall migration 210
Monte–Carlo approach 179, 183

n
nanomaterials, future trends 20
Napierian logarithms 116
National Environmental Policy Act (NEPA) 237
national focal points 28
national food safety strategy 356
national legislation, summary 21–23
national reference laboratories (NRLs) 19
National Veterinary and Food Administration 152
new food contact resins 333
new food contact substance, *see* new food contact resins
New Zealand 369, 373
– agriculture, importance 373
– food standards
–– code 374
–– evolution 369
– regulatory framework, regulations 374
no adverse effect level (NOAEL) 176
no-migration principle 167f., 170, 172
noncarcinogenic food contact substances 230
nonintentionally added substances (NIAS) 171, 197
nontarget analysis 213
no objection letter (NOL) 247
no observed effect level (NOEL) 176f.
Nordic council of ministers 153
Nordic report 151
Norway, legislation in 153
– critical points, control 154
– EK declaration 153
– metals in ceramics, glass, metalwares, and nonceramic materials without enamel 155
– packaging convention 153
– paper and board food contact materials 154
Norwegian Food Safety Authority 154f.
nuclear magnetic resonance (NMR), infrared 199
nylon resins 302
– apparatus, specific specifications 302

o
official feed and food control (OFFC) 17
– regulation (EC) no.882/2004 17f.
oils 342
– olive 201, 216
– packing articles 333
– sun flower 216
optical emission spectrometer (OES) 217
organoleptic testing, packaging 214
overall migration limit (OML) 70, 145, 168

p
packaged drinking water 343
packaging 13–15, 54, 181, 185–187, 189

– approaches 181
– characteristic 14
– containers 329
– – restrictions 313
– contaminants 375
– function 14
– inks, resolution 60
– – chronological development 60
– – documents inventory 60
– – resolution, content 60
– label requirements 342
– materials 118, 214, 248, 256f., 321, 325
– – definition 319
– – hygiene standards 321
– – MERCOSUR GMC resolutions 257
– – metals 118
– – recycling 377
– – regulations, future direction 366
– – rules 321
– – sensory testing 214
– status 54–63
– substance migrating 187
– surveys 186
– type 181, 185, 189
paper and board 91, 92, 112–114, 132–134
– French regulation 91–93
– paper, resolution 56
– – chronological development 56
– – documents inventory 56
– – package content 56
– physicochemical properties 132
– rules 133
– types 112
Partial Agreement 49
– committees 49
– countries 64
per capita approach 185, 189
– advantage 185
Peruvian National Institute 273
petitioner summary data sheet (P-SDS) 34, 42, 44
petroleum waxes 264
Pharmaceutical Affairs Law 291
Pharmacopoea Helvetica 148
phenol resins 300
– apparatus, specific specification 300
physicomathematical diffusion models 188
plastic material 20, 50, 54, 71–74, 88f., 111, 129–131, 199, 258, 350
– additives, reference texts 89
– colorants 54
– – chronological development 54
– – documents inventory 54
– – reference texts 89
– – resolution, content 50, 54
– declaration of compliance 8
– directive 11f., 16
– – 2002/72/EC 163, 165
– – 2002/722/EC 173
– – requirements 11
– dispersions 74
– epoxy plastics 112
– fillers, commodities 80
– food contact materials 16, 20
– – harmonization of rules 20
– – material-specific requirements 16
– food contact packages 263
– – manufacture 7
– – mass transfer 188
– – producer 9
– – specific French legislation 89
– – use of 258
– food-packing container, QS catalog 330–332
– future trends 20
– nonepoxy plastics 111
– overall migration limits (OML) 50
– packaging 251
– qualities 350
– recyclers 251
– recycling process, type 11
– resins, hazardous characteristics 359
– specific migration limits (SML) 50
– standards 350
– surface area, migration limit 258
plastic polymerization 54
– control of aids 54
– – chronological development 55
– – resolution, content 54f.
– documents inventory 54
polycarbonate resins 303
– apparatus, specifications 303
polychlorobiphenyls (PCBs) 133
polycyclic aromatic hydrocarbons (PAHs) 215
polyethylene terephtalate (PET) 10, 261
– apparatus, specifications 302
– resins 302
polylactic acid resins 303
– apparatus, specifications 303
polymeric additive 32f.
polymeric chemical substances, preparations 32, 148
polymeric coatings food 266
polymeric resins 264
polymerization 233
– inhibitors 259
polymers 232, 259

– composition 38
– formation 145
– positive list 259
polymethyl metacrylate resins 302
– apparatus, specifications 302
polymethylpentene resins 302
– apparatus, specifications 302
polypropylene resins 301
– apparatus, specifications 301
polystyrene resins 301
– apparatus, specifications 301
polyvinyl alcohol resins 303
– apparatus, specifications 303
polyvinyl chloride (PVC) 338
– apparatus, specifications 301
– package, use 347
– resins 301
polyvinylidene chloride resins 302
– apparatus, specifications 302
PRC 333
– food hygiene law 333
– import and export commodity inspection law 333
printing inks 359
probabilistic (stochastic) modeling 179, 183
protective coating concept 103
Public Health Committee 20

q

quality assurance system 6
quality control system 5

r

rapid alert system for feed and food (RASFF) 18
raw materials 324
– standard sets 324
– suppliers 170
REACH deadlines 166
REACH regulation (EC) no. 1907/2006 162
ready-to-eat-packages 214
record keeping methods 186
recycled materials 89, 250, 376
– functional barrier, use 250
– plastic materials, manufacturers 250
– process efficacy 251
– recycling process 219
– reference texts 89
–– AFSSA 90
–– CSHPF 89f.
– source control 251
– use limitations 251
reference binding texts, categories 84–86
regenerated cellulose 121, 132
– casings 266
– film 11, 87f., 265
regulatory scheme 245
– enforcement 252
– mandatory requirements 245
– NOL process 247
–– normal submission requirements 248
–– recycled products 250
– polymer resins 247
– result 251
– voluntary requirement 246
Republic of Korea 355
– legislation on food contact materials 355
resins, uses 167
rubber 88, 114–118, 131, 303
– apparatus 303
– categories 114–117
– legislation 131
– products 59
–– manufacturing, scheme 115
–– resolution 59f.
– rules 131

s

safe food, requirements 4
sanitary hazards 355
Scandinavian countries 151
– food contact materials, legislation 151
Scientific Committee on Food (SCF)
– definition 45f.
– EFSA protocols 166
– guidelines 31
self supervision law 141
short-term dietary survey 186
silicone materials 55, 75, 148
– elastomers 75–77, 93
–– French regulation 93
– resolution 55
–– chronological development 55
–– content 55
–– document inventory 55
– rubbers 75–77
silver-plated metals 99
Singapore food regulations 345
specific migration limit (SML) 7, 51, 70, 111, 117, 136, 169, 202, 343
– methods of analysis 343
specific national legislation 19
stainless steel 134
– objects 96
Standards Institution of Israel (SII) 283, 284
– expert forums 284
– General Plastic Committee 285

– marks 286
– technical forums 284
State Administration of Food And Drug (SFDA) 320
state food authority 373
STFI project 193
structure-activity relationships (SARs) 171
substance(s) 32–36, 41f., 50, 166, 182, 327, 324, 366
– antimicrobial substances 39
– authorization procedure 64
– biological properties 366
– categories 166, 324
– chemical structure 41
– classes 32
– evaluation, guidelines 50
– exposure, advantage/disadvantage 182
– information on authorization 36
– information on migration 36
– information on residual content 39
– intended application, information 35
– physical/chemical properties, information 33
– requirements 324
– specifications 33
– specific migration 36
– stability 34
– technological function 35
– toxicological information 40
– – special investigations/additional studies 41
– – toxicological tests, core set 40
– – toxicological tests, reduced set 41
– use 42, 327
summary data sheet (SDS) 42, 64
supplementary French legislation 90–100
Sweden, legislation in 157
– food 158
– voluntary agreement 157
Swedish Food Act 157
Swedish National Food Administration 157
Swiss Federal Office of Public Health (FOPH) 142
– information letters and directives 143
– responsibility 142
Swiss Food Compendium 142, 146
Switzerland 141
– attestation of conformity 143
– Council of Europe 143
– legislative system 141
– – legal texts, availability 142
– – official documents, availability 142
synoptic document 44
– Council of Europe Resolution AP 172
synthetic colorants 359
synthetic resins 309–311, 313

t

temperature-resistant polymer coating systems 80
Thai Industrial Standards (TIS) 353
Thai Industrial Standard Institute (TISI) 353
Thai regulations 348
theoretical exposure level 105
thermographic analysis (TGA) curve 34
thermoplastic cements 267
threshold of regulation (TOR) rule 224, 229, 328
threshold of toxicological concern (TTC) 249
tin
– alloys 95, 98
– free steel, DM 1/6/88 136
– plate 135
tinned iron 95
tissue paper kitchen towels 62
– guidelines 62
– – chronological development 62
– – content 62
– – documents inventory 62
tolerable daily intake (TDI) 7, 44, 51, 176, 182
total reduction factor (TRF) 202
toxicology tests 236
– data, type 237
Treaty of Rome principle 172

u

Unification Italian Committee (UNI) 134
United States of America 36, 272, 275
– Food and Drug Administration (FDA) 38, 179, 184, 223, 376
– – FCN program 328
– – red book 249

v

vinyl chloride monomer (VCM) 161, 199
virtually safe dose (VSD) 226
VWA 111

w

whitened metal, II-K objects 99
wood and cork 121
working group on food contact materials (FCM WG) 29
work package, list 192